T0309762

CLUSTER SECONDARY
ION MASS
SPECTROMETRY

WILEY SERIES ON MASS SPECTROMETRY

A complete list of the titles in this series appears at the end of this volume.

CLUSTER SECONDARY ION MASS SPECTROMETRY

Principles and Applications

Edited by

Christine M. Mahoney

Published by John Wiley & Sons, Inc., Hoboken, New Jersey
Published simultaneously in Canada

For general information on our other products and services or for technical support, please contact our Customer Care Department within the United States at (800) 762-2974, outside the United States at (317) 572-3993 or fax (317) 572-4002.

Wiley also publishes its books in a variety of electronic formats. Some content that appears in print may not be available in electronic formats. For more information about Wiley products, visit our web site at www.wiley.com.

Library of Congress Cataloging-in-Publication Data:

Mahoney, Christine M., 1975-
    Cluster secondary ion mass spectrometry : principles and applications / Christine M. Mahoney.
        pages cm
    Includes bibliographical references and index.
    ISBN 978-0-470-88605-2 (hardback)
    1. Secondary ion mass spectrometry. I. Title.
    QD96.S43M34 2013
    543'.65–dc23

                                                    2012045181

Printed in Singapore

10 9 8 7 6 5 4 3 2 1

CONTENTS

5 MOLECULAR DEPTH PROFILING WITH CLUSTER ION BEAMS

Christine M. Mahoney and Andreas Wucher

Andreas Wucher, Gregory L. Fisher, and Christine M. Mahoney

CONTRIBUTORS

Albert J. Fahey, National Security Directorate, Pacific Northwest National Laboratory, Richland, WA

Andreas Wucher, Department of Physics, University of Duisburg-Essen, Campus Duisburg, Duisburg, Germany

Arnaud Delcorte, Institute of Condensed Matter and Nanosciences, Université catholique de Louvain, Louvain-la-Neuve, Belgium

Bartlomiej Czerwinski, Institute of Condensed Matter and Nanosciences, Université catholique de Louvain, Louvain-la-Neuve, Belgium

Christine M. Mahoney, Environmental Molecular Sciences Laboratory, Pacific Northwest National Laboratory, Richland, WA

Greg Gillen, Surface and Microanalysis Science Division, Materials Measurement Laboratory, National Institute of Standards and Technology, Gaithersburg, MD

Gregory L. Fisher, Physical Electronics, Incorporated, Chanhassen, Minnesota

Joe Bennett, Novati Technologies, Austin, TX

John Vickerman, Surface Analysis Research Centre, Manchester Interdisciplinary Biocentre, School of Chemical Engineering and Analytical Science, The University of Manchester, Manchester, UK

Nick Winograd, Department of Chemistry, Penn State University, University Park, PA

Oscar A. Restrepo, Institute of Condensed Matter and Nanosciences, Université catholique de Louvain, Louvain-la-Neuve, Belgium

Peter Williams, Department of Chemistry & Biochemistry, Arizona State University, Tempe, AZ

ABOUT THE EDITOR

Dr. Christine M. Mahoney is a recognized expert and leader in the field of Secondary Ion Mass Spectrometry (SIMS). Throughout her career she has focused primarily on the application of SIMS to molecular targets, and has played a significant role in the development of cluster SIMS for polymer depth profiling applications. She received her Ph.D in Analytical Chemistry from SUNY Buffalo in 1993. After which, she spent the following eight years at the National Institute of Standards and Technology (NIST), where much of her molecular depth profiling work was performed. Christine is currently employed as a senior research scientist at the Environmental and Molecular Sciences Laboratory (EMSL) at the Pacific Northwest National Laboratory (PNNL) where she continues to lead research in the field of SIMS.

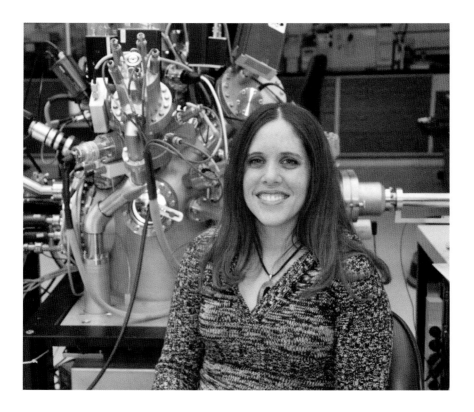

CHAPTER 1

AN INTRODUCTION TO CLUSTER SECONDARY ION MASS SPECTROMETRY (CLUSTER SIMS)*†‡

Christine M. Mahoney and Greg Gillen

Cluster secondary ion mass spectrometry (SIMS) has had a significant impact on the mass spectrometry and surface analysis communities over the past two decades, with its newfound ability to characterize surface and in-depth compositions of molecular species with minimal damage, excellent spatial (100 nm or less) and depth (5 nm) resolutions, and increased sensitivities for bioimaging applications. With the continual development of new cluster ion beam technologies, we are breaking down barriers once thought to be unbreakable, and entering into new fields once labeled as out of reach. Instrument designs are now advancing to account for these new applications, allowing for further improvements in molecular sensitivities, selectivities, and even high throughput analysis. Although we are

*Official contribution of the National Institute of Standards and Technology; not subject to copyright in the United States.

†Commercial equipment and materials are identified in order to adequately specify certain procedures. In no case does such identification imply recommendation or endorsement by the National Institute of Standards and Technology, nor does it imply that the materials or equipment identified are necessarily the best available for the purpose.

‡This document was prepared as an account of work sponsored by an agency of the US Government. Neither the US Government nor any agency thereof, nor Battelle Memorial Institute, nor any of their employees, makes any warranty, express or implied, or assumes any legal liability or responsibility for the accuracy, completeness, or usefulness of any information, apparatus, product, or process disclosed, or represents that its use would not infringe privately owned rights. Reference herein to any specific commercial product, process, or service by trade name, trademark, manufacturer, or otherwise does not necessarily constitute or imply its endorsement, recommendation, or favoring by the US Government or any agency thereof, or Battelle Memorial Institute. The views and opinions of authors expressed herein do not necessarily state or reflect those of the US Government or any agency thereof.

Cluster Secondary Ion Mass Spectrometry: Principles and Applications, First Edition.
Edited by Christine M. Mahoney.
© 2013 John Wiley & Sons, Inc. Published 2013 by John Wiley & Sons, Inc.

only at the beginning of the growth curve toward low damage molecular SIMS, we have come a long way over the past few years, and significant discoveries have been made. This book addresses these new discoveries and describes practical approaches to SIMS analysis of samples using cluster sources, with a focus on soft sample analysis.

1.1 SECONDARY ION MASS SPECTROMETRY IN A NUTSHELL

Before we discuss cluster beam technology, it is appropriate to first review the basics of SIMS. SIMS is a mass spectrometric-based analytical technique, which is used to obtain information about the molecular, elemental, and isotopic composition of a surface. In a conventional SIMS experiment, an energetic primary ion beam, such as Ga^+, Cs^+, or Ar^+ is focused onto a solid sample surface under ultra high vacuum conditions (Fig. 1.1). The interaction of the primary ion beam with the sample results in the sputtering and desorption of secondary ions from the surface of the material. These secondary ions are subsequently extracted into a mass analyzer, resulting in the creation of a mass spectrum that is characteristic of the analyzed surface (Fig. 1.2a), and yielding elemental, isotopic, and molecular information simultaneously, with sensitivities in the parts per million (ppm) to parts per billion (ppb) range. There are three basic types of SIMS instruments that are used most commonly in the field, each employing a different mass analyzer:

1. *Time-of-Flight Secondary Ion Mass Spectrometers (ToF-SIMS).* These spectrometers extract the secondary ions into a field-free drift tube, where the ions are allowed to travel along a known flight path to the detector. As the velocity of a given ion is inversely proportional to its mass, its flight time will vary accordingly, and heavier ions will arrive at the detector later than lighter ions. This type of mass spectrometer allows for simultaneous detection of all secondary ions of a given polarity and has excellent mass resolution. Moreover, because the design utilizes a pulsed ion beam operated at extremely low

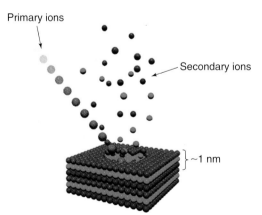

Primary ions

Secondary ions

~1 nm

Figure 1.1 Illustration of the sputtering process in secondary ion mass spectrometry (SIMS).

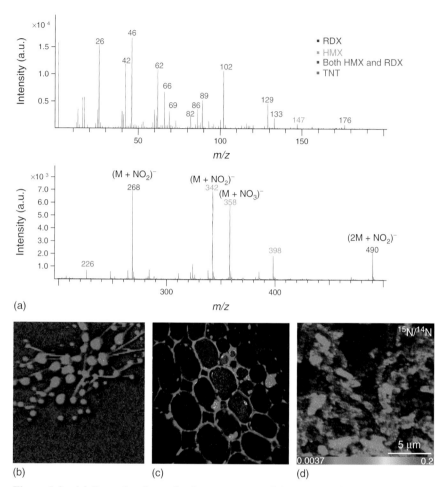

Figure 1.2   (a) Example of negative ion mass spectral data acquired from a sample of composition-4 (C-4) plastic explosive, comprised of poly(isobutylene), RDX explosives, di-isooctylsebacate, and other additives. (b) Example of negative ion molecular imaging (200 × 200 μm) in Semtex plastic explosive, based on RDX explosive, PETN explosive, poly(styrene-*co*-butadiene), and other additives; green = PETN explosive molecules (*m/z* 376), red = binder and oils (*m/z* 25), and blue = SiO_2^- (*m/z* 60) from the Si substrate. (c) Example of positive ion *elemental* mapping of trace elements in plant roots; green = CN^- (*m/z* 26), blue = Si_2^- (*m/z* 28), and red = As (*m/s* 75).[3] (d) Isotopic imaging of bacteria grown in ^{15}N culture medium. Green regions indicate ^{15}N-enriched bacteria, while blue regions indicate more natural isotopic abundances. Hence, the bacteria in the blue regions are not as metabolically active as the green regions.[4]   Figure 1.2c and d recreated from Moore et al.[3] and Kilburn,[4] respectively, with permission from the American Society for Plant Biology and the University of Western Australia.

currents (picoampere range), this mass spectrometer is useful for analysis of surfaces, insulators, and soft materials, which may be prone to ion-induced chemical damage.

2. *Magnetic Sector SIMS instruments.* Magnetic sector SIMS instruments typically use a combination of electrostatic and magnetic sector analyzers for velocity and mass analysis of the sputtered secondary ions. The use of a magnetic field to deflect the ion beam causes lighter ions to be deflected more than the heavier ions, which have a greater momentum. Thus, the ions of differing mass will physically separate into distinct beams. An electrostatic field is also applied to the secondary beam in order to remove any chromatic aberrations. Because of the higher operating currents and continuous beams, these instruments are very useful for depth profiling. However, they are not as ideal for surface analysis and characterization of samples that will charge and/or damage readily.

3. *Quadrupole SIMS Instruments.* These instruments are becoming increasingly rare because of the relatively limited mass resolution attributed to them (unit mass resolution—unable to resolve more than one peak per nominal mass). The quadrupole utilizes a resonating electric field, where only ions with selected masses have stable trajectories through a given oscillating field. Similar to the magnetic sector instruments, these instruments are operated under high primary ion currents and are generally thought of as "dynamic SIMS" instruments (i.e., used for sputter depth profiling and/or bulk analysis of solid samples).

Although these designs are most commonly observed in the SIMS community at present, there are many new exciting designs emerging, which may play a more prominent role in the future.[1,2] These new designs include continuous ion beam designs with multiple mass spectrometers (e.g., quadrupole/ToF for MS-MS analysis) and even a Fourier transform ion cyclotron resonance (FT-ICR) instrument, with mass resolutions approaching 1 million or greater. These new designs will be briefly introduced in Chapters 4 and 8.

1.1.1   SIMS Imaging

In all SIMS instruments, mass spectrometric imaging can be achieved by focusing and rastering the ion beam over a selected area or by using secondary ion optical focusing elements (in the case of magnetic sector instruments), where the secondary ion intensity for a given mass-to-charge ratio (m/z) is monitored as a function of position on the sample. Examples of molecular, elemental, and isotopic mapping of components on surfaces are given in Figure 1.2b–d.[3,4]

1.1.2   SIMS Depth Profiling

SIMS can be utilized for both surface analysis (at low primary ion doses) and in-depth analysis (at high primary ion doses). An example of SIMS depth profiling is shown in Figure 1.3, which depicts the elemental intensities of Cr, Ni, and C,

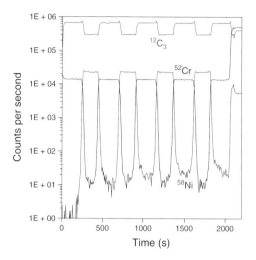

Figure 1.3 Example of SIMS depth profiling in Cr/Ni thin films. Data acquired from NIST SRM 2135a, containing nine alternating layers of Cr and Ni on silicon with nominal layer widths of 53 and 66 nm, respectively.

plotted as a function on increasing primary ion sputtering time in a sample containing Ni/Cr alternating layers. Each Ni and Cr layer is readily observed using SIMS depth profiling, as indicated by the inversely alternating Cr and Ni intensities.

Unlike inorganic samples, organic, polymeric, and biological materials have historically required the use of "static SIMS" analysis conditions, where the primary ion fluence is maintained at or below a critical dose in order to retain the surface in an undamaged state. This critical dose is defined as the "static limit," and is often reported to be at or less than 1×10^{13} ions/cm^2, depending on the sample and the ion beam employed. Unfortunately, this limitation results in decreased sensitivity and precludes compositional depth profiling in soft materials. One potential solution to this limitation is to use cluster or polyatomic primary ion beams (such as C_{60}^+, SF_5^+, or Ar_{700}^+) in place of atomic sources in order to extend the characterization of these samples beyond the static limit.

1.2 BASIC CLUSTER SIMS THEORY

When a cluster ion impacts a surface, the cluster breaks apart and each atom in the cluster retains only a fraction of the initial energy of the ion as described in the relationship shown below in Equation 1.1 (where E_c is the final energy of a constituent atom after collision with the surface, E_0 is the energy of the polyatomic ion before impact, M_c is the mass of the constituent, and M_t is the total mass of the polyatomic ion).[5]

$$E_c = E_0 \left(\frac{M_c}{M_t} \right) \tag{1.1}$$

Since the penetration depth of the ion is proportional to the impact energy of the ion, cluster ion bombardment results in a significant reduction in penetration depth of the ion. This causes surface-localized damage and consequently, preserves

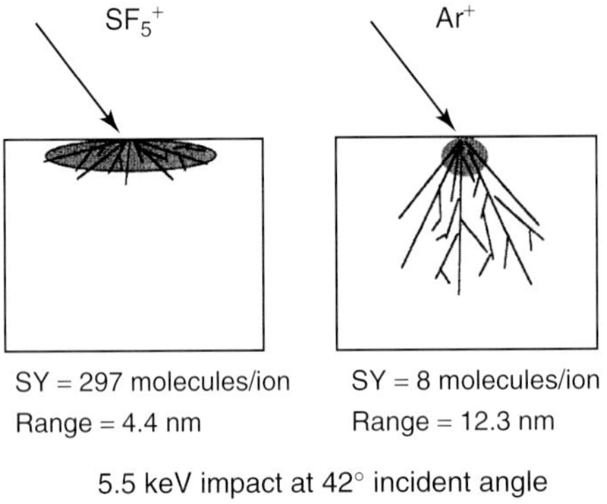

SY = 297 molecules/ion   SY = 8 molecules/ion
Range = 4.4 nm   Range = 12.3 nm

5.5 keV impact at 42° incident angle

Figure 1.4 Graphic illustration suggesting how the high sputter yields and low penetration depths observed with polyatomic ion bombardment may reduce the accumulation of beam-induced damage in an organic thin film. The actual SRIM calculations are indicated below each illustration, where SY represents the calculated sputter yield in a PMMA sample, and the range represents the depth of the projectile into the PMMA sample. Reproduced from Gillen and Roberson[1] with permission from Wiley.

the chemical structure in the subsurface region (Fig. 1.4).[5] Similar energy atomic beams, however, will penetrate deeply, resulting in the breaking of molecular bonds deep into the sample and thus precluding the ability to depth profile in molecular samples. Furthermore, because there are more atoms bombarding the sample simultaneously with cluster ions, the sputtering yield can be significantly enhanced. This is in part because of the increased number of atoms per ion, but is also a result of the formation of a high energy density "collisional spike" regime that is formed with cluster ion bombardment, causing nonlinear sputtering yield enhancements (i.e., sputtering yield of $C_n{}^+ \gg nC^+$).[6]

1.3 CLUSTER SIMS: AN EARLY HISTORY

1.3.1 Nonlinear Sputter Yield Enhancements

The benefits of utilizing polyatomic ions for sputtering was shown as early as 1960, with the observation of nonlinear enhancements in sputtering yields when using polyatomic ions as opposed to atomic ions.[7−10] An example of this nonlinear sputtering effect can be seen in Figure 1.5, which compares the sputtering yield per incoming atom when employing Te^+ ions as compared to $Te_2{}^+$ ions under an identical E_c (Eq. 1.1).[9] It can be seen from Figure 1.5, that the sputtering yield resultant from one $Te_2{}^+$ diatomic ion is greater than the combined sputtering yield from two Te^+ atomic ions of similar E_c.

Although these nonlinear effects were observed much earlier, the benefits of cluster sources (where cluster is defined here as an ion with more than two atoms) for SIMS applications were not realized until the mid to late 1980s. One of the earliest works was published in 1982, in which the authors compared the performance

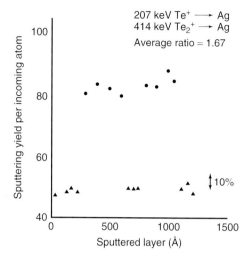

Figure 1.5 Sputtering yield per atom of a polycrystalline silver target using 207 keV Te$^+$ and 414 keV Te$_2$$^+$ ion bombardment as a function of sputtered layer thickness. Nonlinear effects are clearly observed. Reproduced from Anderson and Bay[9] with permission from the American Institute of Physics.

of siloxane molecular ions to Hg$^+$ ions for characterization of oligosaccharides in a glycerol matrix.[11] The results showed a large increase in the ionization of the organic molecules when employing the siloxane cluster source as compared to the atomic Hg$^+$ ion source.

Later, Appelhans et al. used SF$_6$ neutral beams to characterize electrically insulating polymer samples such as polytetrafluoroethylene (PTFE), poly(ethylene terephthalate) (PET), poly(methyl methacrylate) (PMMA), and polyphosphazene, where the authors found that the SF$_6$ cluster beam yielded 3−4 orders of magnitude more intense secondary ion yields from these polymer samples than equivalent energy atomic beams.[12−14] Similar findings were found in the mass spectra of pharmaceutical compounds.[13]

1.3.2 Molecular Depth Profiling

Another unique feature of cluster ion beams as compared to their atomic ion beam counterparts is their ability to retain molecular information as a function of depth in soft materials. The combination of increased sputter yields along with decreased subsurface damage has enabled the SIMS analyst to characterize compositions as a function of depth in organic materials for the first time; a process now referred to as *molecular depth profiling*. Cornett et al. were among the first to demonstrate the feasibility of molecular depth profiling with cluster ion beams, when they discovered that continued bombardment of protein samples with massive glycerol cluster ions yielded constant molecular secondary ion signals with increasing ion fluence, while the same samples irradiated with Xe$^+$ ions yielded the characteristic rapid signal decay that is commonly associated with atomic beams.[15]

An example of molecular depth profiling is demonstrated in Figure 1.6, which shows an early attempt at depth profiling in a thin glutamate film (180 nm) vapor

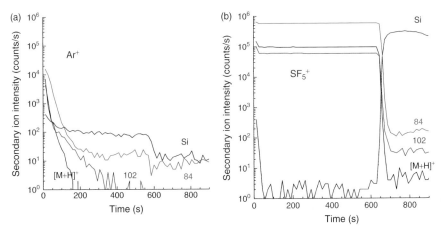

Figure 1.6 Comparison of depth profiles obtained from a 180 nm vapor-deposited glutamate film using (a) Ar^+ and (b) SF_5^+ primary ions under dynamic SIMS conditions. The SF_5^+ primary ion dose required to reach the silicon was 2.4×10^{15} ions/cm^2. Reproduced from Gillen and Roberson[1] with permission from Wiley.

deposited onto a Si substrate.[5] In this example, glutamate molecular ion signal intensities $[M + H]^+$, and fragment ion intensities (m/z 84 and m/z 102) are measured as a function of sputtering time, using both Ar^+ (Fig. 1.6a) and SF_5^+ polyatomic primary ions (Fig. 1.6b). Si^+ ion intensities (m/z 28) were also measured as a function of sputtering time in both examples. When employing the Ar^+ monatomic ions, the molecular signals decay rapidly, as is characteristic of atomic ion bombardment in molecular films. However, when employing polyatomic primary ion sources, the molecular ion and fragment ion intensities of the glutamate remain constant throughout the entire depth of the film. In addition, while the SF_5^+ source was able to profile through the entire film in the 900 s sputter time interval, as indicated by the decreasing molecular ion signal intensities with commensurate increases in the Si, the Ar^+ was unable to sputter through the material during the allotted time interval.

1.4 RECENT DEVELOPMENTS

Since the advent of cluster SIMS, there has been an abundance of work on surface and in-depth characterization of soft materials ranging from simple molecular films and polymers[16,17] to complex biological systems.[2] Cluster primary ion sources such as C_{60}^+, Au_3^+, SF_5^+, Bi_3^+, and $Ar_{(x>500)}^{n+1}$ have resulted in significant improvements (typically >1000-fold) in characteristic molecular secondary ion yields and decreased beam-induced damage. Furthermore, most of these sources have allowed for molecular depth profiling in samples; a feat that was unheard of with previously employed monatomic ion beam sources. With these new cluster sources, beam damage limitations have all but been removed for depth profiling in most organic and polymeric materials. With the increased sensitivity, nanoscale depth resolution (<5 nm), and submicrometer lateral resolution, cluster SIMS is a promising new characterization tool enabling high resolution three-dimensional imaging capabilities for organic and polymeric-based materials (Fig. 1.7 and Fig. 1.8).[16,18]

1.5 ABOUT THIS BOOK

This book will serve as a compendium of knowledge on the topic of cluster SIMS. In this book, in-depth discussions on the various aspects of cluster SIMS and its applications will be presented—from the details of cluster SIMS theory and erosion dynamics, to experimental parameters for optimum depth profiling in molecular samples.

Theoretical discussions regarding cluster ion beam interactions with organic materials will be discussed in Chapter 2, where important aspects of molecular dynamics simulations will be reviewed. This chapter will review the current state of the literature in this field, as well as help one to obtain a better understanding of the physics of cluster ion bombardment in organic, polymeric, and biological samples.

Chapter 3 presents a detailed overview of the myriad of sources that are available, for SIMS, cluster ion beams, or otherwise. This chapter will provide

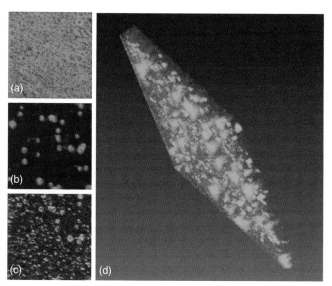

Figure 1.7   Positive secondary ion image maps (100 × 100 µm) of characteristic tetracycline signal (m/z 59) in a PLGA film, acquired using an SF_5^+ sputtering source in conjunction with a Bi_3^+ analysis source. (a) No sputtering, (b) 15 s sputtering with SF_5^+ (~75 nm depth), (c) 75 s sputtering with SF_5^+ (~375 nm depth), and (d) 3D volumetric representation of tetracycline signal (m/z 59) in PLGA film (acquired from approximately the top 2 µm) containing 15% tetracycline; 5 keV SF_5^+ beam energy, operated at 4 nA continuous current and a 500 × 500 µm raster.

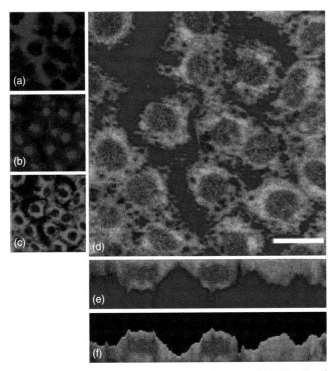

Figure 1.8 (a–d) Two-dimensional (2D) images of NRK cells after the forty-fifth sputter cycle. Summed signals of amino acid fragment ions are represented in red (b), those of phospholipids in green (c), and substrate-derived secondary ions are depicted in blue (a). (d) An overlay of the three images. The scale bar in (d) corresponds to 20 μm. (e) and (f) Vertical xz sections through the sample. Data acquired using C_{60}^{+} sputtering in conjunction with Bi_3^{+} analysis. Reproduced from Breitenstein et al.[18] with permission from Wiley.

information about how these various sources function, what they are used for, and the benefits and disadvantages of each.

Chapters 4 and 5 will provide a comprehensive review of the literature regarding the surface characterization and in-depth analysis of soft materials with cluster SIMS. Chapter 4 will describe the important aspects that need to be considered during any static SIMS experiment employing cluster sources (i.e., the best source, the experimental conditions, etc.). A similar approach will be taken in Chapter 5, which will provide a summary of molecular depth profiling. Both the physics and the chemistry of cluster ion bombardment will be discussed in detail, with the introduction of erosion dynamics theory and a brief description of ion beam irradiation chemistries.

Three-dimensional imaging in soft materials is the ultimate goal in molecular depth profiling. This topic will be introduced in Chapter 6, which will serve as both a review of the literature, and a tutorial for 3D imaging. There are, in particular, many important considerations and corrections that are required in order to obtain accurate representations of 3D SIMS image data. Many of these considerations will be discussed here. Furthermore, this chapter serves as a guide for practical molecular depth profiling and analysis with cluster ion beams, discussing how one should make precise and accurate measurements of depth resolution, damage cross sections and efficiencies, beam conditions, and sputtering rates. The authors will discuss these measurements and more; defining the rules for different scenarios (i.e., organic/organic layers vs organic/inorganic layers), and identifying how and what should be reported in each of these scenarios.

Chapters 7 and 8 will discuss special applications of cluster SIMS for characterization of inorganic materials and biological materials, respectively. Chapter 8 will discuss in detail, the special case of biological samples. Biological materials and cells are particularly challenging and complex, and therefore need special consideration. This chapter will help the reader to better understand the successes and challenges for surface characterization and in-depth analysis of biological samples, and will serve as a detailed review of the field, displaying brilliant 3D molecular images in cells and other biological samples. Finally, all these discussions are wrapped up in Chapter 9, which briefly gives a perspective on what the future holds for the technique of cluster SIMS.

ACKNOWLEDGMENT

The authors would like to acknowledge Kenneth McDermott from the Food and Drug Administration for the provision of samples for analysis in Figure 1.7.

REFERENCES

1. Smith, D. F.; Robinson, E. W.; Tolmachev, A. V.; Heeren, R.; Pasa-Tolic, L. *Anal. Chem.* **2011**, 83, 9552–9556.
2. Fletcher, J. S.; Vickerman, J. C. *Anal. Bioanal. Chem.* **2010**, 396 (1), 85–104.
3. Moore, K. L.; Schroder, M.; Wu, Z.; Martin, B. G. H.; Hawes, C. R.; McGrath, S. P.; Hawkesford, M. J.; Ma, J. F.; Zhao, F. J.; Grovenor, C. R. M. *Plant Physiol.* **2011**, 156 (2), 913–924.
4. Kilburn, M. R. *Isotopic Imaging of Bacteria Grown in 15N Medium*. http://www.ammrf.org.au (accessed Nov 22, 2012).
5. Gillen, G.; Roberson, S. *Rapid Commun. Mass Spectrom.* **1998**, 12, 1303–1312.
6. Sigmund, P.; Claussen, C. *J. Appl. Phys.* **1981**, 52 (2), 990–993.
7. Gronlund, F.; Moore, W. J. *J. Chem. Phys.* **1960**, 32 (5), 1540–1545.
8. Andersen, H. H.; Bay, H. L. *J. Appl. Phys.* **1975**, 46 (6), 2416–2422.
9. Andersen, H. H.; Bay, H. L. *J. Appl. Phys.* **1974**, 45 (2), 953–954.
10. Thompson, D. A.; Johar, S. S. *Appl. Phys. Lett.* **1979**, 34 (5), 342–345.
11. Wong, S. S.; Stoll, R.; Rollgen, F. W. *Zeitschrift fur Naturforschung. A J. Phys. Sci.* **1982**, 37 (7), 718–719.
12. Appelhans, A. D.; Delmore, J. E.; Dahl, D. A. *Anal. Chem.* **1987**, 59 (13), 1685–1691.
13. Appelhans, A. D.; Delmore, J. E. *Anal. Chem.* **1989**, 61 (10), 1087–1093.
14. Appelhans, A. D. *Int. J. Mass Spectrom.* **1989**, 88 (2–3), 161–173.
15. Cornett, D. S.; Lee, T. D.; Mahoney, J. F. *Rapid Commun. Mass Spectrom.* **1994**, 8 (12), 996–1000.
16. Mahoney, C. M. *Mass Spectrom. Rev.* **2010**, 29 (2), 247–293.
17. Wucher, A.; Cheng, J.; Winograd, N. *Appl. Surf. Sci.* **2008**, 255 (4), 959–961.
18. Breitenstein, D.; Rommel, C. E.; Mollers, R.; Wegener, J.; Hagenhoff, B. *Angew. Chem. Int. Ed.* **2007**, 46 (28), 5332–5335.

CHAPTER 2

CLUSTER SIMS OF ORGANIC MATERIALS: THEORETICAL INSIGHTS

Arnaud Delcorte, Oscar A. Restrepo, and Bartlomiej Czerwinski

2.1 INTRODUCTION

> I am ashamed to tell you to how many figures I carried these computations, having no other business at the time.

Sir Isaac Newton

By definition, clusters are "small, multiatom particles." The upper size limit of clusters is reached when the number of atoms is sufficient to reproduce the physical properties of the condensed matter, such as the band structure. Clusters can be made of a collection of atoms or molecules, and of any element in the periodic table—from hydrogen or noble gases to heavy metals. Adding that to the variety of possible surfaces and energy ranges, it becomes clear that an exhaustive theoretical description of energetic cluster-surface interactions constitutes a serious endeavor, apt to mobilize the workforce of generations of researchers. Although the range of cluster sources used for secondary ion mass spectrometry (SIMS) is restricted by technological considerations, the initial domain of commercially available, relatively small heavy metal and light-element cluster sources such as Au_n^+ and C_{60}^+ is soon to be overcome and one reads more and more reports concerning other types of projectiles, such as massive noble gas (Ar), metal (Au), or molecular (water) clusters.[1]

The study of kiloelectronvolt cluster impacts is deeply rooted in classical physics, as far back as Galileo and Newton, who were able to define the notions of

Cluster Secondary Ion Mass Spectrometry: Principles and Applications, First Edition.
Edited by Christine M. Mahoney.
© 2013 John Wiley & Sons, Inc. Published 2013 by John Wiley & Sons, Inc.

momentum and energy and the way these physical quantities can be exchanged or transferred in collisions. Microscopic cluster-surface interactions bear resemblances with macroscopic phenomena, such as meteoric impacts on celestial bodies, bullet penetration in a target, or even rain droplets splashing on the surface of a pond. These analogies stimulate researchers to propose phenomenological models based on similar concepts. For Au clusters impinging on Au surfaces with typical meteoroid velocities ($\sim$22 km/s), the limit between microscopic and macroscopic impact behaviors was identified for cluster nuclearities between 1000 and 10,000.[2]

The state-of-the-art theoretical approaches used for the explanation of cluster interactions with surfaces involve analytical models as well as computer simulations. The analytical formulas resulting from hydrodynamic models are sometimes even coupled with initial molecular dynamics (MD) simulations to predict effects that would be too long to treat with the sole use of MD computer codes (Section 2.3). Nowadays, the models provide a good description of the sputtering (or desorption) process for an ever-increasing number of systems and initial conditions (cluster nature, surface material, projectile energy, and angle). However, one must acknowledge that the detailed understanding of ionization processes of molecules and fragments upon cluster bombardment, and in turn the prediction of ionized fractions, remains out of reach. As was stated in a recent review article on the physics of surface-based organic mass spectrometry, the relatively small number of particles sputtered per impact (10^3 or less), combined with the low measured ion fractions (10^{-3} to 10^{-5}), make any theoretical prediction in that field very difficult.[3] Indeed, assuming that the right physics was in the model, "hundreds of trajectories,* each of weeks to months, would have to be performed in order to make comparisons with experimental distributions."

In this chapter, we will discuss the case of organic and related materials, with a few examples taken from other types of systems when deemed necessary. From the theoretical point of view, cluster-induced sputtering of organic materials offers a particular challenge because one wishes to correctly describe not only the overall dynamics of the bombarded systems, but also predict the fate of each and every sputtered molecule (are they ejected? do they survive or fragment?). Indeed, the ultimate result of an organic SIMS experiment is a complex collection of charged atoms, fragments, intact molecules, recombination products, and ... clusters. And organics are often fragile materials that like to do chemistry. The problem worsens if one needs to model the results of multiple overlapping impacts, in which induced roughness, chemical modification and damage mechanisms, and long-term relaxation effects may play an important role, as is observed upon molecular depth profiling of polymers. The models are not yet able to handle the large complexity of the latter, but forays in the right direction exist.

Keeping the above-mentioned caveats in mind, this chapter will focus on problems of cluster–surface interactions that the theory could successfully tackle. These include several important issues for scientists working in the field of cluster

---

*In MD simulation terminology, a "trajectory" corresponds to the time evolution of the system for a given set of initial conditions, different trajectories corresponding in turn to different sets of initial conditions.

SIMS, such as the difference in behavior between atomic, small cluster and massive cluster projectiles, and the prediction of sputtering yields, the identification of some induced structural and chemical effects, or the first results on repeated cluster bombardment of materials. The main part of the chapter (Section 2.2) describes the methodology and results of MD simulations, while the following section (Section 2.3) bridges the gap with analytical models and highlights some recent hybrid theoretical approaches.

2.2 MOLECULAR DYNAMICS SIMULATIONS OF SPUTTERING WITH CLUSTERS

2.2.1 The Cluster Effect

The first concept that comes to mind concerning cluster impacts on surfaces is nonlinearity. The word has been used many times and with various meanings. Because SIMS is mostly concerned with sputtering and secondary ion yields, nonlinearity of the yield is the phrase that first triggered excitement in the community. Strictly, there is nonlinearity of the sputtering yield Y when the yield generated by a cluster of X_n (with n constituents) is larger than the yield produced by n atoms of the element X, with the same energy per atom (or velocity). There are many reports showing the nonlinearity of the sputtering yields, especially for metal clusters and inorganic surfaces.[4,5] However, this definition of the "cluster effect" is not very tractable for SIMS practitioners, even sometimes confusing, and it is frustrating for physicists because it does not say anything about the fundamentals of the interaction. What the SIMS practitioners usually compare are different beams with sometimes the same energy (and sometimes not), and what they are interested in is to have the highest yield of characteristic ions of the surface. For organic materials, it is either the molecular ion or, in the case of polymers, the fingerprint ions including the repeat unit of the structure. Fortunately, the reports usually indicate that the sputtering yields and the characteristic ion yields of organic materials at a given projectile energy are drastically enhanced by the use of clusters (Au_n, Bi_n, SF_5^+, C_{60}^+, Ar_n^+) in comparison with atomic projectiles (Ga^+, Ar^+, Xe^+, In^+).[6-10] For instance, several reports indicated that at the same acceleration energy, C_{60}^+ ions produce $10^2 - 10^3$ more characteristic ions than Ga^+.[11,12]

For the physicist, nonlinearity may have a different meaning. Indeed, in the original theory of sputtering proposed by Sigmund, a long chapter is devoted to linear collision cascades,[13,14] as opposed to overlapping (nonlinear) cascades or spikes.[15] Linear collision cascades are based on the binary collision approximation, meaning that atoms in the impacted region of the solid interact only two by two and so successive generations of recoil atoms are gently created, with a minimum overlap at a given time. The other situation is one where many atoms move together in a small energized region of the surface, and the description of the interaction must necessarily be many-body. Usually, this situation leads to correlated motion of atoms, pressure waves, and formation of craters.[16] This distinction between linear cascades and collective, nonlinear behavior constitutes a more physical way to describe the cluster effect. An example is shown in the snapshots of Figure 2.1, taken from the MD simulations of the impact of C_{60} on two different substrates, one metallic and one organic, carrying large molecules (polystyrene, PS, with a mass of 7.5 kDa).[3,17] In both cases, the collective nature of the interaction is shown by the huge number of atoms displaced in a correlated manner, inducing the formation of a crater with a spherical geometry around the impact point and the emission of large quantities of matter, including the PS molecules. The characteristic dimension of the disturbed region is about 10 nm.

The initial interactions of the ion beam with a surface provide particularly important information concerning the physics at play and substantiates the distinction made earlier between linear and nonlinear cascades. Figure 2.2, for example, shows the penetration of the projectile (white) and the trajectories of the recoil atoms (colored) propagating in a PS solid with more than 10 eV of kinetic energy over the first 200 fs, which is about 2 orders of magnitude faster than the processes shown in Figure 2.1.[18] The penetration of a 10 keV Ar atom in PS induces a dilute cascade, with well-separated branching points and, essentially, binary interactions, at least in the considered time frame. In contrast, isoenergetic C_{60} creates a comparatively small energized volume where all the atom tracks overlap and many-body interactions are frequent, if not the rule. A few observations must be noted here. First, the Ar atom trajectory is, for the most part, not deflected by the interaction with the target atoms over several nanometers and, therefore, it implants deep in the sample. Second, subcascades are created in depth, which means damage to the underlying PS molecules. Third, some fast recoil atoms intercept the surface plane as far as 6–7 nm from the impact point, which gives an idea of the lateral extension of the sputtering event. In contrast, the carbon atoms of C_{60} stop in the top 2 nm of the surface, with only one recoil atom implanted significantly below the disturbed volume, and "fast" ejection is concentrated in a zone of ∼4 nm around the impact. This early action, with the rapid formation of a very high energy density region, constitutes the "big bang" of the cluster sputtering event and the seed of the correlated radial motions and massive emission yields observed at later times

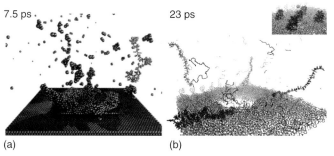

7.5 ps   23 ps

(a)   (b)

Figure 2.1   Fullerene-induced molecular emission. (a) Desorption of a PS 61-mer (red spheres) from a Ag surface (blue spheres) due to 20-keV C_{60} bombardment (7.5 ps).[3] (b) Desorption of 4 PS 61-mers (red) from a polyethylene substrate (blue) bombarded by 10-keV C_{60} (23 ps).[17] Figure 2.1a reprinted from Garrison and Postawa,[3] with permission from Wiley.

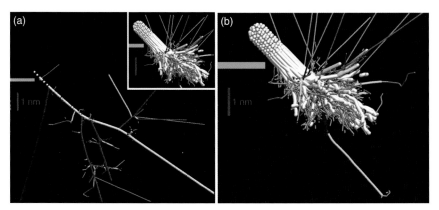

Figure 2.2 Collision cascades in a PS tetramer sample.[18] Side views showing the successive positions of the projectile (white) and recoil atoms (colored) as a function of time up to 200 fs. The atom positions are represented every 25 time steps of the simulation, that is, with time intervals in the range 0.2–2 fs between two positions because of the variable time step. The recoil atoms are color-coded as a function of their kinetic energy: red, greater than 100 eV; yellow, between 50 and 100 eV; blue, between 10 and 50 eV. Except for the recoil atoms with greater than 10 eV of kinetic energy, the atoms of the sample are omitted for clarity. The horizontal position of the sample–vacuum interface before impact is signified by the cyan line segment. (a) 10-keV Ar → PS. (b) 10-keV C_{60} → PS. Recreated from Delcorte and Garrison,[18] with permission from Elsevier.

(Fig. 2.1). To be complete, one must say that if virtually all the cluster impacts lead to "cascade nonlinearity," not all atomic impacts lead to linear cascades. As was shown in several studies, even with atomic projectiles there is a percentage of impacts that lead to clearly overlapping cascades, depending on the nature and energy of the projectile, the nature of the target, and the exact impact point.[19] Therefore, this classification in terms of linearity and nonlinearity must really be used with caution.

Another related difference between atomic and cluster impacts which characterizes the cluster effect concerns the fluctuations of the sputtering yields, originating in atomic SIMS from a variety of recoil track patterns. This effect is illustrated in Figure 2.3 for a series of 15 impacts of Ga and C_{60} on a benzene-covered silver substrate.[3] While the number of sputtered molecules varies between ∼20 and 450 depending on the impact point for Ga bombardment, it is almost constant for C_{60} bombardment. The intuitive explanation is that the buckyball, acting as a unit, does not "see" the detailed structure of the surface anymore because of its large size (7 Å—more than twice the interdistance between the Ag atoms of the substrate and four times the length of a C–C bond). In contrast, for gallium, the collision cascade and the number of sputtered particles are completely different, whether the projectile undergoes a head-on collision in the surface layer or channels deep into the target. Heavy metal clusters, such as Au_3 or Bi_3, fall in between these two extremes. This observation is not trivial because the idea of "blurring" the details of the projectile and target atomistic structures, that is, going

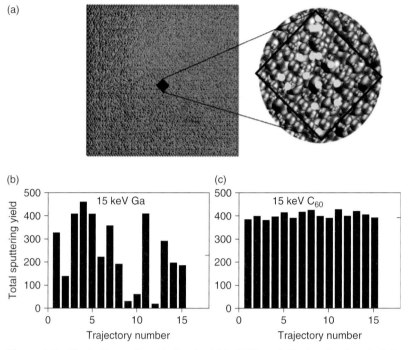

Figure 2.3 Fluctuations of sputtering for 15-keV Ga and C_{60} bombardment of three layers of benzene (red) on Ag (blue).[3] (a) Top view of the sample with a zoom showing the 15 aiming points (yellow). (b, c) Sputtering yield (number of ejected benzene molecules) corresponding to each aiming point for 15 keV Ga (b) and C_{60} bombardment (c). Reprinted from Garrison and Postawa,[3] with permission from Wiley.

from a discrete to a continuous view, has strong implications on the physical reality of the interaction and, in turn, on the choice of models that are appropriate to describe it. For example, one moves from stochastic collision cascades, best modeled as isolated events by Monte-Carlo or MD calculations, to collective motions and shock wave effects that can be described by hydrodynamic and continuum mechanics models (Section 2.3). In the case of MD, this effect allows the theoretician to compute a much lower number of trajectories to obtain a statistically relevant sample of the interactions with clusters than with atomic projectiles.

In the past decade, small cluster interactions with surfaces have been primarily modeled using classical MD. In the usual energy range of SIMS experiments, a variety of targets and projectiles have been described and the picture hidden in this vast jig-saw puzzle begins to unveil. The rest of this section is devoted to MD simulations of these systems, after a brief introduction of the model itself.

2.2.2 Computer Simulations and the Molecular Dynamics "Experiment"

In the energy range used in SIMS, one can usually distinguish two main stages of interaction. The first one (Fig. 2.2), corresponding to the first hundreds of femtoseconds of the projectile–solid interaction, pertains to high energy collisions in which the bond energies are small with respect to the translational energy of the impinging atoms ($\sim$5 eV vs 250 eV for 15-keV C_{60}). In this stage, the atoms of the system behave like marbles or billiard balls and the nature of the bonds is irrelevant. For atomic bombardment, in which the collision cascades are often (but not always) dilute, this stage of the interaction can be reasonably modeled within the binary collision approximation (BCA) using purely repulsive interatomic potentials, as was prescribed in the classical sputtering theory. This is the theoretical framework implemented in simulation codes such as SRIM[20,21] and Marlowe.[22] However, the BCA, as the basis of the physics implemented in these codes, is no longer valid for cluster bombardment, where collective effects including many-body interactions and correlated motions of atoms already dominate in the high energy stage (Fig. 2.2b). For cluster impacts, the BCA-based simulation codes, treating only one-to-one collisions in a sequential manner, must be replaced by another scheme that is able to adequately model interactions between many partners at each time, such as MD simulations. In the late stages of the interaction, when the energy of the projectile and recoil atoms is of the order of magnitude of the bond energies in the solid, the organic sample behaves like the classical beads-and-spring system. Predicting the details of these late events, in which fragments and entire molecules are emitted from organic surfaces, requires a realistic description of many-body interactions for both atomic and cluster bombardment.

A MD simulation constitutes a very well-controlled virtual experiment occurring in the computer. Classical mechanics being completely deterministic, repeating the "experiment" with the same initial conditions will produce the exact same result. It is also an experiment in the sense that the direct observations, for example, quantities of emitted species, amount of damage in the system, size of the formed craters, must be "interpreted" in order to elaborate meaningful physical explanations. In other words, a real understanding can only be attained if the results of the motions of millions of atoms in the simulation can be expressed in simpler rules that relate to known physical processes and/or properties. At variance with real experiments, however, any detail of the interaction that is required to explain the observations is present and can be retrieved from the computer. It is just a matter of time and perseverance.

At each time step of the MD calculation, the forces between the different constituents of the system are calculated from the atomic positions and the interaction potentials. Then Hamilton's equations of motion are integrated to determine the position and velocity of each particle at the following time step.[23,24] In addition to reproducing the behavior of the system for a wide range of energies, from kiloelectronvolt down to subelectronvolt, the MD potentials must provide a correct description of bond scission and bond creation processes. Current algorithms use a sophisticated set of semiempirical and many-body interaction potentials from which the energy and forces in the system are calculated. These potentials usually blend a strongly repulsive low distance wall, accounting for the rigid ball behavior in the beginning of the interaction, with a more complex function including a repulsive and an attractive part at larger distance—the springs or glue mimicking the solid behavior in the subsequent stages of the interaction. For hydrocarbon samples, the C–C, C–H, and H–H interactions are well described by the adaptive intermolecular reactive empirical bond-order (AIREBO) potential.[25] This potential is based on the reactive empirical bond-order (REBO) potential developed by Brenner for hydrocarbon molecules[26–28] and includes nonbonding intermolecular interactions through an adaptive treatment that conserves the reactivity of the REBO potential. It is important to remember that electronic excitations and charge exchange processes are not described by the model, which is classical in nature. Therefore, one should be particularly careful when comparing simulation results to SIMS data. However, the effect of the valence electrons in the bonding chemistry is implicitly taken into account via the interaction potentials.

One important practical issue when performing MD simulations is computer time. Reducing the computer time in order to tackle larger and/or more complex calculations requires a constant effort of computational scientists and, like the memory available for data storage, the computation speed never seems to be sufficient. In sputtering simulations, as much as the specific problem to be addressed, the feasibility of the calculation within a given limit of computer time, commensurable with human patience, often governs more fundamental choices such as the maximum energy of the projectile and the maximum size of the system, the selection of the potentials, or the level of approximation of the model. A discussion of this issue is provided in Reference 3. Several examples of schemes devised to reduce the computational expense will be given along this chapter. Coarse-graining constitutes one useful simplification that was introduced for the description of organic sample sputtering by fullerene projectiles.[29] In these simulations, certain atoms, such as CH, CH_2, and CH_3 residues, are grouped to form united atoms or particles. The advantages of such a coarse-grained approach is that there are fewer particles, the potentials are usually simpler and thus quicker to calculate, and the fast H-vibration is eliminated, which allows for a larger time step to be used in the integration. Of course, all of this is at the expense of the accuracy of the description, especially where the detailed chemistry is a concern, and care must be taken in the interpretation of the results.

2.2.3 Light and Heavy Element Clusters, and the Importance of Mass Matching

The most common cluster projectiles used for SIMS nowadays are either fullerene molecular ions (C_{60}^+), massive Ar_n^+ gas clusters (where $n > 500$), and heavy metal trimers and pentamers (Au_3^+, Bi_3^+, Bi_5^+) or SF_5^+, which constitute an intermediate case. Therefore, it is important to understand the differences between these clusters in terms of sputtering. MD simulations of the bombardment of water ice by 5-keV Au_3 and C_{60} clusters showed that these two projectiles have a very different penetration depth in the solid (Fig. 2.4).[30] The atoms of Au_3 keep a correlated motion over a distance of 5–6 nm inside the ice and they slowly diverge afterward to finally stop at a depth of ∼10 nm under the surface. In contrast, C_{60} explodes in the top 2 nm of the ice target and the carbon atoms do not bury deeper than 4 nm, in agreement with the results presented in Figure 2.2 for a PS sample. The important physical parameter here is not the total energy or the total mass of

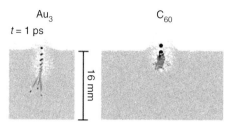

Figure 2.4 Gold cluster and fullerene impacts.[30] Vertical cross-sections (2 nm) of a water ice sample bombarded by Au_3 and C_{60} showing the successive positions of the projectile constituents up to 1 ps. Reprinted from Russo and Garrison,[30] with permission from ACS publications.

the projectile, which are comparable, but rather the energy and the momentum of the constituent atoms. They are respectively ~ 20 and 18 times smaller in the case of C_{60}.

In addition, the energy transfer efficiency in a collision is strongly dependent on the masses of the colliding particles. For instance, the maximum energy transfer factor in a binary collision between two atoms of masses m_1 and m_2 is $4 \times m_2/m_1 \times (1 + m_2/m_1)^{-2}$ which amounts to 0.98 for C impacting O but only 0.28 for Au impacting O (the numbers are similar for all C-, O-, and N-based organic materials). Therefore, heavy Au and Bi atoms can transfer proportionally less energy per collision in the target. As a result of these two factors, larger energy per atom and less effective energy transfer per collision, small heavy metal clusters are almost not deflected in their collisions with the water molecules and then travel particularly far into light-element samples. The results are somewhat different for targets made of heavy elements such as silver or gold.[31] In this case, Au atoms transfer their energy better than C atoms and C atoms impinging on metal atoms are deflected with larger angles (more backscattering).

For water ice and organic targets, the final result is a deeper and narrower crater in the case of Au_3 bombardment, with a sputtering yield that is generally lower. For gold targets bombarded by Bi_3 and C_{60} with energies between 5 and 25 keV, our results indicate that both types of projectiles create hemispherical craters, with slightly larger sizes and higher sputtering yields for Bi_3.[32] When a 2.5 nm film of ice deposited on a Ag crystal is bombarded by 15 keV projectiles, the effects combine, with more H_2O molecules emitted and a wider crater developing in the ice layer for C_{60}, and a crater extending deeper in the Ag substrate with Au_3.[33] The same study clearly showed that C_{60} is more surface sensitive than Au_{1-3}.

2.2.4 Structural Effects in Organic Materials

2.2.4.1 Amorphous Molecular Solids and Polymers

Bulk organic solids and polymers constitute an important field of applications for SIMS analysis and 3D molecular imaging.[1] In addition, it is the family of materials for which the yield enhancement provided by clusters, especially fullerenes, with respect to atomic ions seems to be the largest.[12] Therefore, understanding the physics of cluster bombardment of these materials is a major goal for MD simulations of sputtering.

Figure 2.5 shows a visual rendition of the crater development induced by 5 keV C_{60} in an amorphous polymeric sample made of interlaced chains of 1.4 kDa each.[34] Only the atoms and chain segments that end up above the original surface plane at the end of the trajectory (20 ps) are displayed. The color coding indicates the radial localization of the molecules with respect to the impact point at the beginning of the interaction, cyan being close to the impact site and red being at the periphery. Figure 2.5a–c indicates that the polymer chains initially located in the 5 nm wide impact region (cyan) are ejected by the fullerene. In contrast, many of the chains constituting the purple–red outer ring redeposit around the crater at 20 ps. The movement of radial expansion and redeposition of the chains

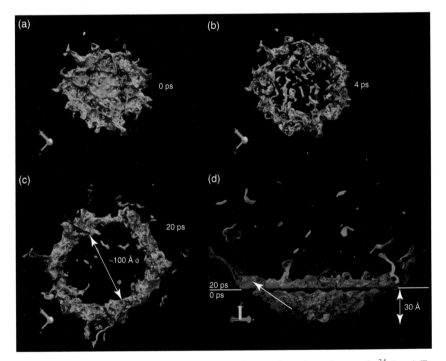

Figure 2.5   Five kiloelectronvolts C_{60} bombardment of a polymeric sample.[34] (a–c) Top views of the bombarded sample showing the time evolution, between 0 and 20 ps, of the material that ends up above the surface plane at the end of the sputtering event (20 ps). This representation provides a direct indication of the origin of the material that is sputtered and/or redeposited in the crater rim. Color coding is from cyan to red from the center (impact point) to the periphery at 0 ps. (d) Side view of the same at 0 ps (below the surface plane) and 20 ps (above the surface plane), showing the upward and outward motion of the bombarded polymer (white arrow). The surface plane is signified by the white line. Reprinted from Delcorte and Garrison,[34] with permission from ACS publications.

is illustrated by the white arrow in the side view of Figure 2.5d. In the end, the blue–purple molecules form the rim of the crater above the red molecules initially present at the periphery. While some entire molecules are emitted, others remain bound to the surface. Although the end crater is 5 nm deep, Figure 2.5d shows that the sputtered and rim material only come from the top 3 nm, while the bottom is compressed into a higher density material. A similar phenomenon has been observed for C_{60} bombardment of water and benzene ices.[3,30]

The nature of the emission process is well described by the velocity vectors of the ejecting molecules and fragments. In Figure 2.6a and b, horizontal and vertical projections of the velocity vectors of the sputtered molecules are presented for a 5 keV fullerene impact on an amorphous icosane sample. The

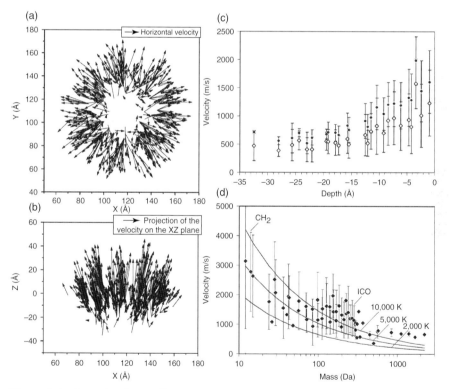

Figure 2.6 Sputtering by C_{60}.[34] Horizontal (a) and vertical (b) projections of the velocity vectors of the molecules ejected upon 5-keV C_{60} bombardment of icosane. The origins of the vectors are placed at the position of the center-of-mass of the molecules before impact. (c) Average total velocity (full diamonds) and vertical component (z) of the velocity vectors (open diamonds) of intact molecules as a function of their depth of origin. (d) Average total velocity of the sputtered species as a function of mass. The gray lines signify the calculated average velocities for Maxwell–Boltzmann distributions (at 2000, 5000, and 10,000 K). Reprinted from Delcorte and Garrison,[34] with permission from ACS publications.

orientation and magnitude of the velocity vectors are correlated with their initial position with respect to the impact point and the motion is essentially upward and outward, in agreement with the crater formation dynamics shown in Figure 2.5. Figure 2.6c also shows that the molecules originating in the depth of the crater have, on average, smaller velocities than those emitted from the topmost layers. A dependence of the axial velocity on the depth of origin was also observed in the sputtering of rare gases[35] and water ice.[36] Finally, for such linear hydrocarbon molecules, the velocities tend to decrease with increasing fragment mass (Fig. 2.6d). Typical velocities for CH_2 species are around 3000 m/s, whereas entire molecules, molecular clusters, and large molecular fragments are slower than 1000 m/s. In Figure 2.6d, the full lines show the average velocities $<v>$ that would be expected for Maxwell–Boltzmann distributions of particles with the corresponding mass and temperature. The very high temperatures indicate that the ejected molecules and fragments arise from a largely superheated nanovolume. Figure 2.5 and Figure 2.6 confirm the mesoscopic nature of the cluster sputtering process for organic materials, with molecules flowing collectively from the surface in an out-of-equilibrium process characterized by short times and high energy gradients.

The evolutions of the crater sizes and the sputtered quantities as a function of energy between 1 and 10 keV are summarized in Table 2.1. The crater volume increases almost linearly in this energy range, with a rate of $\sim$50 nm^3/keV. The variations of the quantities of polymer that are sputtered or redeposited in the crater rim are more complex. Between the onset of sputtering and 2 keV, the sputtered mass increases in a nonlinear manner, while the evolution beyond 2 keV is linear. In contrast, the mass of matter that is above the surface (sputtered + rim) at the end of the trajectories increases almost exponentially. As a result, the ratio of sputtered versus (sputtered + rim) quantities first increases to reach a maximum at 2 keV and decreases slowly afterward. The energy of 2 keV, corresponding to the maximum of this ratio, can therefore be considered as the energy for which the sputtering is

TABLE 2.1   Energy Dependence of the Crater and Sputtering Characteristics of Polymeric Samples Bombarded by C_{60}

C_{60} Energy	1 keV	2 keV	5 keV	10 keV
Crater Ø (Å)	56	76	102	124
Crater + rim Ø (Å)	94	128	161	217
Crater depth (Å)	25	33	49	62
Sputtered mass (Da) [a]	1135	10,910	24,689	55,628
Sputtered + rim (Da) [b]	9674	26,694	68,699	219,697
Ratio [a]/[b]	0.12	0.41	0.36	0.25
Sputtered molecules	0	3	7	19 (2 dimers)

Line 2 (crater + rim Ø) indicates the diameter of the crater including its rim. Line 5 (sputtered + rim) corresponds to the cumulated mass of the sputtered material and the material contained in the crater rim, above the surface.
In all cases, the values have been obtained for a single trajectory.
Reprinted from Delcorte and Garrison,[34] with permission from ACS publications.

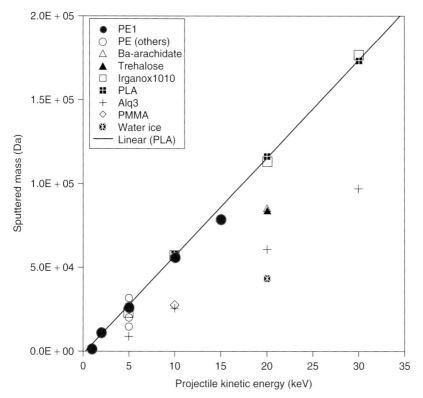

Figure 2.7 Sputtered mass (Da) per C_{60} impact as a function of the projectile energy.[34] The values calculated from MD simulations of hydrocarbon molecular samples are signified by full red ($C_{99}H_{200}$) and open circles ($C_{20}H_{42}$, $C_{249}H_{500}$, and $C_{1000}H_{2002}$). The experimental values are obtained from the literature (see text for details). Reprinted from Delcorte and Garrison,[34] with permission from ACS publications.

the most efficient. Over the considered range of energies, the total sputtered mass always remains inferior to the mass of material displaced to form the rim.

In Figure 2.7, the computed sputtered masses are compared with experimental values obtained from the literature. Between 1 and 15 keV, that is, the energy range investigated in the simulations, the mass sputtered upon C_{60} bombardment of an amorphous sample of $C_{99}H_{200}$ is similar to the experimental values measured by Shard et al. for the polymer additive, Irganox 1010 (1177 Da) and for polylactic acid.[37] The other experimental values are either slightly or significantly under the line, and a detailed discussion is provided in Reference 34. Briefly, the variation of the sputtering yield confirms the influence of the sample nature on the emission process. One important conclusion to be drawn from Figure 2.7 is that the coarse-grained MD constitutes an adequate generic model to describe the sputtering of bulk molecular and polymeric solids by clusters such as C_{60}. It predicts both the linear evolution of the yield with projectile energy beyond a threshold value and even, quantitatively, the total sputtered mass at a given energy.

2.2.4.2   Organic Crystals

In the case of atomic projectiles, the crystallinity of the targets has important consequences on the ion penetration and on the induced sputtering. Energetic atoms may either undergo head-on collisions in the surface or channel between the crystal rows and planes, resulting in significant sputtering yield fluctuations as a function of the impact point (see Fig. 2.3 and Reference 19). Another effect of the target crystallinity is the directionality of the sputtered flux.[38] Owing to their polyatomic nature, clusters should be less affected by effects such as channeling, but a directionality of the sputtered flux commensurate to the main crystal directions might be possible. In organic SIMS, effects of crystallinity on the sputtering yields could have implications for the analysis of liquid crystal-type materials and biological sections, including cells and tissues, where membranes of phospholipid bilayers are a major constituent of the sample. For such targets, an influence of the molecular orientation with respect to the incoming cluster beam on the sputtering process could translate into artifacts in the secondary ion mass spectra and images. Therefore, it constitutes a relevant topic for fundamental investigations.

The simulation of cluster impacts in organic materials began with a crystalline solid, namely *frozen benzene*.[29,39] Effects of crystallinity can be seen, for example, in the patterns of molecular displacements in the bombarded solid as a function of time.[3] Nevertheless, because of the small size, round shape, and perhaps stiffness of benzene, its behavior resembles that of atomic solids such as Ag or amorphous organic hydrocarbons (Section 2.2.4.1). In particular, the craters created by C_{60} in benzene are almost perfectly hemispherical. Some crystallinity effects are also found in the C_{60} bombardment of linear octane and octatetraene.[40] In those simulations, the energy is deposited closer to the surface for the system in which the chains are lying perpendicular to the projectile incidence. However, as was the case for benzene, the effects of crystallinity in these systems made of short molecules are moderate.

In order to further investigate the effects of crystallinity and molecular orientation in cluster SIMS of organic materials, studies of the interaction of cluster projectiles with samples exhibiting more pronounced structural features have been undertaken. The bombardment of model hexacontane crystalline targets[41] and Langmuir–Blodgett multilayers of Ba-arachidate[42] was modeled using MD simulations with a coarse-grain approximation. In the case of hexacontane ($C_{60}H_{122}$; 842 Da), two different orientations of carbon chains, horizontal and vertical, and two cluster projectiles, C_{60} and Au_3, were tested. The results are depicted in Figure 2.8.[43] Irrespective of the projectile, the simulations show that the sputtered mass (sputtering yield) increases dramatically when going from vertically to horizontally oriented hydrocarbon chains, by a factor of 7–8 times. For instance, for 15-keV C_{60} bombardment, the total sputtered mass varies from ~20 kDa for vertical chains (Fig. 2.8b) to >150 kDa for horizontal chains (Fig. 2.8a). With the latter system, large chunks of organic material, including many molecules, are emitted

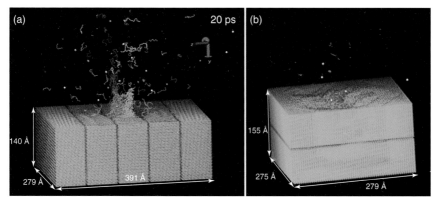

Figure 2.8   Effects of crystallinity.[41,43] Perspective views at 20 ps of the sputtering events induced by 5-keV C_{60} impacts on crystalline hexacontane ($C_{60}H_{122}$) with different orientations: (a) horizontal chain orientation and (b) vertical chain orientation. Reprinted from Delcorte et al., with permission from Wiley.[43]

(beyond 14 kDa). The mechanistic explanation for the observed yield enhancement in Figure 2.8a can be described by two factors. First, the energy deposition by the projectile in the sample surface creates a region with intense upward momentum in the top 2 nm of the sample, which is able to lift up and eventually detach a larger numbers of molecules because they lie parallel to the surface. Second, the reflection of the pressure wave at the interface between two molecular layers tends to confine the energy in the impact region and to reinforce the desorption mechanism. The analysis of the sputtered flux also shows a pronounced directionality of the ejected material with a maximum at a polar angle of ~45° and an azimuth perpendicular to the chain direction, as if the low index molecular planes were gliding on top of each other. In the case of vertical chains (Fig. 2.8b), the energized nanovolume is entirely included in the top molecular layer and the polymeric chains in that volume undergo extensive bond breaking and other stresses. Hence, very few or no intact molecules are sputtered, only fragments. In this case, atomic projectiles may even be more efficient than clusters for sputtering, as suggested by the results obtained with single Au, because they penetrate and break bonds deeper in the sample thereby liberating larger fragments than Au_3 and C_{60} projectiles. These specific behaviors are accompanied by peculiar crater shapes that are far from the hemispherical shapes observed with atomic solids, crystalline molecular solids made of small molecules (benzene), and amorphous polymers. Finally, for high energy bombardment (25 keV), delamination of the top layer is observed in the vertical hexacontane system, which results in the emission of a small number of intact molecules. The delamination process can be explained by the reflection of the energy on the weak interface between the two molecular layers. These predictions of the model remain to be verified by appropriate experiments. In the laboratory, crystals of hexacontane with horizontal molecules have been obtained[44] and ion scattering analyses of hexatriacontane, $C_{36}H_{74}$, with different orientations, have been published.[45] Therefore, the hexacontane model should be easily tested experimentally.

The interaction between C_{60} and a Langmuir–Blodgett film of Ba-arachidate $(CH_3(CH_2)_{18}COOBa; 449 Da)$ was also studied recently with MD simulations. The tilt angle of the Ba-arachidate molecules in the surface layer with respect to the normal is $\sim15°$, that is, the orientation is rather close to the vertical hexacontane described above. Under 15-keV C_{60} bombardment with an incidence close to the alignment of the molecules, the authors report a total sputtered mass of ~70 kDa.[42] This value is much larger than the one mentioned earlier for vertical hexacontane (~20 kDa). The difference can be explained by the smaller size of the arachidate molecules. With the considered energy, fullerene projectiles deposit their energy in the top two Ba-arachidate layers (2.6 nm thickness per layer) and molecules in the top layer receive essentially upward momentum. Therefore, intact molecules from the first and even the second layer are easily desorbed. In contrast, the hexacontane layer is significantly thicker (7.5 nm) and the region of high energy density barely reaches the interface with the second layer at 15 keV. In this case, the molecules can only be broken and not sputtered intact. In spite of these differences, the crater shapes observed in Ba-arachidate are not hemispherical and mirror the sample structure, as was the case for hexacontane.

Finally, although the process is different from atomic channeling, a strong dependence of the final projectile penetration depth on the angle between the incident beam and the molecule orientation is observed with both hexacontane and Ba-arachidate layers. For hexacontane, the penetration depth increases by a factor of ~2 when going from incidence perpendicular to the chains to incidence aligned with the chains. The reason invoked to explain this effect is the opening of the lattice along the chain direction.[42]

2.2.4.3   Thin Organic Layers on Metal Substrates

For cluster bombardment, thin organic layers constitute quite peculiar systems, with observations at variance with bulk organic solids. Experimental data[9,46−48] and computer simulations[49−51] indicate that the use of cluster projectiles for SIMS analysis of organic overlayers does not necessarily provide a benefit in terms of intact organic molecule emission with respect to more traditional monatomic ion beams. The seminal MD simulations of thin organic overlayer bombardment were performed using a silver crystal covered by three layers of benzene.[49] The authors compared the effects of 15-keV Ga and C_{60} projectile impacts in order to highlight differences between atom and cluster bombardment and to identify specific mechanisms induced by fullerene clusters. The obtained results indicate that in the case of 15-keV Ga bombardment, the projectile penetrates easily through the thin benzene overlayer losing, on average, $\sim2.3\%$ of its translational energy in the layer. Direct interactions between the projectile and the benzene molecules induce the formation of energetic fragments that are back reflected by the metal substrate into the organic layer, thereby creating even more fragmented species. When these collisions become less energetic and more molecules are set in motion, the ejection of intact molecules begins. These processes, initiated in the organic layer, result in the sputtering of ~20% of all the intact molecules. After passing the benzene overlayer, the Ga projectile penetrates into the substrate, leaving only a small portion of its energy near the metal surface. Via the collision cascade, upward moving substrate particles initiate the additional ejection of both intact and fragmented organic molecules, some being uplifted via correlated atomic motions.[52] The Ga atom finishes its course in the depth of the silver crystal where it deposits the rest of its energy, as was observed for bare Ag crystals[31] and suggested in Figure 2.2 for PS targets.

A different scenario unfolds under C_{60} bombardment. Because of its larger size, the projectile interacts more strongly with the organic layer. On impact with the silver substrate, the spatial correlation of the carbon atom motions is lost and most of the projectile atoms are back reflected into the organic overlayer. The following events are very similar to those observed for clean silver surfaces.[31] However, the size and depth of the formed crater is significantly smaller as ~60% of the projectile energy is deposited in the benzene overlayer. The energy deposition induces a strong pressure wave that propagates in the plane of the organic layer (Fig. 2.9). The main effect of this pressure wave is the relocation of molecules toward the periphery, rather than their emission. The crater formation process is accompanied by a correlated radial motion of the substrate atoms, which ejects intact benzene molecules in a similar manner as a catapult. In contrast with Ga projectiles, the sputtering induced by fullerenes is characterized by ring-like shapes of ejection sites (Fig. 2.10) and off-normal peaks in the angular distributions of fragmented and intact benzene molecules (Fig. 2.11).

Despite the much larger energy deposition in the organic overlayer region, the use of C_{60} clusters does not result in a yield enhancement comparable with the one observed for bulk organic samples. The sputtering results are summarized in

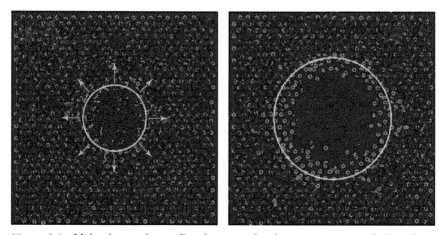

Figure 2.9 Molecular overlayers. Development of a planar pressure wave in three layers of benzene (red–orange) on Ag{111} (blue) upon bombardment by a 15-keV C_{60} projectile: (a) 1.0 ps and (b) 7.0 ps. Color coding of the molecules: red is for the top layer, pink for the intermediate layer, and orange for the bottom layer.

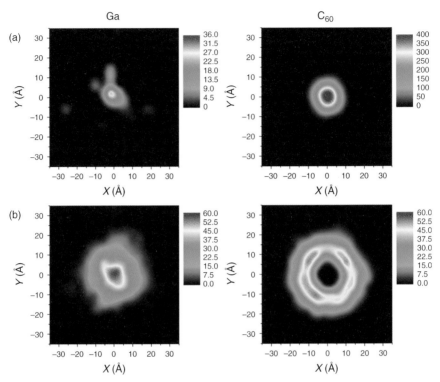

Figure 2.10 Sputtering of benzene overlayers.[49] Spatial distribution of original location of (a) molecular fragments and (b) C_6H_6 molecules sputtered by 15-keV Ga and 15-keV C_{60} impacts at normal incidence. The color represents the number of species emitted per "pixel." Each pixel corresponds to an area of 4×4 Å^2. Reprinted from Postawa et al.,[49] with permission from ACS publications.

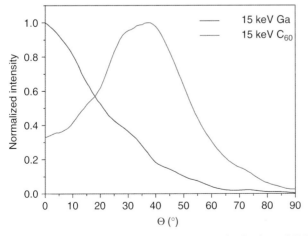

Figure 2.11 Off-normal emission. Angular distribution of C_6H_6 molecules ejected under 15-keV Ga (black) and 15-keV C_{60} (red) bombardment at normal incidence.

TABLE 2.2   Number of Particles Ejected from a Multilayer of C_6H_6 on Ag{111}, a Monolayer of PS4 on Ag{111}, and a Self-Assembled Monolayer of Octanethiol on Au{111} Bombarded by 15-keV Ga and C_{60} Projectiles at Normal Incidence at 15 ps after the Projectile Impact

| | Projectile | | | | | |
| | Benzene | | PS4 | | Octanethiol | |
Particle	Ga	C_{60}	Ga	C_{60}	Ga	C_{60}
Total Ag yield	15	81	14	229		
Total Au yield					19.4	59
Total organic yield	227	398	7.5	12	8.5	70
Intact organic molecules	120	244	5.8	4.7	0.5	0.5
Organic molecules in smaller clusters and fragments	8.5	82	0.1	4.0	4.1	53.5
Organic molecules in larger clusters	99	75	1.2	1.1	3.7	16

The yields for organic species are given in number of ejected benzene, PS4, and octanethiol molecules, respectively. Reprinted from Czerwinski et al.,[49] with permission from Elsevier.

Table 2.2 for three different organic monolayers, benzene, *sec*-butyl-terminated PS tetramers (PS4), and octanethiols.[50] In the case of benzene, the authors identified several phenomena explaining the moderate emission. First, the limited number of available benzene molecules, connected with their low (~0.4 eV) binding energy, makes their ejection possible even via the less energetic processes initiated by the Ga impact. Second, the aforementioned pressure wave induced by C_{60} pushes molecules away from the impact point (Fig. 2.9). Finally, the stronger interaction of C_{60} with the organic layer induces significantly more fragmentation. Combining these explanations, one can conclude that during C_{60} cluster bombardment, small physisorbed molecules will be rather fragmented or pushed away from sputtering region than ejected intact. Additional studies conducted with PS4 physisorbed on Ag{111} and octanethiols chemisorbed on Au{111} provide complementary information. For PS4, the larger molecular size and binding energy, combined with the smaller number of molecules, lead to the decay of the pressure wave and to a significant increase of fragmentation and ejection of substrate particles. In the case of octanethiols, the increased density and thickness of the organic overlayer induce a significant enhancement of organic material emission and a concomitant decrease of substrate material sputtering. However, most of this enhancement is contributed by molecular fragments and not by intact molecule ejection, probably because of the stronger binding to the substrate (Table 2.2).

Finally, the influence of the organic layer−metal interface has also been studied for C_{60} bombardment of Langmuir−Blodgett (LB) multilayers of Ba-arachidate.[42] Figure 2.12 shows the evolution of the total sputtering yield of Ba-arachidate molecules for increasing numbers of LB layers and three projectile energies. The sputtering yield obtained at 15 keV for the monolayer system is

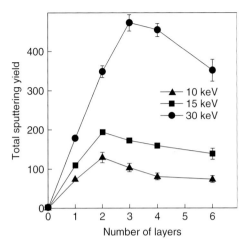

Figure 2.12   Langmuir–Blodgett layers.[42] Dependence of the total sputtering yield of organic material on the Ba-arachidate LB film thickness for three different initial energies of the C_{60} projectile (normal incidence). Reprinted with permission from Paruch et al.,[42] with permission from ACS publications.

quite comparable to the one obtained for octanethiols in Table 2.2, taking into account the weaker binding of Ba-arachidate to the metal substrate. Depending on the projectile energy, the sputtering yield increases with the number of molecular layers in the self-assembly, up to two layers at 10–15 keV and three layers at 30 keV. For thicker assemblies, the yield gradually decreases toward a plateau value, indicating that the substrate does not influence the molecular emission anymore. It is interesting to note that the total sputtered mass at the plateau value (~60 kDa at 15 keV) is also quite comparable to the value obtained for a bulk amorphous polymeric sample (Fig. 2.7). The process of substrate enhancement leading to the maxima observed in Figure 2.12 is explained by the reflection of projectile and recoil carbon atoms by the metal surface, composed of heavier and more densely packed atoms.

2.2.4.4   Hybrid Metal–Organic Samples

On atomic bombardment (Ga^+, In^+), metal nanoparticles condensed on the surface of organic materials induce a strong increase of the analytically significant sample ion yields (metal-assisted secondary ion mass spectrometry or MetA-SIMS).[53] However, with polyatomic projectiles (Bi_3^+, C_{60}^+), a decrease of the sputtered organic ion yield is generally observed.[54,55] These results were quite puzzling at first and MD simulations were conducted using a crystalline polyethylene (PE) sample covered by gold nanoparticles to unravel the effects of the metal overlayer on the dynamics of sputtering.[56,57]

The impacts of 10 keV Ga and C_{60} with a Au_{555} nanoparticle adsorbed on crystalline PE are analyzed in Figure 2.13. Figure 2.13a shows that upon Ga bombardment, the sputtered PE mass increases when the impact point is moved from the bare polymer to the center of the nanoparticle. The large error bars resulting from four different aiming points, calculated for each situation, can be explained by the fluctuations normally observed upon atomic bombardment (Fig. 2.3). For C_{60} bombardment, the trend is reversed, with large ejection yields observed for

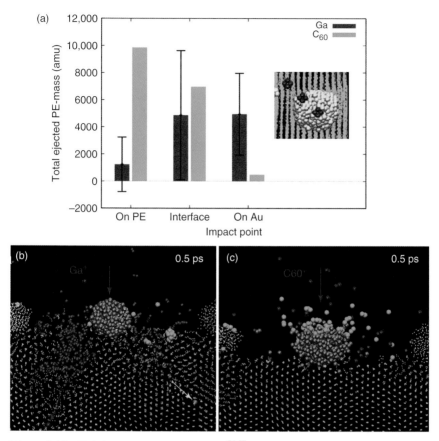

Figure 2.13 Hybrid metal–organic samples.[56,57] Projectile impacts on a crystalline polyethylene covered with gold nanoparticles. (a) Total ejected mass of polyethylene for different impact points and projectiles (Ga and C_{60}). Inset: Top view of the central cluster with three series of aiming points.[57] (b, c) Snapshots of the impacts of 10-keV Ga and C_{60} on the central Au nanoparticle (Au_{555}). Color coding: orange is for the atoms of the central cluster, green and violet are for the Au and PE atoms that are ejected at 30 ps, yellow and cyan are for the remaining Au and PE atoms, and red is for the projectile.[56] Recreated from Restrepo et al.,[56,57] with permission from Wiley.

impacts on the polymer and much smaller yields for impacts on the nanoparticle. The opposite effect of the nanoparticle for Ga and C_{60} bombardment can be explained by the movies of the bombardment event extracted from the simulations. In Figure 2.13b, the Ga projectile interacts with the atoms of the nanoparticle, creating a few energetic Au recoils that, ultimately, induce the formation of a rather dense collision cascade in the PE surface. In this manner, the gold particle helps to confine the atomic projectile energy in the surface and large numbers of PE fragments are emitted. In contrast, when Ga directly hits the PE phase, it often implants deeply and without causing any emission (similar to Fig. 2.2). When C_{60} impinges on the nanoparticle (Fig. 2.13c), it transfers most of its energy to the top of the cluster, causing many Au–Au bond-scissions. However, very few PE fragments are emitted because only slow recoil Au atoms reach the PE substrate. In the process, most of the projectile atoms are backscattered, as was the case with heavy metal substrates. Here, they carry away $\sim$35% of the initial projectile energy. In order to sputter numerous molecular fragments, C_{60} must impact the PE sample directly. In summary, for impacts on the gold cluster, the sputtered mass of PE is about three times larger when Ga is used instead of C_{60}. Conversely, for impacts on the PE surface, the sputtered mass of PE is about one order of magnitude larger when C_{60} is used instead of Ga. Although electronic effects are neglected, the trends set by the simulations qualitatively agree with the measurements.

2.2.5 Induced Chemistry

In many cases, the useful molecular signal in SIMS is dependent on (de)protonation and cationization reactions occurring in the energized nanovolume or in the selvedge region above the surface.[58] These mechanisms are strongly influenced by the molecular environment (matrix effect)[59] but the dependence on the projectile remains largely unknown, although some pioneering studies exist.[60,61] On the other hand, experiments with clusters such as SF_5^+ or C_{60} indicate that the behavior upon continuous bombardment and the ability to retain the molecular information during the depth profile are also strongly sample dependent.[62] Materials such as poly(methyl methacrylate), which tend to depolymerize upon irradiation, can be depth profiled with retention of the molecular information.[63] But others, such as PS,[64] tend to cross-link under the beam, with, at best, a loss of the molecular information and, in the worst case, failure of the sputtering and carbon build up on the surface.[37] Remarkably, beam-induced cross-linking does not seem to occur when large Ar clusters are used for sputtering.[65] In order to control these effects and properly address the related applications, it is very important to develop a better understanding of the chemistry induced by cluster beams. This is a challenge for MD simulations because the accurate prediction of chemical reactions requires far better interaction potentials than the general description of the molecular motions. A few theoretical studies, however, have been devoted to the investigation of reactions induced by cluster beams in the solid, such as ion formation in water ice or free H formation and cross-linking in carbon-based materials.

Hydrocarbons constitute samples of choice for these simulations because of the existence of an appropriate reactive potential (AIREBO)[25] capable of describing ground-state chemistry. Reactions occurring in octane and octatetraene crystals upon C_{60} bombardment were studied by Garrison et al. The results show that the impact of a 10-keV C_{60} can produce about 400 free and reacted H atoms in the energized region.[40] Comparable numbers were obtained upon bombardment of icosane by coronene, small PS molecules, and C_{60}.[66] These numbers may explain the observed increase of protonation reactions upon depth profiling with C_{60}.[59] Another conclusion of the octane/octatetraene study, consistent with previous studies of adsorbed PS oligomers,[51,67] is that C_{60} is able to produce large numbers of molecules with a low internal energy compared to atomic projectiles. However, the previous hypothesis of collisional cooling was discarded in this study. Using a series of samples including the same octane and octatetraene crystals, in addition to benzene ice and fullerite, Webb et al. investigated the effect of the H : C ratio in the sample on the reactions occurring upon C_{60} bombardment.[68] Overall, the number of cross-links, defined as new C–C bonds created between adjacent molecules, decreases with increasing H : C ratio. However, fullerite really stands out of the series. For the three hydrocarbons, a peak of cross-linking around 200–300 fs is rapidly followed by a relaxation of the system so that only a few tens of cross-links remain after 1 or 2 ps (Fig. 2.14). In contrast, for fullerite, the number of cross-links restarts to slowly increase after the peak, up to a value of ~400 at 2 ps. In addition to the absence of "passivating" hydrogen, this is probably due to the very specific nature of fullerite, where each broken molecule naturally exhibits many reactive radical sites that are prone to induce cross-linking. These results explain the failure of molecular depth profiling with C_{60} for fullerite and fullerene-based photovoltaic materials.

Although they do not involve organic materials, two other studies should be mentioned here concerning the simulations of reactions induced by cluster projectiles in solids. First, the effect of carbon building up on silicon surfaces upon low energy bombardment with C_{60} and the transition observed above 10 keV could be elegantly explained by the calculations of Krantzman et al. using the Tersoff potential.[69,70] Second, the ionization produced in water ice by Au_3 and C_{60} was investigated using a water potential with an activation energy of ~17 eV for the

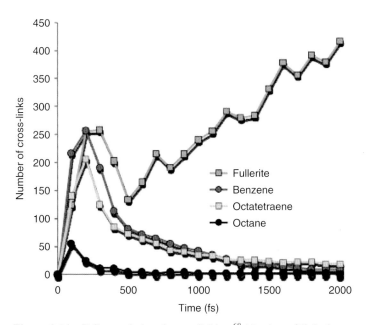

Figure 2.14 Fullerene-induced cross-linking.[68] Number of links between intact molecules as a function of time after the initial impact of a 20 keV C_{60} projectile on different molecular targets. Reprinted from Webb et al.,[68] with permission from Wiley.

dissociation into H^+ and OH^-.[71,72] These simulations show that the energized nanovolume is a cauldron where ground-state atoms coexist with many excited state atoms and ions in a compressed phase for a very short time, and where virtually any reaction is energetically possible. Beyond the specific results, these simulations emphasize the difficulty of properly describing the complex chemistry at work in the first picosecond following cluster impact and, thereby, make us reflect on the limits of the existing models.

2.2.6 Multiple Hits and Depth Profiling

Progress in the modeling of repeated bombardment, concomitant with the development of molecular depth profiling using cluster beams, have been recently made owing to the new "divide and conquer" scheme proposed by the Penn State group and their collaborators.[73,74] In this scheme, the projectile aiming points are randomly chosen over a very large sample and smaller cylindrical volumes, able to contain the effects of a single impact, are defined. When these cylinders do not overlap, the corresponding trajectories are calculated independently, in parallel, and otherwise, sequentially. Then the bombarded cylinders are placed back in the master sample for the next round. This strategy allows to cut the computer cost by dealing with relatively small samples for each impact and to reduce the total time by running several trajectories in parallel. At this point, the methodology has been only applied to inorganic materials such as silver[72] and silicon crystals[75] but, in principle, it should be also applicable to organic materials, provided that the calculation time is decreased.

In the case of silver targets under 20 keV C_{60} bombardment, the results indicate that after the sputtering of a number of atoms equivalent to one crystal layer, only $\sim 20\%$ of the top two layers have been actually removed and that the sputtered material originates from a maximum depth of 5–6 nm, because of the combined effects of roughness development and atomic mixing.[72] This observation provides a limit of depth resolution upon continuous C_{60} cluster bombardment. The average sputtering yield is also significantly lower for the rough surface. Further investigations show that Au_3 produces more corrugated surfaces than C_{60} and that the roughness development is reduced when going from normal to oblique incidence.[73] These effects are illustrated in Figure 2.15, by the fluence dependence of the surface roughness upon C_{60} and Au_3 bombardment, for normal and 70° impacts. Figure 2.15 also indicates that the roughness tends to saturate in all cases after a fluence of $1-2 \times 10^{13}$ ions/cm^2. The results obtained with silicon surfaces indicate similar trends of surface corrugation under ion beam bombardment as well as massive lateral and vertical relocation of atoms in the sample.[75]

The topography of a C_{60} bombarded Si surface is shown in Figure 2.16 after 50 and 200 impacts. Even with 50 impacts, cavities as deep as 5 nm and protrusions as high as 3.5 nm are induced in the surface. However, unlike what was observed in Ag, the topography development in Si did not affect the sputtering yields. Another observation concerns the swelling of the crystal structure, probably due to the incorporation of C atoms in the surface and their reaction with the Si

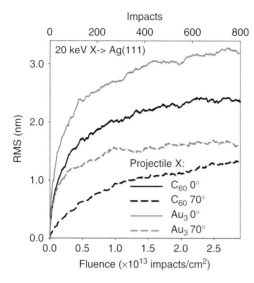

Figure 2.15   Multiple bombardment by clusters.[74] Root-mean-square (RMS) roughness of the entire sample surface versus the ion beam fluence for 20 keV Au_3 and C_{60} bombardment. Reprinted from Paruch et al.,[74] with permission from ACS publications.

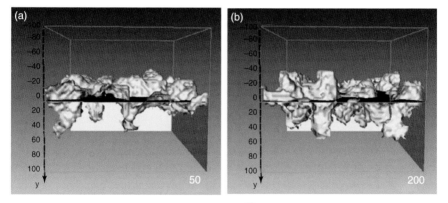

Figure 2.16   Bombardment-induced topography.[75] Isosurface representations of the momentary Si surface after accumulation of (a) 50 and (b) 200 impacts of 20 keV C_{60} projectile (corresponding to projectile ion fluences of 8.8 and 35.2 $\times$ 10^{12} cm^{-2}, respectively). Reprinted from Krantzman and Wucher,[75] with permission from ACS publications.

substrate to form a mixed SiC phase. These results were used to develop a new statistical model of sputtering (SSM). For completeness, it should be also mentioned here that repeated bombardment (up to 20 impacts) of rough Si surfaces by large Ar clusters had been previously modeled by Aoki et al. in order to investigate the smoothing effect experimentally observed with these clusters.[76] Using a Si target with an initial surface defined by a two-dimensional sine function and a corresponding root-mean-square (RMS) roughness of $\sim$10 Å, they showed that the surface roughness was significantly reduced after 20 impacts of large Ar_n clusters

(n = 13,333 and n = 20,000), at an energy of 20 keV. Instead, repeated impacts with smaller clusters were shown to induce an increase of the roughness.

2.2.7 From Small Polyatomic Projectiles to Massive Clusters

2.2.7.1 Light-Element Clusters

The potential of SF_5^+ and C_{60}^+ for higher sensitivity molecular analyses and damageless depth profiling have boosted the investigation of light-element clusters for surface mass spectrometry. In another range of sizes, the interaction of huge molecular clusters with surfaces was also extensively studied experimentally. In particular, glycerol and H_2O nano and microdroplets have been used in massive cluster impact (MCI),[77] electrospray droplet impact (EDI),[78,79] impact desolvation of electrosprayed microdroplets (IDEM),[80] and desorption electrospray ionization (DESI).[81,82]

Building on the results obtained with C_{60}, a series of theoretical studies have been devoted to the comparison of larger light-element clusters, essentially fullerenes, impinging on a bulk organic sample, benzene.[29,83−85] At 5 keV, a maximum molecular emission is observed between C_{20} and C_{60} and the yields decrease slightly for larger clusters up to C_{180} (2.2 kDa). For higher kinetic energies, the maximum shifts to larger cluster sizes.[84] The results are explained by the size of the ejection cone derived from the distribution of the projectile energy in the surface shortly after impact [mesoscopic energy deposition footprint (MEDF) model].[30,86] In all cases, small clusters such as C_6H_6 and C_6H_8 provide slightly lower yields of benzene molecules.

The influence of the projectile size was also studied using a series of hydrocarbon projectiles with masses ranging from 0.3 to 110 kDa.[66,87] Large molecular clusters or "nanodrops" with a total energy of 10 keV create hemispherical craters with a diameter of >10 nm, larger than those induced by isoenergetic fullerenes, but the quantity of sputtered material is less. The sputtered masses per nucleon of all the tested projectiles fit on one line when plotted as a function of energy per nucleon in the projectile (Fig. 2.17c). The change of slope around 1 eV/nucleon corresponds to a physical change in the interaction process, with the sputtering yield depending on nuclearity below 1 eV/nucleon and not anymore beyond that value (the realm of SF_5^+, C_{60}, and coronene). Unlike relatively small clusters such as C_{60} or coronene,[88] large molecular projectiles allow us to probe this range of energies, below 1 eV/nucleon, in which fragmentation of the sample is minimal with yet significant molecular emission yields (Fig. 2.17a and b). For 2D mass spectrometric analysis, the suppression or strong reduction of sample fragmentation is desirable, even with the trade-off of reduced emission yields. However, the simulations also show that these organic projectiles made of large molecules require higher kinetic energies to overcome the regime of net deposition, an issue which could be avoided with inert projectiles such as noble gas clusters or organic clusters made of smaller, volatile molecules, such as toluene.[89] For this reason, the molecular clusters described in Reference 66 would not appear to be good candidates for molecular depth profiling and 3D analysis of organic materials. Another remarkable result of

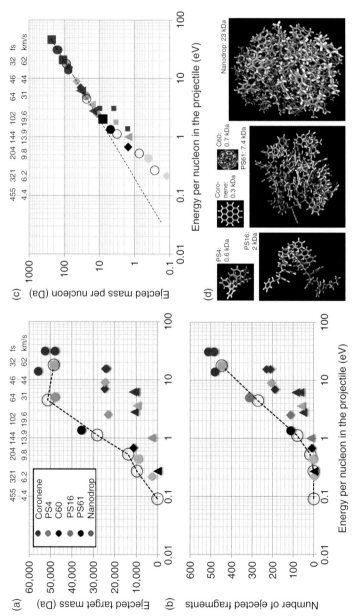

Figure 2.17 From small to massive carbon-based clusters.[66] (a) Sputtered target mass, as a function of the energy per nucleon in the projectile. (b) Number of sputtered fragments. (c) Sputtered target mass per nucleon. The color coding determines the projectile. The symbol shapes refer to the total energy of the projectiles: 1 keV, small squares; 2 keV, triangles; 5 keV, diamonds; 10 keV, circles; 15 keV, large squares. The large open circles connected with a broken line represent $(PS_4)_n$ molecular clusters with n = 1–197. The upper scales indicate the projectile velocities in kilometer per second and the corresponding time to travel 2 nm. (d) Molecular projectiles used in the simulations. Reproduced from Delcorte et al.,[66] with permission from ACS publications.

these simulations is that because of their large momentum, massive organic clusters under normal incidence also induce more molecular mixing in soft materials than fullerenes. This molecular mixing is reduced when going to off-normal angles.[17]

The ability of hydrocarbon nanodroplets to desorb large molecules was further tested using 110 kDa $(PS4)_{197}$ molecular clusters, with 10 keV of incident energy.[17] The samples were PS oligomers of 7.5 kDa adsorbed either on a Au{111} surface or on a PE substrate. A series of incidence angles (from $0°$ to $60°$ with respect to the surface normal) and impact points (from 0 to 10 nm from the molecule) were tested. The results indicate that $(PS4)_{197}$ nanodroplets can desorb a large molecule from a gold substrate by two distinct processes, a "spring" mechanism resulting from the compression and release of the molecule under direct impact (Fig. 2.18a and b) and a "washing" mechanism caused by the splashing of the projectile on the gold substrate. This latter mechanism is also efficient for PE substrates but only in the case of oblique incidence (Fig. 2.18c). Although the internal energies of the desorbed molecules vary with the mechanism, the substrate, and the details of the impact conditions, they are emitted without fragmentation in all the tested cases. The situation of soft desorption sampled with light-element nanodroplets (below 1 eV/nucleon, reminiscent of EDI, and MCI[90,91]) constitutes, therefore, an interesting alternative to the use of smaller clusters.

The interaction of massive H_2O clusters with surfaces has been modeled in a series of studies by Tomsic et al.[92-94] The cluster sizes were around $(H_2O)_{1000}$ and the velocities up to a few kilometers per second . For impacts on a purely repulsive surface, the water clusters undergo asymmetric fragmentation, starting in the leading part of the cluster, with a fragment size distribution following a negative power law. The fragments are mostly reflected with grazing angles. They also observed reflected monomers with a very high kinetic energy. The variety of interactions occurring upon normal impact of frozen water clusters with a metal surface is illustrated in Figure 2.19, with 1–6 km/s $(H_2O)_{13892}$ ($\sim$10 nm in diameter).[95] With a velocity of 1 km/s, the 10 nm nanodroplet is simply deformed by the impact and bounces back toward the vacuum (Fig. 2.19a). This energy/molecule is in the range sampled by EDI. With a velocity of 2 km/s, the projectile forms daughter droplets upon impact (Fig. 2.19b). Finally, at 6 km/s ($\sim$6 eV/molecule), the water nanodroplet disintegrates completely upon impact (Fig. 2.19c). Remarkably, these simulations show that the behavior upon impact of the water nanocluster depends on other parameters such as its size and its temperature. If the temperature of the projectile is raised from 0 to 450 K, the 1 km/s impact becomes sufficient to induce its fragmentation in daughter droplets (Fig. 2.19d). In the same manner, an increase of the droplet size results in a lowering of the velocity needed for fragmentation as a result of the change of balance between the surface and bulk forces. These observations explain why the relatively slow (100 m/s) microdroplets used in DESI also fragment upon impact. Impacts of 1–4 km/s $(H_2O)_{3330}$ clusters with surfaces were also modeled by Urbassek et al. in the context of the development of impact desolvation of IDEM.[96-98] They found that for projectile velocities below 3 km/s, an echistatin protein embedded in the water cluster could keep a solvation shell and preserve its conformation upon impact. At this point in time, desorption of

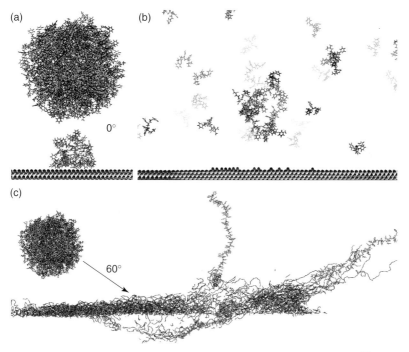

Figure 2.18   Large molecule emission by massive light-element clusters.[17] Desorption of a PS 61-mer (red) from a Au{111} substrate (blue). Projectile: 10 keV $(PS4)_{197}$ organic nanodrop (gray); direct hit, $0°$ incidence, (a) 0 and (b) 12 ps. (c) Bombardment of PS 61-mers adsorbed on a polymeric substrate (blue); side hit, $60°$ incidence, the projectile atoms are omitted for clarity. Reprinted from Delcorte and Garrison,[17] with permission from Elsevier.

organic molecules by massive H_2O (or similar) clusters has not been investigated. Therefore, one must rely on the result obtained with hydrocarbon clusters of similar constituent masses in order to understand those aspects.

2.2.7.2   Large Argon Clusters

The results of recent time-of-flight (ToF)-SIMS studies show that argon gas cluster ion beams are excellent candidates to replace C_{60}^+ beams for surface analysis, depth profiling, and 3D imaging of organic materials.[99] For instance, large Ar clusters perform very well where C_{60} beams fail, for example, for the molecular depth profiling of cross-linkable materials.[65] In addition, experimental studies show that the low fragmentation situation predicted in the simulations involving massive hydrocarbon clusters (Fig. 2.17) is also attained with large noble gas clusters.[100] For thick arginine, leucine, and triglycine films, fragmentation is strongly reduced when going from Ar_{300} to $Ar_{1500-2000}$ with the same acceleration energy (10 or 15 keV).

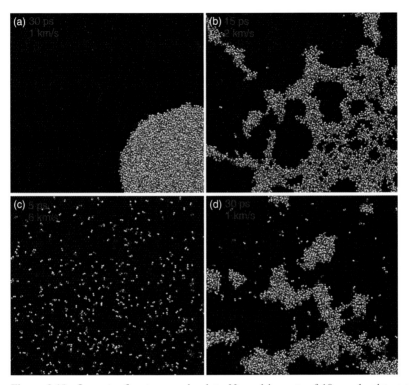

Figure 2.19 Impacts of water nanodroplets. Normal impacts of 10 nm droplets on a rigid Au{111} surface. (a) 1 km/s, 0 K; (b) 2 km/s, 0 K; (c) 6 km/s, 0 K; (d) 1 km/s, 450 K. The indicated velocities and temperatures refer to the droplets before impact.

MD computer simulations have been performed to explain the processes that take place during argon cluster bombardment of different materials. The objects of the first studies were inorganic materials. The results obtained upon bombardment of silicon {100} surfaces[101–103] clarified the relationship between the parameters of the interaction, such as the projectile size, its total kinetic energy and the energy per atom, and the end results of the sputtering process. Depending on the chosen combination of parameters, one may bombard the Si{100} surface without any significant displacement of silicon atoms or inducing severe damage in the surface. The maximum amount of displacements and, thus, the maximum emission of substrate atoms for a selected initial energy of the cluster are directly connected to the cluster size (number of atoms).[102] Additional studies showed that the initial surface structure has a significant influence on its final roughness[104] and that bombardment with large Ar clusters at glancing angles can be used for surface smoothing.[105,106]

A second field of theoretical studies concerns the bombardment of thin organic layers. As was the case in Section 2.2.4.3, Ag{111} substrates covered by a monolayer of *sec*-butyl-terminated PS[107] or by a monolayer of benzene[108] were chosen as model systems. The results reveal that argon clusters are much more

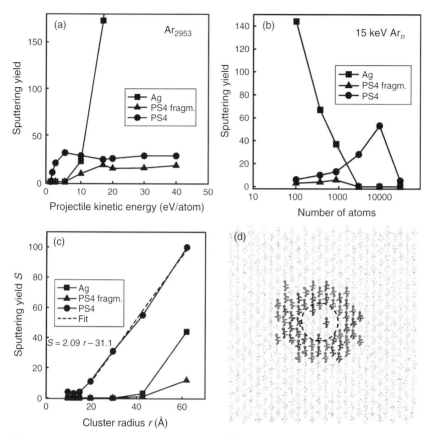

Figure 2.20 Sputtering by large Ar clusters.[107] Dependence of the sputtering yields of silver atoms (squares), fragmented (triangles), and intact (circles) PS4 molecules ejected from a PS tetramer/Ag{111} system as a function of (a) the kinetic energy of the Ar_{2953} projectile for normal incidence bombardment, (b) the number of atoms composing the cluster projectiles (15 keV), (c) the radius of the projectile (kinetic energy of 5 eV/atom; the dashed line is a linear fit to the data for molecules composed of more than 200 atoms). (d) Initial locations of PS tetramers ejected with a kinetic energy higher than 30 eV (red), lower than 30 but greater than 2 eV (green), and lower than 2 eV (blue) for 15-keV Ar_{9000} bombardment at normal incidence. Reprinted from Rzeznik et al.,[107] with permission from ACS publications.

effective at removing intact molecules from thin organic layers than fullerenes. For instance, bombardment by 15 keV Ar_{2953} leads to the emission of 33 PS4 molecules while C_{60} clusters with the same energy are only able to desorb four molecules.[39] Here again, several trends are observed as a function of the initial parameters. If the number of atoms in the projectile (size, radius) is kept constant, the ejection yield of intact PS4 molecules increases with increasing energy up to a saturation value. The saturation point for intact molecular emission also corresponds to the threshold for the ejection of fragments and substrate atoms (Fig. 2.20a). When the initial energy of the projectile is constant, the ejection of substrate atoms and molecular fragments decreases with increasing cluster size. However, the yield of intact molecules has a nonmonotonic behavior with a maximum depending on the cluster energy (Fig. 2.20b). At constant energy per atom, one observes an almost linear dependence of the sputtering yield of intact PS4 molecules on the argon cluster radius (Fig. 2.20c). There are two possible ejection mechanisms of organic molecules upon impact of argon clusters at the saturation energy. The first mechanism is one in which the ejection of organic molecules is initiated by the flux of Ar atoms collectively pushing the molecules toward their first neighbors. These neighboring molecules play the role of takeoff board for the molecules washed away by the cluster. The same process occurs in cascade between successive rings of molecules. This chain ejection pattern lasts as long as the kinetic energy is sufficient to lift up molecules.[107] As a result, the molecules with the highest energies come from the first molecular ring around the impact (Fig. 2.20d). Another mechanism is responsible for the ejection of low energy molecules initially located below the projectile. These molecules are sputtered by collective interactions with the substrate atoms recovering from the impact. In contrast with the "spring" mechanism described in Section 2.2.7.1 for larger polymeric adsorbates, in this case it is the compressed substrate that releases enough energy to lift up the molecules. It acts like a trampoline hurling molecules into the vacuum.[107] Because of the predominant role of the first ejection mechanism, the angular distribution of sputtered intact molecules has a maximum at an oblique angle ($\sim 65°$). The kinetic energy distribution is also different than for C_{60} bombardment. The characteristic features of this distribution are three local maxima that are associated with different times of ejection of organic material. Similar trends are observed for the argon cluster bombardment of benzene monolayers. However, the shape of the ejection region at the saturation energy is no longer a ring but becomes a full circle, as a consequence of the smaller size and binding energy of benzene.[108]

The Ar_{872} sputtering of LB overlayers of Ba-arachidate molecules reveals some similarities with the PS4 and C_6H_6 cases.[109] In the case of single layers, the increase of the initial energy per atom leads to a correlated increase in the ejection of intact molecules. The characteristic energy threshold for emission of molecular fragments and substrate atoms is also observed. However, an important difference exists in the angular distribution of intact molecules. At variance with the PS4 case, the ejection maximum is located at $\sim 15°$. One also notices the sputtering yield enhancement due to argon backscattering in the substrate, as was the case for C_{60} bombardment (Fig. 2.12). These observations suggest that the process of organic material ejection of LB layers is closer to the case of C_{60} bombardment than to the argon cluster bombardment of thin PS4 and benzene layers.

Finally, preliminary results obtained with a solid benzene crystal[3] reveal that the size of the crater created by a 5-keV Ar_{872} impact is larger than the one induced by a fullerene with the same energy. However, the larger crater is not accompanied by a higher sputtering yield but by a more pronounced compression of the material, probably due to the larger momentum of the projectile.[30] The total sputtering yield for 5-keV C_{60} bombardment is 317 molecules while barely 97 molecules are emitted by 5-keV Ar_{872}. Additional studies show that the efficiency of the sputtering process strongly depends on the size of the Ar cluster and its impact angle.[110] Small clusters such as Ar_{366} behave in a quite similar manner as do C_n fullerenes, while for large argon clusters such as Ar_{2953}, a nonmonotonic angular dependence is observed with the maximum at $\sim 45-55°$.[39,111]

2.2.7.3 Massive Gold Clusters

Massive gold clusters (Au_{400}) have been studied for a number of years and their usefulness for sputtering and analyses of organic surfaces were demonstrated in the last decade.[112,113] However, the energies used in the experiments are often significantly higher than the usual energy range of SIMS (>100 keV) and direct comparisons with other projectiles are therefore difficult to establish. These high energies also constitute an issue for modeling because the required sample sizes and, in turn, computation times become prohibitive in the case of organic targets.

Urbassek and coworkers studied the penetration of energetic Au_n clusters, including massive Au_{402}, in various materials including van der Waals solids[114], graphite, and gold.[115] Their results demonstrate the "clearing-the-way" effect of clusters, that is, the fact that clusters penetrate deeper than single atoms with the same velocity. The dependence of the range on the cluster size follows a power law for all the considered materials. They also investigated the influence of the target cohesive energy on the cluster range by artificially varying the potential well depth for the van der Waals system and showed that this parameter was only influential at relatively low energy, below 100 eV/atom.[114]

Recently, two articles reported on the sputtering of a polymeric material by Au_{400} projectiles with 10–40 keV of kinetic energy.[116,117] As an example, Figure 2.21a shows a vertical cross-section of the polymer bombarded by 40-keV

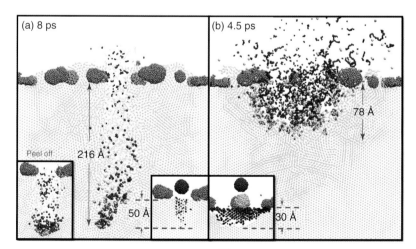

Figure 2.21 Massive Au cluster impacts.[117] Cross-sections of the target at the time of maximum penetration for impacts at 40 keV on (a) polyethylene and (b) a supported gold nanoparticle (Au_{555}). The central insets show the origin of the PE fragments (purple). The left inset shows the "peeling off" of the projectile at 1.5 ps (see text). Reprinted from Restrepo and Delcorte,[117] with permission from ACS publications.

Au_{400} (the polymer chains are perpendicular to the plane of the paper).[116] The Au_{400} cluster implants deeply in the sample, depositing its energy in a narrow cylindrical track where the material is damaged. The color coding (red for the projectile outer atomic shell) suggests that the front shell atoms clear the way for the core atoms (yellow) and that the projectile is "peeled off" as it penetrates deeper in the target. After relaxation of the target, the final penetration depth of 40-keV Au_{400} in PE is ~14 nm. This value is twice larger than the ~7 nm obtained by Anders and Urbassek for graphite, a carbon target that is also two times denser and has a larger cohesive energy than the polymer.[115] For comparison, the calculated range of 40-keV Au_{400} in frozen argon is close to 30 nm.[114] In Figure 2.21a, bond-scissions in the polymer occur only in the track of the projectile and, as a result, all the ejected fragments originate in the top 5 nm of the track (purple atoms in the inset).

In contrast, when the Au_{400} cluster impinges on an adsorbed Au nanoparticle (metal-assisted SIMS configuration), it is immediately stopped by the heavy metal and transfers all of its energy in the surface region of the polymer (Fig. 2.21b). Then, the crater is hemispherical, very similar to the case of light-element clusters bombarding organic materials. Hemispherical craters are also observed for Au_n clusters impinging on flat Au(111) surfaces.[118] In Figure 2.21b, the disintegration of the Au_{400} cluster in the surface induces numerous bond-scissions in the polymer, causing the emission of a large number of fragments from a nanovolume of >10 nm in width and ~3 nm in depth. As a result, the predicted sputtered mass of polymer is one order of magnitude larger for impacts on the nanoparticle than for direct impacts on the polymer.

2.3 OTHER MODELS

2.3.1 Analytical Models: From Linear Collision Cascades to Fluid Dynamics

For kiloelectronvolt atomic bombardment, the linear collision cascade model, developed by Thompson[119] and Sigmund,[13,14] was the most widely accepted theory describing energy transfer and secondary particle emission. Indeed, in many cases, recoil atom trajectories do not overlap, as shown in Figure 2.2a. Sigmund used Boltzmann's transport theory to calculate the statistical function $f(r,v,t)$ that gives the average state of particles with time and, in turn, to predict the angular and energy distributions of ejected atoms. One of the successes of the theory was its predicting the E^{-2} dependence of the recoil atom distribution in the solid and the maximum of the ejected atom distribution at half the binding (sublimation) energy of the material. As was abundantly discussed and illustrated earlier (Fig. 2.5, Fig. 2.6, Fig. 2.8, Fig. 2.9, and Fig. 2.11), in the case of cluster bombardment, many-body collective and correlated motions dominate, indicating that hydrodynamic or continuum mechanics models might be more appropriate. For the case of overlapping cascades and high energy densities, Sigmund also developed the thermal spike concept in which the excited volume behaves like a dense gas at high temperature.[120] Other approaches that seem adequate to describe the physics of cluster bombardment involve macroscopic pressure (shock) waves resulting from the projectile impact. The interest in shock waves and material cratering resulting from high velocity impacts of small particles stemmed from space research, where spacecraft are struck by swift micrometeorites. Those concepts were readily applied to massive cluster impacts[121,122] and, maybe less naturally, to the domains of megaelectronvolts (and kiloelectronvolts) ion–surface interactions.[123,124] The common ground of these models is that the high energy density deposited within the vicinity of the projectile track leads to the propagation of a supersonic mechanical disturbance through the medium resulting in the mechanical ablation of material at the surface.[122,125] After the release of the compression energy, thermal energy is left in the sample which can also cause the vaporization of material from the excited region (thermal desorption). In a related ("pressure pulse") model,[126] the kinetic energy directly or indirectly transferred along the primary ion path propagates by diffusion because of the initial kinetic energy gradient, and the momentum acquired by the molecules is proportional to the time integral of this energy density gradient. Desorption occurs when the energy corresponding to this momentum exceeds the surface binding energy.

According to Bitenski, the probability of spike events apt to induce shock waves is not negligible in kiloelectronvolt atomic ion bombardment.[127] Shock waves were also detected in early MD simulations of 5-keV Ar bombardment of a copper target.[128] With a shift of paradigm, Urbassek described the emission caused by kiloelectronvolt ions in low cohesive energy, van der Waals solids in terms of "gas flow," in accordance with early MD simulations.[129] Related concepts have been used to explain the ejection of intact molecules in fast atom bombardment (FAB) with liquid matrices (spraying process).[130] In a more recent model based on the thermal spike concept and using the formalism of fluid dynamics, Jakas et al. propose that the high pressure built up in the projectile track induces a rapid lateral and vertical expansion of the material and they calculate the volume of the ejection cone on the basis of the material properties.[131,132] This model was successfully adapted by the Garrison group to describe the sputtering of materials by kiloelectronvolt C_{60} projectiles (Section 2.3.2). In this large family of models, the intact ejection of kilodalton molecules, commonly observed in organic SIMS, is naturally explained by the correlated momentum of atoms directed toward the vacuum,[133] a feature also present in all the MD simulations of cluster bombardment.

2.3.2 Recent Developments and Hybrid Approaches

The fluid dynamics equations established by Jakas et al. were adapted to the case of small cluster bombardment (Au_3, C_{60}) of ice in the MEDF model. In this model, the initial quantities needed for the fluid dynamics calculations, which are the track radius (R_{cyl}) and the average energy density ($\bar{E}$), are derived from short time MD simulations (Fig. 2.22a).[30] The calculated volumes of the ejection cones agree very well with the predictions of full MD calculations between 5 and 40 keV and, because this hybrid model only requires the simulation of relatively small volumes of material for very short times, predictions can be made for impact energies that are out of reach of full MD calculations (>100 keV), as illustrated in Figure 2.22b.[86] Interestingly, the model indicates that in order to be productive for sputtering, the

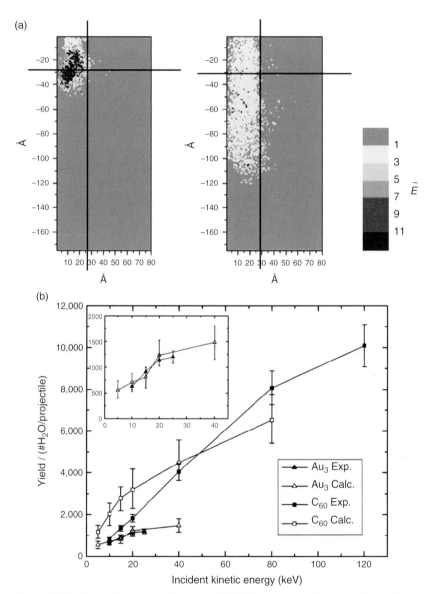

Figure 2.22 Deposited energy density and MEDF model. (a) Contour plots of the excitation energy $\bar{E}$ for 20 keV C_{60} and Au_3 bombardment. The black lines represent the position of R_{cyl}. (b) Measured and predicted total sputtering yields as a function of incident kinetic energy. The inset highlights the Au_3 experimental and calculated data. Reprinted from Russo et al.,[86] with permission from ACS publications.

energy of the projectile must be deposited within a depth equal to R_{cyl}, which is the radius of the energized track. For 5-keV bombardment, R_{cyl} is close to 2 nm for both Au_3 and C_{60}, but in contrast with C_{60}, Au_3 transfers most of its energy much deeper than R_{cyl} resulting in a lower average energy density $\bar{E}$ near the surface. Because the sputtering yield is proportional to $(R_{cyl}^3 \times \bar{E})$, Au_3 ejects significantly less material. The MEDF model was successfully applied to the description of the sputtering of Si, SiC, diamond, and graphite by fullerenes.[134,135]

Recently, MD simulations helped to develop a statistical sputtering model that explains the erosion of Si surfaces under continuous C_{60} bombardment (recall Section 2.2.6).[75] The Penn State group also used MD simulations to propose an analytical model describing the penetration of fullerene clusters in molecular solids. The simulations are consistent with an analytical model that considers the projectile as a single entity submitted to friction while it penetrates the organic medium.[84] This quadratic friction model (the friction force is proportional to v^2, like a turbulent flow) describes the exponential decay of the projectile velocity with time via a simple expression depending essentially on the mass and size of the cluster and on the density of the substrate. The model confirms that the motions of 20-keV fullerenes in benzene are adequately described as those of single particles.

As was already mentioned, atomistic MD simulations are limited in terms of system size and interaction times. At best, the required computer time increases linearly with those parameters and, therefore, an increase by one order of magnitude of one of them often makes the simulation intractable. For instance, the impact of a 25 nm water droplet (similar calculations as those described in Fig. 2.19) takes several months to calculate on an Intel Xeon E5450 (3 GHz) processor and considering much larger droplets and surfaces within this scheme would be unrealistic. Coarse-graining and parallelization of the calculations are helpful but in some cases insufficient. With the trade-off of giving up the detailed microscopic view of the events, other computer simulation schemes were proposed for specific applications. For instance, the breathing sphere model, where each molecule is represented by one entity with the ability to store and release internal energy, was successfully applied to describe the dynamics of laser ablation experiments.[136] Another interesting case concerns the simulation of droplet impacts on surfaces, aiming to explain the dynamic processes occurring in DESI mass spectrometry.[137,138] The authors treated the system using a multiphase numerical fluid mechanics approach, with a mesh of 0.12 μm, allowing them to describe the impact of a ~4 μm droplet on a thin water film, the situation that is believed to occur in DESI after a few droplet impacts on the sample. Figure 2.23 shows two snapshots taken from these simulations, after 0.4 and 0.8 μs, for an incidence angle of 55°, which is customary in DESI experiments. With this approach, the authors could analyze the impact dynamics, including the number, size, and momentum distributions of the daughter droplets as a function of the initial velocity and incidence angle, and correlate their results to existing experiments.

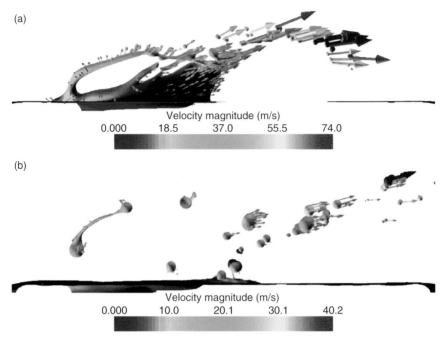

(a)

Velocity magnitude (m/s)
0.000   18.5   37.0   55.5   74.0

(b)

Velocity magnitude (m/s)
0.000   10.0   20.1   30.1   40.2

Figure 2.23   Microdroplet impact on a water surface.[137] Side views of the simulation at
(a) 0.4 µs and (b) 0.8 µs (3.7 µm droplet, 120 m/s, 55° incidence). The arrows and color
coding indicate relative velocity: blue is fluid at rest and red is fluid at the maximum
velocity in the simulation.   Reprinted from Costa and Cooks,[138] with permission
from Elsevier.

2.4   CONCLUSIONS

Many types of clusters, from a few atoms to massive nanoparticles and microdroplets, made of light (C) or heavy (Au) elements and with velocities ranging from
hundreds of meters per second to hundreds of kilometers per second are used as
probes for molecular surface analysis today. This huge variety of primary beams has
considerably broadened the scope of ion beam-based mass spectrometries and the
domain of the underlying physical interactions relevant to these methods. Because
of this diversity, cluster surface interactions cover a range of situations and ter-
minologies encompassing collision cascades, spikes, shock waves, hydrodynamic
flow, track formation, or cratering that initially pertained to microscopic as well as
macroscopic physics. By varying parameters such as the cluster size or the cluster
velocity, one can move from one range of interactions to another, as was shown for
light-element clusters, gold particles, or water droplets. This chapter demonstrates
that MD computer simulations constitute one of the most powerful and versatile
methods to unravel the multifaceted problem of cluster–surface interactions. With
them, we gain an outstanding microscopic view of the interaction of clusters with
complex surfaces, including detailed predictions of damage and molecular emission.

They provide a theoretical framework for understanding experimental observations and also make predictions that should help narrowing down the field of relevant experimental parameters.

Despite the successes of the MD simulations of cluster-induced sputtering, there are still limitations to overcome, either because of the required computer time and resources or, more fundamentally, because of the approximations used in the models. Simplified models or hybrid schemes may help in the first case but the second issue is far more difficult to deal with. For instance, the model can at best provide an approximate description of the state of the matter in the first stage of the cluster impact, the "witches brew" that is the high energy density region created in an organic material by a C_{60} projectile, where "the concept of discussing the reaction dynamics as a series of simple reaction events is not possible."[72] Another limitation is that only ground-state chemistry can be realistically approached even with the best potentials, discarding the electronic effects required to explain and predict ionization in SIMS. To address ionization, implementing some ion formation scenarios in the simulations is possible but probably insufficient given the complex nature of the energized region. For example, to explain ionization under massive cluster impact, Hiraoka et al. propose a mechanism based on electron–phonon coupling after the coherent phonon excitation induced by the subpicosecond compression of the target material.[78] Nevertheless, recent theoretical developments involving repeated bombardment, hybrid or multiscale models are encouraging and the description of more complex systems should be tractable in the future, partly because of conceptually new approaches and partly because computers always get faster. With the help of technological development and creativity, we can therefore hope to "measure what is measurable, and make measurable what is not so," as Galileo put it about four centuries ago.

ACKNOWLEDGMENTS

The authors wish to thank Prof. Barbara Garrison whose research achievements have shaped the world of sputtering simulations for continued collaboration and support over the years. OR acknowledges the financial support of the French Community of Belgium via the Concerted Research Action program (ARC NANHYMO: convention 07/12-003). BC is supported by a postdoctoral fellowship of the Belgian Fonds National pour la Recherche Scientifique (FNRS). AD is a Research Associate of the FNRS.

REFERENCES

1. Gardella Jr, J. A.; Clarke, S. *Proceedings of the Seventeenth International Conference on Secondary Ion Mass Spectrometry (SIMS XVII)*; Toronto, Ontario, Canada, **2011**.
2. Samela, J.; Nordlund, K. *Phys. Rev. Lett.* **2008**, 101 (2), 27601.
3. Garrison, B. J.; Postawa, Z. *Mass Spectrom. Rev.* **2008**, 27 (4), 289–315.
4. Andersen, H. H.; Brunelle, A.; Della-Negra, S.; Depauw, J.; Jacquet, D.; Le Beyec, Y.; Chaumont, J.; Bernas, H. *Phys. Rev. Lett.* **1998**, 80 (24), 5433–5436.
5. Morozov, S. N.; Rasulev, U. K. *Nucl. Instrum. Methods Phys. Res. B.* **2003**, 203, 192–197.
6. Appelhans, A. D.; Delmore, J. E. *Anal. Chem.* **1989**, 61 (10), 1087–1093.

7. Blain, M. G.; Della-Negra, S.; Joret, H.; Le Beyec, Y.; Schweikert, E. A. *Phys. Rev. Lett.* **1989**, 63 (15), 1625–1628.

8. Mahoney, J. F.; Perel, J.; Ruatta, S. A.; Martino, P. A.; Husain, S.; Cook, K.; Lee, T. D. *Rapid Commun. Mass Spectrom.* **1991**, 5 (10), 441–445.

9. Kotter, F.; Benninghoven, A. *Appl. Surf. Sci.* **1998**, 133 (1–2), 47–57.

10. Gillen, G.; Roberson, S. *Rapid Commun. Mass Spectrom.* **1998**, 12, 1303–1312.

11. Weibel, D.; Wong, S.; Lockyer, N.; Blenkinsopp, P.; Hill, R.; Vickerman, J. C. *Anal. Chem.* **2003**, 75 (7), 1754–1764.

12. Kollmer, F. *Appl. Surf. Sci.* **2004**, 231, 153–158.

13. Sigmund, P. *Phys. Rev.* **1969**, 184 (2), 383.

14. Sigmund, P. In *Sputtering by Particle Bombardment I*, Behrisch, R. Ed., Springer Verlag: Berlin, **1981**. pp. 9.

15. Sigmund, P.; Claussen, C. *J. Appl. Phys.* **1981**, 52 (2), 990–993.

16. Jakas, M. M.; Harrison, D. E., Jr. *Phys. Rev. Lett.* **1985**, 55 (17), 1782–1785.

17. Delcorte, A.; Garrison, B. J. *Nucl. Instrum. Methods Phys. Res. B.* **2011**, 269 (14), 1572–1577.

18. Delcorte, A.; Garrison, B. J. *Nucl. Instrum. Methods Phys. Res. B.* **2007**, 255 (1), 223–228.

19. Delcorte, A.; Garrison, B. J. *J. Phys. Chem. B.* **2000**, 104 (29), 6785–6800

20. Ziegler, J. F.; Biersack, J. P.; Littmark, U. *The Stopping and Ranges of Ions in Solids*; **1985**; Vol. 1; Pergamon Press, New York.

21. *Stopping Range of Ions in Matter*. http://www.srim.org/ (accessed Dec 3, 2012).

22. Robinson, M. T.; Behrisch, R. In *Sputtering by Particle Bombardment I*; Behrisch, R., Ed.; Springer Verlag, Berlin, **1981**, p 73.

23. Harrison Jr, D. E. *Crit. Rev. Solid State Mater. Sci.* **1988**, 14 (S1), 1–78.

24. Winograd, N.; Garrison, B. J. *Ion Spectroscopies for Surface Analysis*. Plenum Press: New York, **1991**.

25. Stuart, S. J.; Tutein, A. B.; Harrison, J. A. *J Chem Phys.* **2000**, 112, 6472.

26. Brenner, D. W. *Phys. Rev. B.* **1990**, 42 (15), 9458.

27. Brenner, D. W.; Harrison, J. A.; White, C. T.; Colton, R. J. *Thin Solid Films* **1991**, 206 (1–2), 220–223.

28. Brenner, D. W.; Shenderova, O. A.; Harrison, J. A.; Stuart, S. J.; Ni, B.; Sinnott, S. B. *J. Phys. Condens. Matter* **2002**, 14, 783.

29. Smiley, E. J.; Postawa, Z.; Wojciechowski, I. A.; Winograd, N.; Garrison, B. J. *Appl. Surf. Sci.* **2006**, 252 (19), 6436–6439.

30. Russo Jr, M. F.; Garrison, B. J. *Anal. Chem.* **2006**, 78 (20), 7206–7210.

31. Postawa, Z.; Czerwinski, B.; Szewczyk, M.; Smiley, E. J.; Winograd, N.; Garrison, B. J. *Anal. Chem.* **2003**, 75 (17), 4402–4407.

32. Delcorte, A.; Leblanc, C.; Pleunis, C.; Hamraoui, K. *J. Phys. Chem. C*. **2013**. In Press.

33. Szakal, C.; Kozole, J.; Russo Jr, M. F.; Garrison, B. J.; Winograd, N. *Phys. Rev. Lett.* **2006**, 96 (21), 216104.

34. Delcorte, A.; Garrison, B. J. *J. Phys. Chem. C* **2007**, (111), 15312–15324.

35. Urbassek, H. M.; Waldeer, K. T. *Phys. Rev. Lett.* **1991**, 67 (1), 105.

36. Wojciechowski, I. A.; Garrison, B. J. *J. Phys. Chem. A.* **2006**, 110 (4), 1389–1392.

37. Shard, A. G.; Brewer, P. J.; Green, F. M.; Gilmore, I. S. *Surf. Interface Anal.* **2007**, 39 (4), 294–298.

38. He, C.; Rosencrance, S. W.; Postawa, Z.; Xu, C.; Chatterjee, R.; Riederer, D. E.; Garrison, B. J.; Winograd, N. *Nucl. Instrum. Methods Phys. Res. B* **1995**, 100 (2), 209–212.

39. Czerwinski, B.; Rzeznik, L.; Paruch, R.; Garrison, B. J.; Postawa, Z. *Vacuum* **2009**, 83, S95–S98.

40. Garrison, B. J.; Postawa, Z.; Ryan, K. E.; Vickerman, J. C.; Webb, R. P.; Winograd, N. *Anal. Chem.* **2009**, 81, 2260–2267

41. Hamraoui, K.; Delcorte, A. *J. Phys. Chem. C* **2010**, 114 (12), 5458–5467.

42. Paruch, R.; Rzeznik, L.; Czerwinski, B.; Garrison, B. J.; Winograd, N.; Postawa, Z. *J. Phys. Chem. C* **2009**, 113 (14), 5641–5648.

43. Delcorte, A.; Bertrand, P.; Garrison, B. J.; Hamraoui, K.; Mouhib, T.; Restrepo, O. A.; Santos, C. N.; Yunus, S. *Surf. Interface Anal.* **2010**, 42 (8), 1380–1386.

44. Magonov, S. N.; Yeung, E. S. *Langmuir* **2003**, 19 (3), 500–504.

45. Bertrand, P.; Bu, H.; Rabalais, J. W. *J. Phys Chem.* **1993**, 97 (51), 13788–13791.
46. Stapel, D.; Brox, O.; Benninghoven, A. *Appl. Surf. Sci.* **1999**, 140 (1–2), 156–167.
47. Diehnelt, C. W.; Van Stipdonk, M. J.; Schweikert, E. A. *Int. J. Mass Spectrom.* **2001**, 207 (1–2), 111–122.
48. Weibel, D. E.; Lockyer, N.; Vickerman, J. C. *Appl. Surf. Sci.* **2004**, 231–232, 146–152.
49. Postawa, Z.; Czerwinski, B.; Winograd, N.; Garrison, B. J. *J. Phys. Chem. B.* **2005**, 109 (24), 11973–11979.
50. Czerwinski, B.; Samson, R.; Garrison, B. J.; Winograd, N.; Postawa, Z. *Vacuum* **2006**, 81 (2), 167–173.
51. Czerwinski, B.; Delcorte, A.; Garrison, B. J.; Samson, R.; Winograd, N.; Postawa, Z. *Appl. Surf. Sci.* **2006**, 252 (19), 6419–6412.
52. Garrison, B. J.; Delcorte, A.; Krantzman, K. D. *Acc. Chem. Res.* **2000**, 33 (2), 69–77.
53. Delcorte, A.; Medard, N.; Bertrand, P. *Anal. Chem.* **2002**, 74 (19), 4955–4968.
54. Delcorte, A.; Yunus, S.; Wehbe, N.; Nieuwjaer, N.; Poleunis, C.; Felten, A.; Houssiau, L.; Pireaux, J. J.; Bertrand, P. *Anal. Chem.* **2007**, 79 (10), 3673–3689.
55. Wehbe, N.; Heile, A.; Arlinghaus, H.-F.; Bertrand, P.; Delcorte, A. *Anal. Chem.* **2008**, 80 (16), 6235–6244.
56. Restrepo, O. A.; Delcorte, A. *Surf. Interface Anal.* **2011**, 43 (1–2), 70–73.
57. Restrepo, O. A.; Prabhakaran, A.; Hamraoui, K.; Wehbe, N.; Yunus, S.; Bertrand, P.; Delcorte, A. *Surf. Interface Anal.* **2011**, 42 (6–7), 1030–1034.
58. Delcorte, A. Fundamental Aspects of Organic SIMS. In *ToF-SIMS: Surface Analysis by Mass Spectrometry*; Vickerman, J. C., Briggs, D., Eds.; Surface Spectra/IMPublications: Chichester, UK, **2001**.
59. Delcorte, A. *Appl. Surf. Sci.* **2006**, 252 (19), 6582–6587.
60. Conlan, X. A.; Lockyer, N. P.; Vickerman, J. C. *Rapid Commun. Mass Spectrom.* **2006**, 20 (8), 1327–1334.
61. Conlan, X. A.; Biddulph, G. X.; Lockyer, N. P.; Vickerman, J. C. *Appl. Surf. Sci.* **2006**, 252 (19), 6506–6508.
62. Mahoney, C. M. *Mass Spectrom. Rev.* **2010**, 29 (2), 247–293.
63. Moellers, R.; Tuccitto, N.; Torrisi, V.; Niehuis, E.; Licciardello, A. *Appl. Surf. Sci.* **2006**, 252 (19), 6509–6512.
64. Nieuwjaer, N.; Poleunis, C.; Delcorte, A.; Bertrand, P. *Surf. Interface Anal.* **2009**, 41 (1), 6–10.
65. Ninomiya, S.; Ichiki, K.; Yamada, H.; Nakata, Y.; Seki, T.; Aoki, T.; Matsuo, J. *Rapid Commun. Mass Spectrom.* **2009**, 23 (11), 1601.
66. Delcorte, A.; Garrison, B. J.; Hamraoui, K. *Anal. Chem.* **2009**, 81 (16), 6676–6686.
67. Delcorte, A.; Poleunis, C.; Bertrand, P. *Appl. Surf. Sci.* **2006**, 252 (19), 6542–6546.
68. Webb, R. P.; Garrison, B. J.; Vickerman, J. C. *Surf. Interface Anal.* **2010**, 43 (1–2), 116–119.
69. Krantzman, K. D.; Kingsbury, D. B.; Garrison, B. J. *Nucl. Instrum. Methods Phys. Res. B* **2007**, 255 (1), 238–241.
70. Krantzman, K. D.; Kingsbury, D. B.; Garrison, B. J. *Appl. Surf. Sci.* **2006**, 252 (19), 6463–6465.
71. Wojciechowski, I. A.; Garrison, B. J. *J. Phys. Chem. B* **2005**, 109 (7), 2894–2898.
72. Ryan, K. E.; Wojciechowski, I. A.; Garrison, B. J. *J. Phys. Chem. C* **2007**, 111 (34), 12822–12826.
73. Russo Jr, M. F.; Postawa, Z.; Garrison, B. J. *J. Phys. Chem. C* **2009**, 113 (8), 3270–3276.
74. Paruch, R.; Rzeznick, L.; Russo, M. F.; Garrison, B. J.; Postawa, Z. *J. Phys. Chem. C* **2010**, 114 (12), 5532–5539.
75. Krantzman, K. D.; Wucher, A. *J. Phys. Chem. C* **2010**, 114 (12), 5480–5490.
76. Aoki, T.; Matsuo, J. *Nucl. Instrum. Methods Phys. Res. B* **2005**, 228 (S1), 46–50.
77. Cornett, D. S.; Lee, T. D.; Mahoney, J. F. *Rapid Commun. Mass Spectrom.* **1994**, 8 (12), 996–1000.
78. Hiraoka, K.; Mori, K.; Asakawa, D. *J. Mass Spectrom.* **2006**, 41 (7), 894–902.
79. Asakawa, D.; Fujimaki, S.; Hashimoto, Y.; Mori, K.; Hiraoka, K. *Rapid Commun. Mass Spectrom.* **2007**, 21 (10), 1579–1586.
80. Aksyonov, S. A.; Williams, P. *Rapid Commun. Mass Spectrom.* **2001**, 15 (21), 2001–2006.
81. Takats, Z.; Wiseman, J. M.; Cooks, R. G. *J. Mass Spectrom.* **2005**, (40), 1261–1275.
82. Kauppila, T. J.; Wiseman, J. M.; Ketola, R. A.; Kotiaho, T.; Cooks, R. G.; Kostiainen, R. *Rapid Commun. Mass Spectrom.* **2006**, 20 (3), 387–392.

83. Smiley, E. J.; Winograd, N.; Garrison, B. J. *Anal. Chem.* **2007**, 79 (2), 494–499.

84. Garrison, B. J.; Ryan, K. E.; Russo Jr, M. F.; Smiley, E. J.; Postawa, Z. *J. Phys. Chem. C (letters)*. **2007**, 111 (28), 10135–10136.

85. Ryan, K. E.; Garrison, B. J. *Anal. Chem.* **2008**, 80 (17), 6666–6670.

86. Russo Jr, M. F.; Szakal, C.; Kozole, J.; Winograd, N.; Garrison, B. J. *Anal. Chem.* **2007**, 79 (12), 4493–4498.

87. Delcorte, A.; Garrison, B. J.; Hamraoui, K. *Surf. Interface Anal.* **2011**, 43 (1–2), 16–19.

88. Biddulph, G. X.; Piwowar, A. M.; Fletcher, J. S.; Lockyer, N. P.; Vickerman, J. C. *Anal. Chem.* **2007**, 79 (19), 7529–7266.

89. Chan, C.-C.; Bogar, M. S.; Miller, S. A.; Attygalle, A. B. *J. Am. Soc. Mass Spectrom.* **2010**, 21 (9), 1554–1560.

90. Hiraoka, K.; Asakawa, D.; Fujimaki, S.; Takamizawa, A.; Mori, K. *Eur. Phys. J. D* **2006**, 38 (1), 225–229.

91. Moritani, K.; Houzumi, S.; Takeshima, K.; Toyoda, N.; Mochiji, K. *J. Phys. Chem. C* **2008**, 112 (30), 11357–11362.

92. Tomsic, A.; Schroeder, H.; Kompa, K. L.; Gebhardt, C. R. *J. Chem. Phys.* **2003**, 119 (12), 6314–6323.

93. Tomsic, A.; Gebhardt, C. R. *Chem. Phys. Lett.* **2004**, 386 (1–3), 55–59.

94. Tomsic, A.; Gebhardt, C. R. *J. Chem. Phys.* **2005**, 123 (6), 064704.

95. Delcorte, A.; Garrison, B.J. *Nucl. Instr. Meth. B*. **2012**, In press.

96. Thirumuruganandham, S. P.; Urbassek, H. M. *Int. J. Mass Spectrom.* **2010**, 289 (2–3), 119–127.

97. Thirumuruganandham, S. P.; Urbassek, H. M. *Rapid Commun. Mass Spectrom.* **2010**, 24 (3), 349–354.

98. Sun, S. N.; Urbassek, H. M. *J. Phys. Chem. B* **2011**, 115 (45), 13280–13286.

99. Rabbani, S.; Barber, A. M.; Fletcher, J. S.; Lockyer, N. P.; Vickerman, J. C. *Anal. Chem.* **2011**, 83 (10), 3793–3800.

100. Ninomiya, S.; Nakata, Y.; Ichiki, K.; Seki, T.; Aoki, T.; Matsuo, J. *Nucl. Instrum. Methods Phys. Res. B* **2007**, 256 (1), 493–496.

101. Aoki, T.; Matsuo, J.; Insepov, Z.; Yamada, I. *Nucl. Instrum. Methods Phys. Res. B* **1997**, 121 (1–4), 49–52.

102. Aoki, T.; Matsuo, J.; Takaoka, G. *Nucl. Instrum. Methods Phys. Res. B* **2003**, 202, 278–282.

103. Aoki, T.; Matsuo, J.; Takaoka, G.; Yamada, I. *Nucl. Instrum. Methods Phys. Res. B* **2003**, 206, 861–865.

104. Aoki, T.; Matsuo, J. *Nucl. Instrum. Methods Phys. Res. B* **2004**, 216, 185–190.

105. Aoki, T.; Matsuo, J. *Nucl. Instrum. Methods Phys. Res. B* **2007**, 255 (1), 265–268.

106. Aoki, T.; Matsuo, J. *Nucl. Instrum. Methods Phys. Res. B* **2007**, 261 (1–2), 639–642.

107. Rzeznik, L.; Czerwinski, B.; Garrison, B. J.; Winograd, N.; Postawa, Z. *J. Phys. Chem. C* **2008**, 112 (2), 521–531.

108. Rzeznick, L.; Czerwinski, B.; Paruch, R.; Garrison, B. J.; Postawa, Z. *Nucl. Instrum. Methods Phys. Res. B* **2009**, 267 (8–9), 1436–1439.

109. Rzeznik, L.; Paruch, R.; Czerwinski, B.; Garrison, B. J.; Postawa, Z. *Vacuum* **2006**, 83, S155–S158.

110. Czerwinski, B.; Rzeznick, L.; Paruch, R.; Garrison, B. J.; Postawa, Z. *Nucl. Instrum. Methods Phys. Res. B* **2011**, 269 (14), 1578–1581.

111. Czerwinski, B.; Rzeznik, L.; Paruch, R.; Garrison, B. J.; Postawa, Z. *Nucl. Instrum. Methods Phys. Res. B* **2009**, 267 (8–9), 1440–1443.

112. Tempez, A.; Schultz, J. A.; Della Negra, S.; Depauw, J.; Jacquet, D.; Novikov, A.; Lebeyec, Y.; Pautrat, M.; Caroff, M.; Ugarov, M. *Rapid Commun. Mass Spectrom.* **2004**, 18 (4), 371–376.

113. Novikov, A.; Della-Negra, S.; Fallavier, M.; Le Beyec, Y.; Pautrat, M.; Schultz, J. A.; Tempez, A.; Woods, A. *Rapid Commun. Mass Spectrom.* **2005**, 19 (13), 1851–1857.

114. Anders, C.; Urbassek, H. M. *Nucl. Instrum. Methods Phys. Res. B.* **2008**, 266 (1), 44–48.

115. Anders, C.; Urbassek, H. M. *Nucl. Instrum. Methods Phys. Res. B.* **2005**, 228 (S1), 57–63.

116. Restrepo, O. A.; Prabhakaran, A.; Delcorte, A. *Nucl. Instrum. Methods Phys. Res. B.* **2011**, 269 (14), 1595–1599.

117. Restrepo, O. A.; Delcorte, A. *J. Phys. Chem. C* **2011**, 115 (26), 12751–12759.

118. Zimmerman, S.; Urbassek, H. M. *Nucl. Instrum. Methods Phys. Res. B.* **2007**, 255 (1), 208–213.
119. Thompson, M. W. *Philos. Mag.* **1968**, 18 (152), 377.
120. Pachuta, S. J.; Cooks, R. G. *Chem. Rev.* **1987**, 87 (3), 647–669.
121. Beuhler, R.; Friedman, L. *Chem. Rev.* **1986**, 86 (3), 521–537.
122. Mahoney, J. F.; Perel, J.; Lee, T. D.; Martino, P. A.; Williams P. *J. Am. Soc. Mass Spectrom.* **1992**, 3 (4), 311–317.
123. Carter, G. *Radiat. Eff. Lett.* **1979**, 43 (4–5), 193–199.
124. Reimann, C. T. *Fundatmental Process in Sputtering of Atoms and Molecules*. 43 ed.; Sigmund, P., Ed.; Matematisk-fysiske Meddelelser: Copenhagen, **1993**.
125. Bitensky, I. S.; Parilis, E. S. *Nucl. Instrum. Methods Phys. Res. B* **1987**, 21 (1–4), 26–36.
126. Johnson, R. E.; Sundqvist, B. U. R.; Hedin, A.; Fenyo, D. *Phys. Rev. B* **1989**, 40 (1), 49–53.
127. Bitensky, I. S. *Nucl. Instrum. Methods Phys. Res. B* **1993**, 83 (1), 110–116.
128. Webb, R. P.; Harrison Jr, D. E. *Appl. Phys. Lett.* **1981**, 39 (4), 311–312.
129. Urbassek, H. M.; Michl, J. *Nucl. Instrum. Methods Phys. Res. B* **1987**, 22 (4), 480–490.
130. Wong, S. S.; Rollgen, F. W. *Nucl. Instrum. Methods Phys. Res. B* **1986**, 14 (4–6), 436–447.
131. Jakas, M. M.; Bringa, E. M. *Phys. Rev. B* **2000**, 62 (2), 824–830.
132. Jakas, M. M.; Bringa, E. M.; Johnson, R. E. *Phys. Rev. B* **2002**, 65 (16), 165425.
133. Sundqvist, B. U. R. *Nucl. Instrum. Methods Phys. Res. B* **1990**, 48 (1–4), 517–524.
134. Krantzman, K. D.; Webb, R. P.; Garrison, B. J. *Appl. Surf. Sci.* **2008**, 225 (4), 837–840.
135. Krantzman, K. D.; Garrison, B. J. *J. Phys. Chem. C* **2009**, 113 (8), 3239–3245.
136. Zhigilei, L. V.; Kodali, P.; Garrison, B. J. *J. Phys. Chem. B* **1997**, 101 (11), 2028–2037.
137. Costa, A. B.; Cooks, R. G. *Chem. Commun.* **2007**, 38, 3915–3917.
138. Costa, A. B.; Cooks, R. G. *Chem. Phys. Lett.* **2008**, 464 (1–3), 1–8.

CHAPTER 3

ION SOURCES USED FOR SECONDARY ION MASS SPECTROMETRY

Albert J. Fahey

3.1 INTRODUCTION

It has been approximately 100 years now since Dunoyer[1] produced the first molecular beam showing that atoms in a vacuum move along straight lines. Since then, the applications of molecular beams have been numerous throughout the twentieth- and into the twenty-first century, laying the groundwork for significant technological advances as well as fundamental research. The use of molecular beams for sputtering has been but one of their applications.

Although the phenomenon of sputtering had been known since the mid-nineteenth century it was not until the late 1940s when advancements in vacuum technology led to the first experiments to demonstrate sputtering of secondary ions by Herzog and Viehbock.[2] Subsequent experimental work by Laegreid and Wehner[3] suggested that ions produced during sputtering could be used to produce beams of secondary ions. Later, in the late 1960s, others such as Nelson[4] and Sigmund[5] investigated the theoretical mechanisms of sputtering, leading to a reasonable understanding of this phenomenon. All of these investigators concerned themselves almost exclusively with beams of single atomic ions.

Currently, secondary ion mass spectrometry (SIMS) is a well-developed field having significant applications in the measurement of implanted dopants and thin-film compositions for semiconductor manufacturing and for other inorganic materials and in the measurement of isotope ratios and trace elements for geochemistry and cosmochemistry, just to name a few. In addition, it has found significant application in the analysis of organic and biological materials, as evidenced in this

Cluster Secondary Ion Mass Spectrometry: Principles and Applications, First Edition.
Edited by Christine M. Mahoney.
© 2013 John Wiley & Sons, Inc. Published 2013 by John Wiley & Sons, Inc.

book. The use of SIMS to study organic materials is currently enjoying significant advancements and growth largely because of the development of what are referred to as *cluster ion beams* that can be used for sputtering. However, referring to all such ion beams that have been developed is somewhat of a misnomer. For the purpose of clarity, in this chapter, we will divide the discussion of primary beams used for SIMS and their sources into (i) atomic, which refers to sources that produce single atom ions; (ii) molecular, referring to sources that produce ion beams of atoms that are chemically bound to each other immediately before being ionized (i.e., in the gas phase); and (iii) clusters, which are usually made up of atoms of a single element that would not, in equilibrium, normally chemically bond in the gas phase (e.g., a noble gas). However, these divisions are not clear cut, especially as a single source design can also be used to produce different ion beams. Each of these requires different mechanisms for their production and also use or need different ionization mechanisms to produce the final ionic species.

The use and implementation of each source type and its efficacy to address the problem at hand is significantly influenced by the ion optical column to which it is attached, especially in cases where it is desired to select specific very large clusters from a mixed-mass beam. The significance of column design is recognized but is beyond the scope of this chapter. Features of the column design will only be mentioned when key to the utilization of a particular source and where those features may not be obvious or common.

3.2 RESEARCH NEEDS THAT HAVE INFLUENCED THE DEVELOPMENT OF PRIMARY ION SOURCES FOR SPUTTERING

SIMS is the only microanalytical method that promises the potential to determine the molecular and/or atomic composition of a material not only on its surface but throughout its depth in three dimensions in the analyzed region. This is because it is the only method that can remove an atom or molecule from a material for analysis in a mass spectrometer, with a lateral spatial resolution limited only by the focal size of the primary beam or the chromatic aberrations of the ion microscope system. Chromatic aberrations of ion microscopes can be reduced to increase the spatial resolution of the imaged area but this comes at the expense of secondary ion signal and much lower efficiency. Field-emission ion sources now produce nanometer-focused beams of Ga able to mill 100 nm sections of semiconductors and other materials for later analysis by methods such as TEM (Transmission Electron Microscopy) or AEM (Analytical Electron Microscopy). However, commercial incarnations of SIMS instruments, although significantly capable and applicable to a wide range of problems, have fallen short of this goal. Part of the reason for this is that current primary ion beam designs lack the necessary characteristics to accomplish it.

Researchers have sought to produce ion beams imbued with all of the necessary properties to accomplish the potential promise of SIMS. They have been successful at accomplishing some of this, although not yet all. Some of the initial impetus to produce new ion beams was to improve the secondary ion emission of species of interest, initially from elements such as boron, oxygen, and phosphorus. The production of finer-scale semiconductors pushed the need to measure dopant levels in very shallow depth ranges. The recognition that high depth resolution and minimal layer intermixing in depth profiling was influenced by the incident beam energy gave rise to the notion that not only could directly reducing the energy of the incident atomic ion beam improve depth resolution, but also reducing the energy per atom by using molecular, or cluster, beams that spread the energy laterally across the surface of the material being analyzed.[6,7]

Some advancement came easily, as with the microbeam Cs ion source produced by CAMECA on their "ims" series instruments and the use of very low energies, especially with Cs, to analyze ultrashallow implants for semiconductor applications. Others, such as the use of molecular or cluster beams had fewer acceptances by commercial manufacturers despite their good performance and relative ease with which they could be produced. The ion beams used for SIMS, up until about 2003, were primarily atomic, or beams of small molecules (e.g., O_2^+). This made it very difficult or impossible to perform a depth profile of any organic, polymer, or biological material to any significant depth (e.g., see review by Mahoney[8]). Atomic, or small-molecule-high-energy, ion beams effectively destroyed organic samples during analysis, especially when profiling, cleaving molecular bonds, and changing the material under analysis more radically than in inorganic samples.

The development of cluster and molecular ion beams has opened up a wide range of applications of SIMS to organic and biological samples.[9] As yet, however, the ionization yields for molecular species are still relatively poor. The lateral resolution, limited by the beam size for nonstigmatic SIMS instruments, does not approach that achievable with Ga field-emission sources that have been used in applications apart from SIMS.

3.3 FUNCTIONAL ASPECTS OF VARIOUS ION SOURCES

Basic operational parameters can be set forth to rate the efficacy of ion beams for use as primary ion sources with SIMS. These include obvious parameters as well as more subtle parameters that can be significant.

3.3.1 Energy Spread in the Beam

The lower the energy spread in the ion beam as it leaves the ionization region, the more tightly the beam can be focused through a series of lenses onto the sample surface. The chromatic aberration can be written as $d_c = K_c r_a \Delta V / V_0$ where d_c is the diameter of a focused ion beam (FIB), initially of radius r_a passing through an einzel lens with a characteristic dimensionless factor K_c (typically 1–5), and $\Delta V / V_0$ is the relative energy spread in the beam. The spectral-width of ion energies produces chromatic aberrations in the focusing thus limiting beam diameters and the effectiveness of focusing elements. Low energy spreads can be generally viewed as coupled to the axial depth of the region of ions extracted from the source to enter the beam. In the case of a plasma source, for example, the deeper into the plasma the extraction field removes ions, the broader the energy distribution of ions in the final beam.

3.3.2 Point-Source Ionization

As the refraction of ions passing through an einzel lens is slightly greater than proportional to their distance from the axis, the ion paths do not cross at a single point but form an envelope whose narrowest cross section is given by: $d_s = K_s r_a^3 / D^2$, where r_a is the radius of the beam entering the lens, K_s is a dimensionless factor (typically ranging from 1 to 5), and D is a scaling factor related to the axial extent of the field in the lens.[10] Clearly, the benefit of having a spatially small source diameter is significant as the final focused beam depends on the cube of the input radius.

3.3.3 Stable Emission

Ion emission must be stable over short and long time periods with different constraints on the level of stability depending on the application. Typically, short-term stability is required where ratios of secondary species will be formed and compared. For these measurements, the stability must be high enough such that secondary signal variations are small during the course of a cycle through the masses for a quadrupole or magnetic sector instrument.

When performing a depth profile, or large area map, signal variations due to changes in the primary ion intensity can be troubling. For example, liquid metal field-emission sources can become unstable as they age and simply cease emission, although these are not the only types of sources that can be problematic. Some of the most stable sources involve thermal ionization, such as the Cs^+ ion source. Emission is stabilized simply by the thermal mass of the source tip as long as there is enough reservoir material to ionize.[11]

3.3.4 Ion Reactivity

Numerous publications exist that discuss the benefits derived from sputtering with reactive ion beams as opposed to nonreactive beams, such as noble gas ions. The most commonly known and widely studied are the use of O_2^+ or O^- for the generation of positive secondary ions and Cs^+ for the production of negative secondary ions. Both of these have been widely studied for use with inorganic matrices, primarily those of interest to the semiconductor industry.[12] The effects of sputter rate and ionization efficiency have been separated for several systems using these ion beams. Primarily, the use of such ions enhances the secondary ion yield during sputtering, making reactive species such as O and Cs desirable for sputtering.

3.3.5 Source Lifetime

The time between source-rebuilds varies considerably for the sources currently used for SIMS analysis. Not surprisingly, there is a trade-off between the benefits of using a particular source and ion combination (often along with the selected operating regime) and the amount of time and money consumed for maintenance and repair. The precise parameters for source lifetime are not consistently documented for the commercially available sources, much less for the ion sources discussed as one-off designs for novel ion beams and their application.

In general, sources are not widely adopted if their lifetimes are less than ~50 h, or 1 week's worth of run time. The sources provided on commercial instruments have run times ranging from several hundred to one thousand hours when used as specified and within conservative operating parameters. There are several ways in which a source lifetime can be defined; the most common are that the stability degrades, making it effectively unusable, or that no ions are generated at all. The lifetimes of sources can generally be extended with the proper selection of materials, operating parameters, and construction geometry; however, commercial suppliers often do not have the resources or compelling markets to make such changes.

3.3.6 Penetration Depth and Surface Energy Spread of the Projectile

One of the original reasons for examining the use of cluster or molecular ion beams was to reduce the interlayer mixing for inorganic depth profiling, thus increasing the depth resolution of SIMS analysis. This is specifically important for use in advanced semiconductor device manufacture and development. Additionally, it was documented[13] that the use of cluster or molecular ion beams reduces the damage to samples, especially organic systems, thus preserving molecular signals throughout the analysis allowing depth-profile analysis of organic systems. Currently, it is well understood that the use of cluster and molecular ion beams, at least partly because of the shallow penetration depth, is the preferred method for analysis and profiling of organic systems.[7,8,14]

3.4 ATOMIC ION SOURCES

There are a considerable number of ways to produce beams of atomic ions for use as a primary ion source. This section reviews the ion sources that are used to produce these atomic ion beams. Although these are monatomic and cannot in any way be considered "cluster" sources, the production of these atomic ion beams is similar to and has often preceded their molecular or cluster cousins, lending conceptual notions from one to the other. These include field emission, typically from tips of a low melting point metal or metal eutectic that can form Taylor cones with slight heating and an applied field,[15] radio frequency (RF) ionization in a plasma,[16,17] electron-impact sources that are typically used to produce beams of positive ions from gases,[18] thermal ionization,[11] DC-glow discharges,[19,20] and sputtering.[21]

3.4.1 Field Emission

A widely used ion beam produced by field emission is Ga^+. Such ion beams are typically produced for use in focused ion milling instruments (often referred to as *focused ion beams* or FIBs) and can be focused into beams with nanometer

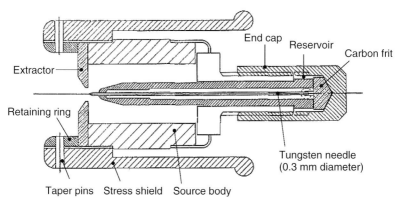

Figure 3.1   Schematic of Ga field-emission source tip. A thin film of liquid metal covers a tungsten needle. The liquid metal film is subjected to a high electric field, whereby a Taylor cone is formed (distortion of the liquid metal in the form of a conical protrusion). Electric fields near the tip of the Taylor cone are sufficiently high to cause emission of ions. Reproduced from Prewett and Jefferies,[15] with permission from the American Institute of Physics (AIP).

diameters.[15] The ion beams focus very well because of their low energy spread and near point-source emission characteristics (see section 3.3). Emission stability of Ga$^+$ from field-emission tips is reasonably stable but must be controlled through feedback that adjusts the voltage on an extraction electrode to stabilize the emission current (typically $1-2$ μA). If ion emission does not occur spontaneously when the extraction field has been raised, the tip can be heated to mobilize Ga from the tip-reservoir and to supposedly reform the Taylor cone so that the geometry is more conducive for emission. An example field-emission source diagram from Prewett and Jefferies[15] is shown in Figure 3.1.

The promise of high brightness achieved with field-emission sources has encouraged researchers to investigate the application of this type of ion source to the production of atomic ion beams from metals other than Ga (e.g., Au, In, Bi), from gases, and for the production of cluster and molecular beams as discussed subsequently. Unfortunately, Ga offers no significant chemical enhancement of ion yields. Other elements have been used in field-emission sources but have yet to demonstrate the small beam diameters available with Ga.

3.4.2   Radio Frequency (RF) Ionization

The ionization of gases through the application of RF electromagnetic fields has become a staple of molecular beam generation. A simple illustration of the components of an RF source is shown in Figure 3.2. Early applications of RF dissociation and ionization are described by King and Zacharias.[22] One of the primary advantages of the use of RF discharges is the lack of electrodes in the plasma to contaminate the beam or to react with it.

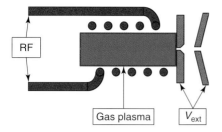

Figure 3.2 Cartoon showing the basic components of an RF plasma ion source. The gas plasma (blue) is contained within a glass or quartz vessel. The end cap and extractor (orange) are biased appropriately so that ions are accelerated out of the plasma. The RF coils (red) are external to the plasma and are often cooled along with the quartz vessel.

RF-driven plasma sources have been developed and used by Ji et al.[16] for the production of noble gas atomic ion beams with FIB applications. Recently, Smith et al.[23] have demonstrated a commercial version of an RF ion source. Claims of very high brightness have been made by Smith et al. but as yet have not been confirmed when compared with other ion sources under similar operating conditions.

3.4.3 Electron Impact

Electron-impact ion sources have been used primarily with Ar^+ as sputtering sources in time-of-flight SIMS instruments that remove layers of material during depth-profile analysis. The ion source design used on the ION-TOF instruments[24] is a simple ion box design not dissimilar to sources that have been designed since 1947[18] This type of ion source is common and can be purchased by a number of third-party supplies for various applications of sputtering, including cleaning, and for other surface analysis methods such as Auger electron spectroscopy and X-ray photoelectron spectroscopy. A cartoon of an electron-impact ion source is shown in Figure 3.3.

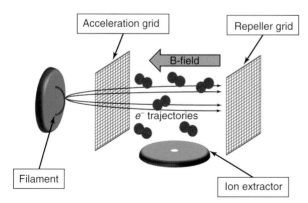

Figure 3.3 Cartoon of the geometry and operation of a typical electron-impact ion source. A heated filament creates electrons that pass through an acceleration grid. Each grid is 80–90% transparent to electrons so that electrons that do not suffer a collision between the grids and can reenter the ionization region between them. The electrons ionize gas (e.g., Ar) that is leaked into the system. The ions created are then extracted and subsequently focused.

The basic elements of an electron-impact source are (i) a source of electrons, usually a thermionic source (e.g., a heated filament) and (ii) acceleration grid and repeller grid that, together, produce an electric field-free region where electrons can impact the atomic or molecular species to be ionized. Often, the repeller electrode is a grid, as shown, so that electrons that make it through the field-free ionization region can be recycled into the field-free region to have another chance at producing an ion. Often, magnetic fields are added to perturb the electron trajectories into helical orbits around the magnetic field lines, increasing the chance of ionizing a target. Finally, (iii) an extraction electrode with an aperture is properly placed and biased to remove ionized species from the field-free region and to form them into an ion beam.

3.4.4 Thermal Ionization

There are few sources that use thermal ionization. The most commonly known ion beam produced by thermal ionization is the Cs^+ ion beam.[11] One of the advantages of this type of source is that the stability of the ion beam is very high and the energy spread in the beam is low. In addition, the Cs^+ ion sources generally have long lifetimes. Figure 3.4 shows a schematic of the first Cs ion source used for SIMS.

The Cs charge was held in the reservoir where it could be heated to produce Cs vapor. The vapor would make its way down the steel tube to the porous ionizer that was heated to produce thermal ionization of the Cs atoms. Typically, the reservoir heater was not used for SIMS applications because the General Ionex source design was capable of providing far more beam current than could reasonably be used for SIMS analysis, and upon source shutdown the reservoir cooler was used to quickly reduce the temperature of the Cs so that it did not clog the porous ionizer.

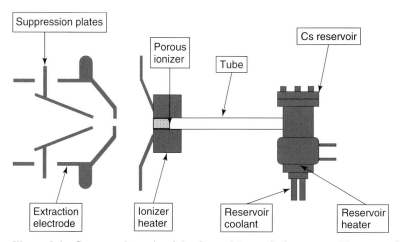

Figure 3.4 Cartoon schematic of the General Ionex Cs ion source. The reservoir accommodated a full 2 g ampule of Cs metal, which lasted many hundreds of hours. The porous ionizer was typically made of sintered tungsten pressed into stainless steel tubing.

A subsequent version of the Cs^+ ion source called the *microbeam Cs source* was used by Cameca on the "ims-N series" of SIMS instruments as well as on the nanoSIMS. At normal incidence, it is capable of focusing to a 50–100 nm spot with picoamp currents.

3.4.5 DC-Glow Discharge

There have been numerous atomic ion beams produced by duoplasmatrons, a type of DC-glow discharge ion source.[19,20,25–27] In all DC-glow discharge systems a cathode, whether cold and emitting electrons by field emission, or hot and emitting them by thermionic emission, provides electrons to ionize a gas between the cathode and anode. Typically, the anode is biased positively to attract and accelerate electrons away from the cathode, producing impacts with atoms that ionize them along the way. A recent publication by Pillatsch et al.[25] discusses the generation of several negative ion beams by simple manipulation of the gases used in the duoplasmatron including F-, Cl-, Br-, and I- (see also Liebl and Harrison[26]). Two papers by LeJeune[19,20] discuss the detailed operating parameters of the duoplasmatron ion source that has been the mainstay of magnetic sector SIMS for more than 30 years. This duoplasmatron source is by far the best described source for SIMS in the open literature. Figure 3.5 shows a cutaway diagram of a duoplasmatron ion source. A duoplasmatron consists of a hallow cathode through which a gas (often O_2) is leaked. This gas is then bombarded with electrons and a high energy plasma is created. A magnetic coil is also used to pinch the plasma in the center for increased brightness. This venerable ion source design for SIMS has produced both positive and negative ions for countless studies.

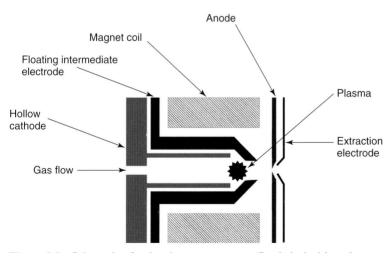

Figure 3.5 Schematic of a duoplasmatron source. Gas is leaked into the system via a needle valve. The cathode then bombards the gas with electrons, creating a high energy plasma. The ions from the plasma are then extracted and subsequently focused to form an ion beam.

3.4.6   Sputtering

The only other source of negative ions, besides the duoplasmatron, has been sputter primary ion sources. Yurimoto et al.[28] produced negative Cu and Au ion beams for analysis of oxygen isotopes in insulating samples. The BLAKE-V source that they developed was the first application of a sputter ion source to SIMS. Figure 3.6 depicts the basic geometry and components of the source.

This source used ionized Xe to sputter the target material, employing Cs vapor that coated the target material surface to aid in the production of negative ions. Other sputter sources (e.g., Alton et al. shown in Figure 3.7)[21] produced Cs ion beams inside the source that directly sputtered the target material configured in such a way as to extract negative ions from the source aperture.

Although apparently elegant in design and producing microampere beam currents, such sources were never adapted for routine use in SIMS instruments to produce negative atomic primary ions.

3.5   MOLECULAR ION SOURCES

There are several true molecular ion sources that have been used for SIMS. These are O_2^+ and O_2^- generated in a cold-cathode discharge duoplasmatron, also, O_2^+ generated by RF discharge and electron-impact. There are two incarnations of SF_5^+ ion sources, an electron-impact source made by ION-TOF, GmbH, and a tri-plasmatron source used by Gillen and Fahey.[29] C_{60} is a popular molecular source that works by heating solid C_{60} until the vapor pressure is high enough to produce a significant beam ionized by electron impact to form C_{60}^+ (as well as C_{60}^{2+}). Similarly, $Os(CO)_n^+$ beams were generated by heating followed by electron-impact ionization. Beyond this no other molecular ion beam sources (save for the standard O_2^+ produced in a duoplasmatron) have been employed in any significant number of SIMS studies. Although SF_5^+ and C_{60}^+ ion beams are commonly referred to in the literature as "cluster" ion beams they are better categorized as molecular ion beams.

3.5.1   Field Emission

Field-emission style molecular ion beams have been explored primarily with "room temperature ionic liquids" (RTILs). These ions beams have been produced at the tips of W needles as well as Taylor cones developed by liquids passed through capillaries.[30] Electrospray ionization methods also fit into this category.[31] Generally these ion beams have the most intense peaks with the molecular mass of the cationic or anionic monomer of the liquid; however they can produce clusters of more than one monomer in the beam, thus blurring the distinction between molecular and true cluster beam sources.

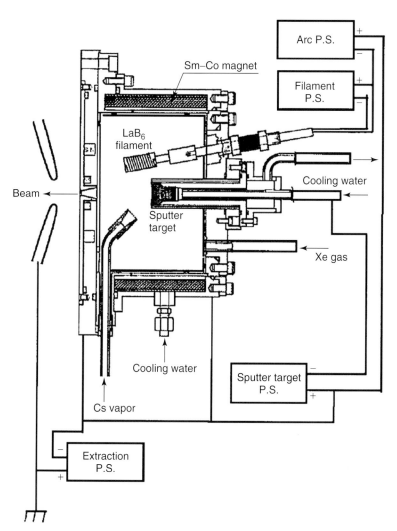

Figure 3.6 Cutaway drawing showing the Blake-V source used by Yurimoto et al.[28] to produce a Au$^-$ beam to be used for the measurement of O isotopes. Xe gas is leaked into the chamber. LaB$_6$ emits electrons that ionize the Xe, which subsequently sputters the Au target. Cs vapor is used to enhance ionization. Reproduced from Yurimoto et al.,[28] with permission from the American Institute of Physics (AIP).

3.5.2 Radio Frequency Discharge

O$_2^+$ ions have been produced by RF discharge. A commercial ion source is being produced and has been described by Smith et al.[23] Figure 3.2 shows the basic geometry that is common to all inductively coupled systems.

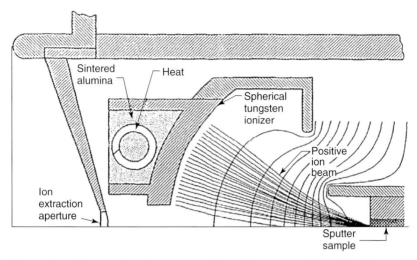

Figure 3.7 Cross section of the sputter primary ion source designed by Alton et al.[21] Field lines and ion trajectories drawn in the figure show how the spherical tungsten ionizer functions to direct Cs ions at the target material. The target material is then sputtered, creating negative ions that are then extracted and subsequently focused. Reproduced from Alton et al.,[21] with permission from Elsevier.

3.5.3 Electron Impact

The electron-impact source used in the ION-TOF designs, as described earlier, has been used to generate Ar^+, as well as O_2^+ and SF_5^+ ions with the appropriate filaments (either W or Ir) and minor materials changes to increase the operating lifetime of the source with SF_5^+. The beam size at the sample could be focused to ~ 2 µm and the source lifetime was several hundred hours. The maximum current depends on the accelerating voltage but could be as high as 30 nA at 10 keV.

By far, the ion source that has produced the most interest and shown the most significant advancements for organic molecular analysis is the C_{60} ion source. Although it is commonly referred to as a "cluster" ion source it is more accurately described as a molecular ion source that uses electron impact as its ionization mechanism. In this source, solid C_{60} is heated ($\sim 500\ °C$) to produce enough vapor[32] so that electron-impact ionization can be effective and produce a reasonably intense ion beam.

Weibel et al.[33] describe the operation of the source shown in Figure 3.8. The source produces a 2 nA beam of C_{60}^+ and can focus to a spot size of approximately 3 µm* at 0.1 nA. This design is configured to produce pulsed ion beams for use with a time-of-flight mass spectrometer.

Fujiwara et al.[34] produced ion beams of $Os_3(CO)_n^+$ and $Ir_4(CO)_7^+$ with an electron-impact source. While no references could be found as to the application

---

*Current designs have spatial resolutions on the order of 1 µm.

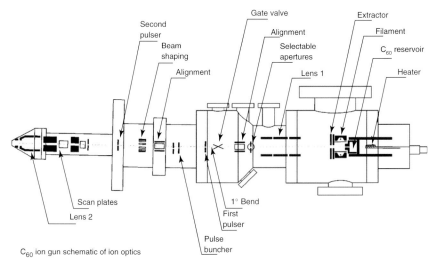

C$_{60}$ ion gun schematic of ion optics

Figure 3.8   Cutaway illustration of C$_{60}$ ion source design successfully used in numerous SIMS studies. A heated reservoir containing C$_{60}$ produces a vapor that is subsequently ionized via electron impact. Reproduced from Weibel et al.,[33] with permission from ACS publications.

of Os$_3$(CO)$_n^+$ to SIMS analysis, Ir$_4$(CO)$_7^+$ was successfully utilized for SIMS applications.[35]

3.5.4   DC-Glow Discharge

The duoplasmatron, as mentioned earlier, produced not only atomic ions but molecular ions as well, most notably O$_2^+$. Other molecular species have been extracted but by far the most prevalent is O$_2^+$. Although other geometries could produce ion beams, the duoplasmatron has been the brightest ion source available and has been employed since the 1960s.

3.5.5   Sputtering

Clusters of negative ions have been successfully produced in several geometries by sputtering of target materials with Cs$^+$ ions.[36] The appellation molecular or cluster ion source is not clearly differentiated with sources that produce ion beams by sputtering as atoms in the resultant ion beams clearly were present together in the solids from which they were removed. Experiments with C-clusters of up to C$_{10}$ showed significant advantages for the secondary ion signals from organic materials. The ion beams produced with the source described by Gillen et al. (via direct sputtering of the target with a Cs ion beam), were able to be focused into a spot of a few micrometers; however, sputtering of the target material in the source produced degradation of the beam intensity over the course of several hours, resulting in a daily replacement of the source target material.[36]

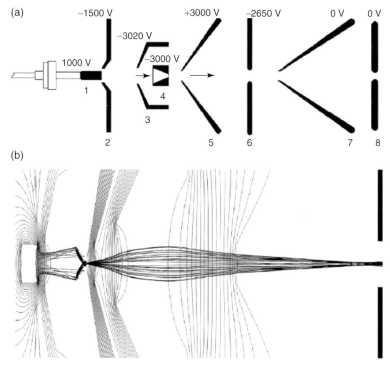

Figure 3.9 A schematic diagram of the sputter ion source from Belykh, used to produce Si_n^- and CU_n^- clusters, showing: (a) (1) the Cs^+ ion gun; (2) the extraction electrode; (3) the shield electrode; (4) the sputter cone target (Si or Au); (5–7) the three-electrode immersion lens, and (8) the entrance aperture. (b) A computer model of the sputter ion source in two dimensions is shown with ion trajectories of sputtered negative ions in the "sputter cone target–entrance aperture" space. Reprinted from Belykh et al.,[37] with permission from the American Institute of Physics.

A sputter-based primary ion source with a novel geometry was produced by Belykh et al.[37] and used on a Cameca ims-4f to produce negative secondary ions. This source is shown in Figure 3.9. This source managed to produce several nanoamps of Si_n^- and Cu_n^- ions that could be focused into an ~60 μm spot. The source lifetime was reported to be ~150 h or more.

3.6 CLUSTER ION SOURCES

There are in total four types of sources that have been used to produce cluster ion beams for SIMS: massive gas clusters produced via supersonic expansions (e.g., jets), sputtering-produced negative ion clusters, field-emission liquid metal ion sources that produce small clusters of atoms (such as Bi_3^+), and electrospray[31] and ionic liquids[30].

3.6.1 Jets and Electron Impact (Massive Gas Clusters)

The production of large numbers of truly massive clusters can only be achieved through the use of supersonic jets where the hydrodynamic conditions at the nozzle produce large clusters, essentially by condensation. An extensive treatment of the production of these types of ion beams can be found in the chapters of the compendia assembled by Pauly.[38] The shape of the nozzle is critical for the production of numerous clusters. The largest number and highest average cluster size, shown by Hagen and Obert[39] and later confirmed by Kappes et al.,[40] are produced with relatively simple divergent nozzles. The large achievable cluster sizes ($n > 700$) makes this source ideal for analysis of organic and biological materials by SIMS.

A disadvantage to using these beams is that the clusters have a broad energy distribution. This makes them difficult to focus and makes it more difficult to select the cluster size. Typical spot sizes for this source are in the $20-30$ μm range, much larger than is typical for SIMS analysis. Although cooling of the clusters can be achieved, thus reducing their energy spread, this comes at the expense of additional complications in the source.[38]

A simple schematic of the geometry of a massive gas cluster source is shown in Figure 3.10. In a massive gas cluster source, the clusters are not generally ionized when produced. Ionization of the clusters has been done by simple electron impact in configurations similar to that shown in Figure 3.10 (e.g., see Henkes et al.[41]). These sources require multiple stages of differential pumping in order to direct the final ion beam into a high vacuum region where the sample and analyzer are located. Generally this results in large columns with a turbomolecular pump for each stage of pumping. Selection of cluster size (charge-to-mass ratio) is typically done with a Wien filter, although some other methods have been applied for this such as the orthogonal time-of-flight method used by Von Issendorff and Palmer.[42]

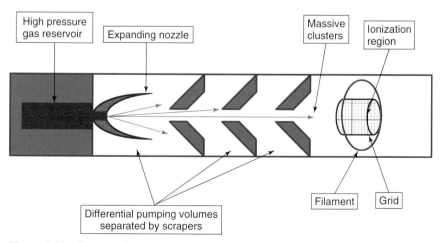

Figure 3.10 Cartoon of the components of a massive gas cluster source showing the expansion nozzle, accommodation for differential pumping, and the ionizer. Not shown is the mechanism for selection of cluster size, or charge-to-mass ratio.

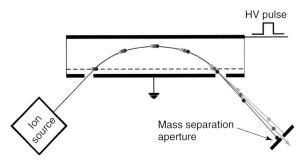

Figure 3.11 Illustration showing additional mass separation of massive gas cluster sources through momentum deflection.[46] After the ions are formed they undergo an initial gross filtering with a Wien filter. They then enter a 90° pulsing unit, allowing for a pulsed primary ion beam for analysis, and also allowing for additional mass separation. Only a narrow range of cluster sizes will fit through the mass separation aperture. Courtesy of Ewald Niehuis from ION-TOF.[46]

Matsuo et al.[43] have described in detail the operation and development of this type of source for SIMS applications producing ion beam currents of several milliamps. Mochiji et al.[44] reported the use of an Ar cluster source with SIMS showing impressive results. Recently, Rading et al.[45,46] have described the operation of an Ar cluster source on an ION-TOF ToF-SIMS V instrument, where they have employed an additional mass separation method via a 90° pulsing system to further enhance mass separation as illustrated in Figure 3.11. They claim that the cluster size distribution can be narrowed such that $(n/\Delta n)$ is approximately 60–100. Another source design has been developed based on the C_{60} source design shown previously in Figure 3.8. This design utilizes a combination of a Wien filter with double pulsing to achieve a cluster size distribution similar to that reported for ION-TOF designs.

3.6.2 Field Emission

Clusters produced by field-emission sources have been, thus far, limited to two types: clusters from ionic liquids of $n > 1$ monomers of the ionic liquid (e.g., massive water clusters or electrospray)[47] and groups of metal ions from metals such as Au and Bi, although the intensity of these metal clusters is lower than the monomer atomic ions. Electrospray droplet impact sources (EDIs) create large water droplets (mixed with dilute acetic acid) that are transported into a quadrupole ion guide via a 400 μm orifice. The droplets are accelerated into the vacuum and toward the sample of interest.

Emission of metal clusters appears largely to be dependent on the ability of the metal, whether it be a eutectic or elemental, to flow into a Taylor cone at reasonable temperatures. It is not clear why Ga does not form significant clusters on field emission given the similarity of the source geometry to Bi and Au field emitters. Bi and Au, with their relatively low melting points as monoisotopic elements, are perfect candidates to use. The significant improvement in secondary ion yield of molecules simply with a cluster of a few Au or Bi atoms is impressive (e.g., Nagy and Walker[48]). When used for analysis with time-of-flight mass spectrometers the beams must be separated by mass and charge. This is typically done via a flight-time method. Metal-cluster field-emission sources have similar properties to Ga-based LMIG (Liquid Metal Ion Gun) sources but typically must be run at a raised temperature and apparently have poorer spatial resolution, in addition the life of the source may be shorter. The spatial resolution of Bi cluster ions produced by field emission has been demonstrated to be of the order of 100 nm or less.

Very large metal ion clusters, containing as many as 400 atoms, have been made by field emission from a gold–silicon eutectic by Bouneau et al.[49] The effects of their impact on secondary ion emission have been reported by Novikov et al.[50] and by Benguerba et al.[51] No information was presented in either of these works regarding the focusing characteristics of the beams; however, one gets the impression that the beam size is of order 1 mm or less.

3.7 SUMMARY

The application of very large projectile ions as primary ions for SIMS has given rise to hope that SIMS will be able to produce nanometer-scale molecular information in organic and biological materials, both in lateral dimension and in depth. Such information, if attainable, will yield a wealth of understanding of basic biochemical processes at the cellular and hopefully subcellular level. In addition, the availability of such ion sources gives hope of fulfilling the potential that SIMS has to offer as a probe of molecular and atomic composition in organic and inorganic materials.

Although one could argue that the development of massive gas cluster ion sources depends, to some extent, on having modern turbomolecular pumps and the high speed, high voltage pulsing electronics needed for differential pumping and cluster mass, respectively, it is surprising that it has taken 10–15 years for them to be considered for use with SIMS. With the development of and experimentation with the appropriate ion optics, the focusing of massive gas cluster ion beams as well as other molecular beams will become more controlled and, with any luck, we will have highly FIBs that will be able to desorb large intact-molecules as ions to be detected in mass spectrometers.

What is also surprising is the dearth of selection of ion beams of small molecules and atoms that are available for any application, not just SIMS alone. The use of negative primary ions greatly reduces, although does not eliminate, the effects of charging on insulating samples; yet we currently have only O^- available to us on commercial SIMS instruments. Given the breakthrough measurements made possible by molecular ion beams such as SF_5^+, C_{60}^+, and by clusters such as of Bi, Au, and now Ar, one cannot help but imagine what other improvements might exist with the application of these and other species as primary ions and to what level these would enable us to exploit the promise that SIMS has to offer.

REFERENCES

1. Dunoyer, L. *J. Radium* **1911**, 8, 142.
2. Herzog, R. F. K.; Viehbock, F. P. *Phys. Rev.* **1949**, 76 (6), 855.
3. Laegreid, N.; Wehner, G. K. *J Appl. Phys.* **1961**, 32 (3), 365–369.
4. Nelson, R. S. *Philos. Mag.* **1965**, 11 (110), 291–302.
5. Sigmund, P. *Phys. Rev.* **1969**, 184 (2), 383.
6. Delcorte, A.; Garrison, B. *J. Nucl. Instrum. Methods Phys. Res. B* **2007**, 255 (1), 223–228.
7. Townes, J. A.; White, A. K.; Wiggins, E. N.; Krantzman, K. D.; Garrison, B. J.; Winograd, N. *J. Phys. Chem. A* **1999**, 103 (24), 4587–4589.
8. Mahoney, C. M. *Mass Spectrom. Rev.* **2010**, 29 (2), 247–293.
9. Winograd, N. *Anal. Chem.* **2005**, 77 (7), 142–149.
10. Liebl, H. *Applied Charged Particle Optics*, 1st ed.; Springer-Verlag: Berlin, **2010**; pp 90–91.
11. Williams, P.; Lewis, R. K.; Evans Jr, C. A.; Hanley, P. R. *Anal. Chem.* **1977**, 49 (9), 1399–1403.
12. Wilson, R. G.; Stevie, F. A.; Magee, C. W. *Secondary Ion Mass Spectrometry: A Practical Handbook for Depth Profiling and Bulk Impurity Analysis*; John Wiley & Sons, Inc.: New York, **1989**.
13. Gillen, G.; Fahey, A. *Appl. Surf. Sci.* **2003**, 203, 209–213.
14. Delcorte, A.; Garrison B. *Nucl. Instrum. Methods Phys. Res. B* **2007**, 255 (1), 223–228.
15. Prewett, P. D.; Jefferies, D. K. *J. Phys. D Appl. Phys.* **1980**, 13, 1747.
16. Ji, Q.; Jiang, X.; King, T. J.; Leung, K. N.; Standiford, K.; Wilde, S. B. *J. Vac. Sci. Technol. B* **2002**, 20 (6), 2717–2720.
17. Jiang, X.; Ji, Q.; Chang, A.; Leung, K. N. *Rev. Sci. Instrum.* **2003**, 74, 2288.
18. Nier, A. O. *Rev. Sci. Instrum.* **1947**, 18 (6), 398–411.
19. Lejeune, C. *Nucl. Instrum. Methods Phys. Res. B* **1974**, 116 (3), 417–428.
20. Lejeune, C. *Nucl. Instrum. Methods Phys. Res. B* **1974**, 116 (3), 429–443.
21. Alton, G. D.; McConnell, J. W.; Tajima, S.; Nelson, G. J. *Nucl. Instrum. Methods Phys. Res. B* **1987**, 24, 826–833.
22. King, J. G.; Zacharias, J. R. *Adv. Electron. Electron Phys.* **1956**, 8, 1–88.
23. Smith, N.; Tesch, P.; Martin, N.; Boswell, R. *Microsc. Microanal.* **2009**, 15 (S2), 312–313.
24. Schwieters, J.; Cramer, H. G.; Heller, T.; Jurgens, U.; Niehuis, E.; Zehnpfenning, J.; Benninghoven, A. *J. Vac. Sci. Technol. A* **1991**, 9 (6), 2864–2871.
25. Pillatsch, L.; Wirtz, T.; Migeon, H. N.; Scherrer, H. *Nucl. Instrum. Methods Phys. Res. B* **2011**, 269 (9), 1036–1040.
26. Liebl, H.; Harrison, W. W. *Int. J. Mass Spectrom.* **1976**, 22 (3), 237–246.
27. Coath, C. D.; Long, J. V. P. *Rev. Sci. Instrum.* **1995**, 66 (2), 1018–1023.
28. Yurimoto, H.; Mori, Y.; Yamamoto, H. *Rev. Sci. Instrum.* **1993**, 64 (5), 1146–1149.
29. Gillen, G.; Fahey, A. *Appl. Surf. Sci.* **2003**, 203, 209–213.
30. Larriba, C.; Castro, S.; de la Mora, J. F.; Lozano, P. *J. Appl. Phys.* **2007**, 101, 084303.
31. Hiraoka, K.; Asakawa, D.; Fujimaki, S.; Takamizawa, A.; Mori, K. *Eur. Phys. J. D* **2006**, 38 (1), 225–229.
32. Diky, V. V.; Kabo, G. J. *Russ. Chem. Rev.* **2000**, 69, 95.
33. Weibel, D.; Wong, S.; Lockyer, N.; Blenkinsopp, P.; Hill, R.; Vickerman, J. C. *Anal. Chem.* **2003**, 75 (7), 1754–1764.
34. Fujiwara, Y.; Kondou, K.; Teranishi, Y.; Nonaka, H.; Fujimoto, T.; Kurokawa, A.; Ichimura, S.; Tomita, M. *Surf. Interface Anal.* **2006**, 38, 1539–1544.
35. Fujiwara, Y.; Kondou, K.; Teranishi, Y.; Nonaka, H.; Fujimoto, T.; Kurokawa, A.; Ichimura, S.; Tomita, M. *J. Appl. Phys.* **2006**, 100, 043305.
36. Gillen, G.; King, L.; Freibaum, B.; Lareau, R.; Bennett, J.; Chmara, F. *J. Vac. Sci. Technol. A* **2001**, 19, 568.
37. Belykh, S. F.; Palitsin, V. V.; Veryovkin, I. V.; Kovarsky, A. P.; Chang, R. J. H.; Adriaens, A.; Dowsett, M. G.; Adams, F. *Rev. Sci. Instrum.* **2007**, 78, 085101.
38. Pauly, H. *Atom, Molecule, and Cluster Beams: Cluster Beams, Fast and Slow Beams, Accessory Equipment, and Applications*; 2nd ed.; Springer Verlag: Berlin, Heidelberg, New York **2000**.
39. Hagena, O. F.; Obert, W. *J. Chem. Phys.* Berlin, Heidelberg, New York; **1972**, 56, 1793.

40. Kappes, M. M.; Schar, M.; Schumacher, E. *J. Phys. Chem.* **1987**, 91 (3), 658–663.

41. Henkes, W.; Hoffman, V.; Mikosch, F. 1*Rev. Sci. Instrum.* **1977**, 48 (6), 675–681.

42. Von Issendorff, B.; Palmer, R. E. *Rev. Sci. Instrum.* **1999**, 70, 4497.

43. Matsuo, J.; Okubo, C.; Seki, T.; Aoki, T.; Toyoda, N.; Yamada, I. *Nucl. Instrum. Methods Phys. Res. B* **2004**, 219, 463–467.

44. Mochiji, K.; Hashinokuchi, M.; Moritani, K.; Toyoda, N. *Rapid Commun. Mass Spectrom.* **2009**, 23 (5), 648–652.

45. Rading, D.; Moellers, R.; Cramer, H. G.; Niehuis, E. Dual beam depth profiling of polymer materials: comparison of C60 and AR cluster ion beams for sputtering. *Surf. Interface Anal.*, 43 (1–2), 198–200.

46. Kayser, S.; Rading, D.; Moellers, R.; Kollmer, F.; Niehuis, E. Surface spectrometry using large argon clusters. *Surf. Interface Anal.* **2013**, 45 (1), 133–133.

47. Hiraoka, K.; Mori, K.; Asakawa, D. *J. Mass Spectrom.* **2006**, 41 (7), 894–902.

48. Nagy, G.; Walker, A. V. *Int. J. Mass Spectrom.* **2007**, 262 (1–2), 144–153.

49. Bouneau, S.; Della-Negra, S.; Depauw, J.; Jacquet, D.; Le Beyec, Y.; Mouffron, J. P.; Novikov, A.; Pautrat, M. *Nucl. Instrum. Methods Phys. Res. B* **2004**, 225 (4), 579–589.

50. Novikov, A.; Caroff, M.; Della-Negra, S.; Depauw, J.; Fallavier, M.; Le Beyec, Y.; Pautrat, M.; Schultz, J. A.; Tempez, A.; Woods, A. S. *Rapid Commun. Mass Spectrom.* **2005**, 19 (13), 1851–1857.

51. Benguerba, M.; Brunelle, A.; Della-Negra, S.; Depauw, J.; Joret, H.; Le Beyec, Y.; Blain, M. G.; Schweikert, E. A.; Assayag, G. B.; Sudraud, P. *Nucl. Instrum. Methods Phys. Res. B* **1991**, 62 (1), 8–22.

CHAPTER 4

SURFACE ANALYSIS OF ORGANIC MATERIALS WITH POLYATOMIC PRIMARY ION SOURCES*†‡

Christine M. Mahoney

4.1 INTRODUCTION

Cluster ion sources are continually proving themselves to be superior tools for surface analysis and imaging applications, allowing for significant enhancements in molecular sensitivities and improved overall secondary ion yields for organic and biological secondary ion mass spectrometry (SIMS) applications.[1] The combination of enhanced molecular yields, high spatial resolution imaging, and the potential for molecular depth profiling has led to a major paradigm shift, particularly in the biological SIMS community.[2−4]

---

*Official contribution of the National Institute of Standards and Technology; not subject to copyright in the United States.

†Commercial equipment and materials are identified in order to adequately specify certain procedures. In no case does such identification imply recommendation or endorsement by the National Institute of Standards and Technology, nor does it imply that the materials or equipment identified are necessarily the best available for the purpose.

‡This document was prepared as an account of work sponsored by an agency of the US Government. Neither the US Government nor any agency thereof, nor Battelle Memorial Institute, nor any of their employees, makes any warranty, express or implied, or assumes any legal liability or responsibility for the accuracy, completeness, or usefulness of any information, apparatus, product, or process disclosed, or represents that its use would not infringe privately owned rights. Reference herein to any specific commercial product, process, or service by trade name, trademark, manufacturer, or otherwise does not necessarily constitute or imply its endorsement, recommendation, or favoring by the US Government or any agency thereof, or Battelle Memorial Institute. The views and opinions of authors expressed herein do not necessarily state or reflect those of the US Government or any agency thereof.

Cluster Secondary Ion Mass Spectrometry: Principles and Applications, First Edition.
Edited by Christine M. Mahoney.
© 2013 John Wiley & Sons, Inc. Published 2013 by John Wiley & Sons, Inc.

The benefits of utilizing cluster sources were shown as early as in 1960, with the observation of nonlinear enhancements in sputtering yields when using polyatomic beams as opposed to atomic beams.[5−8] However, cluster sources were not employed for SIMS applications until much later. One of the earliest works was published in 1982, where the authors compared the performance of siloxane primary ions to Hg^+ ions for characterization of oligosaccharides in a glycerol matrix.[9] The results indicated that there was a large increase in the ionization of the sugars when employing the siloxane cluster source as compared to the atomic Hg^+ ion source.

Later, Appelhans et al.[10] used SF_6 neutral beams to characterize electrically insulating polymer samples, and discovered that the SF_6 cluster beam yielded an increase in secondary ion yield intensities by up to 3–4 orders of magnitude as compared to equivalent energy atomic beams. Similar findings were discovered in the mass spectra of pharmaceutical compounds.[11]

Benninghoven and coworkers performed a series of systematic studies on the molecular secondary ion emission from organic and polymeric surfaces under 10 keV Ar^+, Xe^+, and SF_5^+ bombardment (where Xe^+ and SF_5^+ have similar masses: m/z 131 and m/z 127, respectively).[12−15] For all investigated surfaces, there was an increase in the secondary ion yields in the order of $Ar^+ < Xe^+ < SF_5^+$ primary ion bombardment up to a factor of 1000. The ion yield enhancement observed was more prominent in the high mass region when using SF_5^+.

Since then, countless examples of signal enhancements using cluster sources for organic, polymeric, and biological systems have been reported and a review of this material is warranted.[14−28] This chapter addresses the application of cluster ion beams for "static" SIMS (surface analysis) of soft materials. The advantages and disadvantages, as well as important experimental parameters for surface analysis and/or imaging with cluster ion sources are described.

4.2 CLUSTER SOURCES IN STATIC SIMS

4.2.1 A Brief Introduction to Static SIMS

One of the inherent problems with employing atomic ion beams for SIMS analysis of organic samples and other types of radiation sensitive materials is that the ion beam creates extensive damage to the sample in the form of molecular bond breaking and/or cross-linking, particularly at the site of the ion beam track (see Chapters 1 and 2).[29] This damage may present itself in the form of cross-linking and eventual graphitization of the material, resulting in a complete loss of signal characteristic of the molecular structure with concurrent increases in C- and graphitic-type peaks. An early sign of damage will be an increase in the number of polycyclic aromatic hydrocarbons (PAH) in the mass spectrum, as has been shown by numerous authors.[1,30−32] For more detailed information about the damage processes in organic samples perturbed by ion beams, see Chapter 5, Section 5.5.4.

In order to minimize the damage created by atomic beams, the ion doses are kept low such that the probability of a primary ion hitting the same spot twice is minimal, a term dubbed the *static SIMS limit* by Benninghoven and coworkers.[33−36] Obtaining data below this limit, typically reported to be around 1×10^{13} ions/cm², will ensure that one is characterizing *only the surface* of a material, and/or that very *little to no molecular damage* is occurring to the sample during the course of the experiment. However, limiting the primary ion dose in such a manner will also decrease the sensitivity of the method.

4.2.2   Analysis beyond the Static Limit

Significant reductions in beam-induced damage accumulation are observed when employing cluster ion sources as primary ion beams for SIMS. Because the damage created by cluster beams is surface localized, any damage to subsurface layers is minimized (see Chapter 2). The decreased beam-induced damage to subsurface layers thus allows for molecular depth profiling in organic and polymeric materials; an exciting prospect that will be discussed in greater detail in Chapters 5 and 6.

With reference to surface analysis applications, a decrease in beam-induced damage allows for longer signal averaging times—"beyond the static limit" that is required for atomic beams. This means we are now able to maintain molecular signal for much higher primary ion doses, as exemplified in Figure 4.1, which shows the molecular and fragment ion intensities plotted as a function of sputter time with Ar^+ (Fig. 4.1a) and SF_5^+ (Fig. 4.1b) primary ion bombardment.[24] One can see that the signal can be maintained for a much longer time period, beyond the static limit, when employing the cluster source. While increased signal longevity greatly enhances the sensitivity of the method, one now needs to consider the implications of moving beyond the static limit. In other words, are we still looking at the surface or is the resultant data becoming more representative of the bulk?

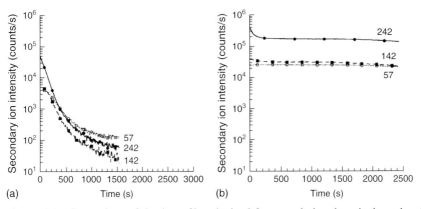

Figure 4.1   Comparison of depth profiles obtained from a solution-deposited tetrabutyl ammonium bromide thin film using Ar^+ (a) and SF_5^+ (b) primary ions under dynamic SIMS conditions. Reprinted from Gillen and Roberson,[24] with permission from Wiley.

4.2.3 Increased Ion Yields

Because multiple ions are bombarding the sample simultaneously when employing cluster sources, significant molecular signal enhancements are often observed as compared to atomic ion beams (see Chapter 2 for mechanistic details). An example is given in Figure 4.2, which shows representative negative cholesterol dimer images (256 μm × 256 μm) obtained from a rat brain cross section.[28] The image represents the interfacial region between lipid-rich gray matter (a) and cholesterol-rich white matter (b). It is evident on comparison of the images, that there are significant signal enhancements when employing Bi_3^+ cluster ions as opposed to Bi^+ atomic ions of the same dose.

Although these signal enhancements result in increased sensitivity for trace constituents, one needs to keep in mind that there is often a corresponding increase in the background in the mass spectrum. This increased background can be particularly problematic, as it may lead to interferences. One potential solution is the application of high mass resolution spectrometers (orbitrap, ion-cyclotron resonance), which are capable of separating out such interferences. Some examples of prototype instruments will be discussed in greater detail in Section 4.6.

Cluster SIMS is particularly useful for the analysis of polymeric materials. Figure 4.3, for example, shows mass spectra obtained from thin films of poly(ethylene glycol) (PEG) using both cluster ion and atomic ion beams. The PEG [average molecular weight of 1000 g/mol (PEG 1000)] was deposited onto Si wafers, and spectra were acquired using both Ga^+ and Au_3^+ ion beams. Figure 4.3a and b depicts the resulting mass spectra acquired with Ga^+ and Au_3^+ ions respectively.[20] There is a clear enhancement in molecular secondary ion signals when using Au_3^+ as indicated by the scaling of the two spectra. More importantly, while PEG molecular weight distributions (labeled peaks in Figure 4.3) are well defined in the mass spectrum acquired with Au_3^+, the

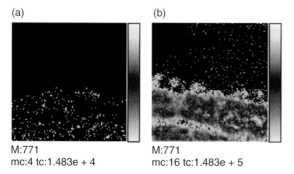

(a)

(b)

M:771
mc:4 tc:1.483e + 4

M:771
mc:16 tc:1.483e + 5

Figure 4.2 Negative secondary ion images (256 μm × 256 μm) of the cholesterol dimer ion $(2M + H)^-$ at m/z 771, obtained from a rat brain sample under irradiation with (a) Bi_1^+ and (b) Bi_3^+ primary ions, at a primary ion dose density of 10^{12} ions/cm^2. The corpus callosum is located at the bottom part of the image, and is colocated with the cholesterol signal. Reprinted from Touboul et al.,[28] with permission from Elsevier.

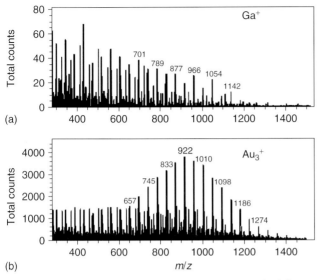

(a)

(b)

Figure 4.3 Positive secondary ion mass spectra acquired from a poly(ethylene glycol) (PEG 1000) sample, using (a) Ga^+ and (b) Au_3^+ primary ions. Labeled peaks are identified as $[M + Na]^+$ peaks with the number of monomer units varying from 9 to 26. Recreated from Aimoto et al.,[20] with permission from Elsevier.

corresponding fragment ion distributions (unlabeled peaks at lower masses) are much less intense as compared to what is observed in the Ga^+ spectrum. This indicates that the cluster source causes decreased fragmentation in polymers. Similar examples of decreased fragmentation in polymers and enhancements in molecular weight distributions can be found in the literature for polystyrene (PS), poly(ethylene terephthalate) (PET), poly(methyl methacrylate) (PMMA), and polytetrafluoroethylene (PTFE) with C_{60}^+.[21,37,38]

It should be noted that these results are not always typical, and in some cases, there is actually an increase in fragmentation with cluster beams as compared to atomic beams, depending on the sample and the beam conditions.[39,40] Overall, the extent of fragmentation is dependent on the type of sample being analyzed, its molecular weight, its thickness, the type of ion beam employed, and other experimental conditions as are discussed later in the chapter.

4.2.4 Decreased Charging

Cluster sources have been known to alleviate charging.[10,41−44] With some cluster beams, charge compensation is not required at all, even in the dynamic SIMS mode.[43,44] This effect can be attributed to the significant increases in secondary ion sputtering yields.[43] As we know that sample charging results from a net imbalance of primary ion charges entering the sample and secondary ion charges being removed from the sample via sputtering, an increased flux of secondary ions is likely to offset the charge carried by the primary ion beam.

4.2.5 Surface Cleaning

One of the greatest benefits of cluster ion sources is their ability to sputter clean organic surfaces that have become contaminated or to restore surfaces that have been highly damaged by ion beam processing. Some common examples are given in Figure 4.4 and Figure 4.5. In Figure 4.4, mass spectra were acquired before (a) and after (b) sputtering of a thin PMMA film using SF_5^+, where the sample was found to be contaminated with some form of silicone.[24] In Figure 4.4a, much of the PMMA signal is masked by the silicone contaminant. After sputtering (Fig. 4.4b), however, this contaminant was completely removed, and peaks characteristic of the PMMA polymer m/z 59 and m/z 69, for example, became more prevalent in the spectrum. Also shown in the mass spectra are peaks consistent with PAH, such as m/z 77 and m/z 91, and Cs (m/z 133). It is not stated in the paper[24] what the source of this Cs is (likely contamination from Cs^+ source, also used on this instrument), but it is clear that the PAH peaks are a result of minor beam-induced damage of the PMMA structure resultant from SF_5^+ sputtering.[30,45,46]

Figure 4.5 shows an example of the utility of surface cleaning in real-world systems. The sample analyzed is a single granule of smokeless gunpowder, cross-sectioned and subsequently pressed into indium foil.[47] In the figure, NO_2^- was imaged in order to determine the location of the explosive nitroglycerin constituents. Very little signal is observed at the surface of the cross section before sputter cleaning (Fig. 4.5a). However, the signal was easily recovered after sputter cleaning with an SF_5^+ source (Fig. 4.5b).

Finally, cluster ion beams can be utilized to recover molecular signal from previously ion beam-damaged regions.[24,48,49] This ability has particular implications for analysis of focused ion beam (FIB) cross sections that are created by a high energy Ga ion beam for subsequent analysis of buried structures within

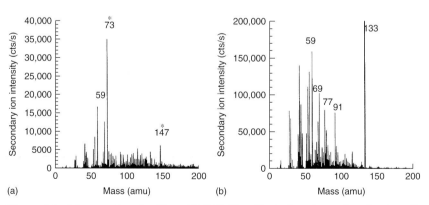

Figure 4.4 Mass spectra of a 50 nm PMMA film, illustrating the sputter cleaning of contaminants from the surface using SF_5^+. (a) before sputter cleaning and (b) after sputter cleaning. The peaks labeled with * represent peaks characteristic of a silicone contaminant, commonly observed during static SIMS experiments. Recreated from Gillen and Roberson,[24] with permission from Wiley.

(a) (b)

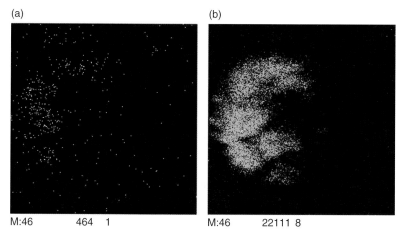

M:46   464   1   M:46   22111 8

Figure 4.5   Negative secondary ion images (500 μm × 500 μm) acquired from a smokeless powder sample before (a) and after (b) 5 keV SF_5^+ sputtering at an approximate dose of 7.5×10^{13} ions/cm^2. The nitrate signal at m/z 46 (NO_2^-), indicates the location of the explosive constituents in the powder. The sample was cross-sectioned with a razor blade and subsequently pressed into indium foil.

a sample. FIB sectioning is commonly used for secondary electron microscopy (SEM) and transmission electron microscopy (TEM) imaging. More recently, such methods are being developed for SIMS applications (see Chapter 6).

Creating cross sections in this manner inherently generates ion beam damage in the surface region of organic materials. This damage layer will need to be removed before successful molecular analysis of these cross sections can be performed with SIMS. Cluster ion beams are capable of removing this damage layer and some examples will be given in Chapters 5 and 6.

4.3   EXPERIMENTAL CONSIDERATIONS

4.3.1   When to Employ Cluster Sources as Opposed to Atomic Sources

Although cluster sources are turning out to be the beam of choice for organic analysis, there are cases when atomic ion beams are still preferred. The benefits of cluster ion beams, for example, are less obvious for thin organic overlayers on dense inorganic substrates.[13−15,21,24,50−53] Stapel et al.,[14] for example, performed several studies on series of poly(methyl acrylate) (PMA) Langmuir–Blodgett (LB) layers, with the number of PMA layers (n) ranging from $n = 1$ to $n = 9$. The secondary ion yield enhancement was shown to be significantly pronounced for multilayer coverage.[14] Furthermore, the disappearance cross sections (defined as the exponential decay in signal as a function of ion fluence) were also found to decrease for multilayer films bombarded with molecular beams. Weibel et al.[21] also measured marked decreases in disappearance cross sections in thicker films when employing cluster sources, where the authors characterized both thin monolayer films and thick films of PS (MW = 2000) with both Ga^+ and C_{60}^+ ions. It was also noted in this latter study that for the thin films, the values for disappearance cross sections were very similar for the two primary ions, whereas for the thicker films, the disappearance cross sections were much reduced with C_{60}^+ as compared to Ga^+.

There are also examples where increasing the ion beam polyatomicity in organic samples will actually *speed up* cross-linking damage processes, because of the resultant increased localized energy density of the ion beam at the center of the impact zone. This is true particularly in the case of cross-linking materials, such as PS.[1] In Chapter 5, an example of this is given where employing Bi^+ versus Bi_3^+ as an analysis source for molecular depth profiling resulted in a significant change in the damage characteristics of the profile.[54,55]

Overall, the benefits of cluster ion beams for SIMS analysis of organic materials outweigh the drawbacks, and the community has, by and large, switched to cluster beams for analysis of soft samples. But the question is, what about inorganic materials? There have also been some benefits observed for characterization of inorganic materials such as metals and semiconductors.[56−64] However, cluster sources are not currently used for inorganic materials analysis owing to the already well-established protocols for depth profiling and analysis of these materials. In some cases, there can even be a disadvantage in using certain cluster sources (e.g., C_{60} sputtering of Si).[65] It is unclear at this point in time, if the benefit of using cluster sources for these types of materials is significant enough to warrant a change. This subject will be discussed in greater detail in Chapter 7, which is devoted to analysis of inorganic materials with cluster beams.

4.3.2   Type of Cluster Source Used

Nowadays, there are multiple source designs to choose from (see Chapter 3 for more information on source designs). This can be confusing at times, particularly when deciding upon which source to purchase. Here, a brief overview of each type of cluster ion source and their potential applications in surface analysis will be presented.

4.3.2.1   Liquid Metal Ion Gun (LMIG)

The liquid metal ion gun (LMIG) sources, such as Au_n^+ or Bi_n^+, remain unparalleled in their ability to obtain high spatial resolution molecular information from samples. The increased signal associated with LMIG cluster sources, combined with the exquisite spatial resolution, has led to a considerable amount of research on biological- and tissue-imaging applications.[2−4,28,66−71]

Au (and Bi) cluster sources can achieve spatial resolutions as low as 25 nm for inorganic materials. However, for organic materials, spatial resolution is more realistically on the order of 100–200 nm, due to beam damage constraints. This higher spatial resolution value, which we name *useful lateral resolution*, has been defined by Kollmer et al.[72] (see Appendix A). The authors report on useful lateral resolutions from an organic pigment color filter array. The values reported range from ~1300 nm for Ga^+, 400 nm for Au^+ and Bi^+, to <150 nm for Au_3^+ and Bi_3^+. Although the beam efficiency is higher for C_{60}^+, and would therefore reduce this value even further, C_{60}^+ has a much larger spot size than an LMIG (1–2 μm).

It should be noted that although LMIG sources are ideal for imaging applications, they tend to accumulate damage more rapidly than other sources, such as SF_5^+, C_{60}^+, and $Ar_{n>500}$, even when utilized as an analysis source during depth profiling experiments.[54,55] Therefore, for higher dose experiments, where the high spatial resolution of an LMIG is not required, it may be more beneficial to use a different cluster source. Employing magnetic- sector-based SIMS ion microscope instrumentation, which focuses the secondary ions via optical columns, is a viable alternative approach to this problem, allowing for the application of large defocused beams for imaging applications. Surprisingly, other than at the National Institute of Standards and Technology (NIST), very few sources have been employed with this platform.[24]

4.3.2.2 C_{60}^+ for Mass Spectral Analysis and Imaging Applications

C_{60}^+ ion beams, although not as useful for high spatial resolution imaging as are LMIG clusters, can be utilized for low damage sputtering and analysis of organic materials under higher primary ion doses. Thus an inherent increase in the sensitivity can be observed when longer signal-averaging times are employed. Furthermore, in most cases, there are increased molecular yields associated with C_{60}^{n+} as compared to LMIG clusters.[72] Figure 4.6 plots the relative secondary ion yields (Y) (Fig. 4.6a), or the number of ions detected per incident primary ion, the corresponding disappearance or damage cross sections (σ) (Fig. 4.6b), and efficiencies ($E = Y/\sigma$) (Fig. 4.6c) of various atomic and cluster primary ion sources bombarding an Irganox film on polyethylene.[72] From the yield plots in Figure 4.6a, one can clearly see that C_{60}^+ significantly enhances the Irganox molecular ion yields as

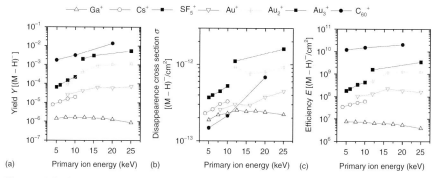

Figure 4.6 Secondary ion yield (a), disappearance cross section (b), and efficiency (c) of the quasi-molecular ion $[M - H]^-$ from a submonolayer sample of Irganox 1010 on low density polyethylene (LDPE) plotted as a function of the primary ion energy and species. Reprinted from Kollmer et al.,[72] with permission from Elsevier.

compared to other sources. These values are closely followed by SF_5^+ and Au_3^+. Note that Bi_3^+ is not included in this plot. It is likely that there are even greater enhancements with Bi_3^+, as it is known to have increased yields relative to Au_3^+. More recent work has indicated that Bi clusters with a nuclearity greater than 3, perform similarly to C_{60}^+ in terms of efficiency.[73]

Plots of disappearance cross section are shown in Figure 4.6b. These plots show that the least amount of damage is imparted to the sample by C_{60}^+ as compared to the other clusters, at least at lower beam energies (10 keV and lower). Also note that, in most cases, the cluster sources impart a much larger damage cross section than do the atomic ion beams. However, when employed at low beam energies, the damage cross sections of C_{60}^+ are even lower than that observed for the atomic ion sources.

The ratio of these parameters gives us the secondary ion formation efficiency, E, plotted in Figure 4.6c. As can be seen, the efficiency is an order of magnitude greater for C_{60}^{n+} than for that observed with Au_3^+ or SF_5^+ sources. Hence, if spot size is not a deterrent, C_{60}^+ appears to be preferable over LMIG sources for analysis of organic materials.

4.3.2.3   The Gas Cluster Ion Beam (GCIB)

The Ar cluster source, or gas cluster ion beam (GCIB) source, is showing particular promise as a tool for surface analysis. Spectra acquired using massive Ar gas clusters have decreased fragmentation and increased molecular yields as compared to C_{60}^{n+}. Figure 4.7, for example, shows representative mass spectra acquired from thin angiotensin III films using 20 keV Ar_{2000} (Fig. 4.7a), and 20 keV C_{60} (Fig. 4.7b).[74] As can be seen, the spectrum acquired using C_{60} has increased fragmentation associated with it when compared to Ar_{2000}. Moreover, the mass spectra acquired with 20 keV Ar_{60} was found to be very similar to that of C_{60} (not shown). The benefit observed with the massive gas cluster source, therefore, is attributed to the decreased energy per constituent atom that comes along with the larger cluster size (see Chapter 2).

While GCIB sources are proving to be superior for molecular depth profiling applications (see Chapter 5), and are proving to be useful as a soft ionization source for SIMS, the spatial resolution of the gas cluster source (30 µm) is severely limited, as compared to what is typically observed with other cluster sources. It may be that there will be improvements to this design in the future, as this source has only recently been developed for commercial applications.

4.3.2.4 Au_{400}^{4+}

Au_{400}^{4+} cluster ions are also being employed for surface analysis applications, to distinguish mass spectral features on a 10 nm spatial scale. Au_{400}^{4+} cluster ions were first introduced by Tempez et al.[75] In this work, both small (Au_n^+, where $n < 10$) and large (Au_{400}^{4+}) gold clusters were utilized to characterize peptide samples. Significant signal enhancements and decreased damage cross sections were observed with Au_{400}^{4+} as compared to Au_5^+ and Au_9^+ ions. Au_{400}^{4+} ions

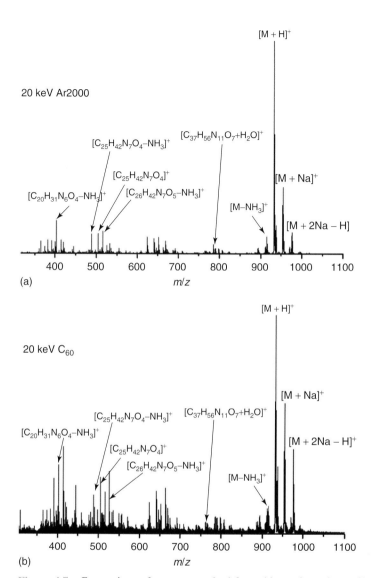

Figure 4.7 Comparison of spectra acquired from thin angiotensin III films acquired with C_{60} (a) and Ar_{2000} (b) at an ion fluence of 2×10^{12} ions/cm^2. Reprinted from Rabbani et al.,[74] with permission from the ACS publications.

were also shown to further enhance the signal by creating Au–analyte adducts, which are not observed with smaller gold clusters such as $Au_5{}^+$.[76]

Schweikert et al. further developed this source for single-impact events, where instead of a focused beam of ions, a single $Au_{400}{}^{4+}$ projectile impacts the surface, resulting in secondary ion emission from nanoscale regions.[77–79] The authors compared the performance of $Au_{400}{}^{4+}$ to that of $C_{60}{}^+$ (also operated in the single impact, or event-by-event mode), where it was found that the Au_{400}^{4+} projectile induced abundant multi-ion emission (e.g., the average number of detected ions emitted per event from glycine was 12.5 ions as opposed 0.15 ions per event using C_{60}^{+}).[79] This single-impact event design was based on an earlier design, originally developed by Van Stipdonk et al.[80] for C_{60}^{+}, $(CsI)_nCs^{+}$, and Ga^{+} beams.

Though this source is not yet designed for imaging experiments (currently, it cannot be determined where the beam is actually sampling), this type of source may have major implications for imaging in the future, as the spatial resolution will only be limited by the lateral extent of the surface area influenced by the single-impact event, and not the focusing elements of the source. Indeed, it has been proven to determine nanoscale heterogeneity within single-impact sites in samples of PS cast on alumina whiskers.[81]

4.3.2.5   Other Sources

Metal cluster complex ions, such as $Ir_4(CO)_{12}$ and $Os_3(CO)_{12}$ are also being explored as possible cluster sources.[59-63,82] These sources are comprised of chemically synthesized organometallic compounds, which consist of a metallic framework of several metal atoms and surrounding ligands.[62] These clusters have much higher molecular weights as compared to more conventional cluster beams such as C_{60} and SF_5^{+}, and are therefore expected to be much more useful for analysis of inorganic materials (see Chapter 2 for a discussion on the importance of mass matching). Thus far, the metal cluster complex ion beam has been shown to significantly improve depth resolutions for inorganic depth profiling applications over more conventional approaches (e.g., O_2 depth profiling).

Very little research has been performed using metal cluster complex sources for analysis of organic systems. One such work has focused on the characterization of PMMA samples, where the authors show that the metal cluster complex source enhanced signal by an order of magnitude over O_2 bombardment, with a particularly strong enhancement in the high mass region.[59]

Other cluster sources that have been developed for SIMS applications include ionic projectiles such as CsI^{+}, Cs_2I^{+} and Cs_3I_2,[50,51,83] CF_3^{+},[84,85] O_3^{+},[86] ReO_4,[87-93] $Ir_4(CO)_7^{+}$,[59] and $(Bi_2O_3)_n$ BiO^{+}.[51] For more information regarding source designs, see Chapter 3.

4.3.3   Cluster Size Considerations

The relationship of cluster size to signal enhancement and damage characteristics has been investigated in detail, particularly for small clusters.[14,15,17-19,40,94-96] All results indicate a saturation effect whereby the yields, damage cross sections, and efficiencies become constant with further increases in size. An example is given in Figure 4.8, which is reprinted from Stapel and coworkers,[14] where the secondary ion yields are plotted as a function of primary ion mass (b) and nuclearity (a) for PMA LB-layers on silicon.[14] The primary ions employed in the study, including Ne^{+}, Ar^{+}, Xe^{+}, O_2^{+}, CO_2^{+}, SF_5^{+}, $C_7H_7^{+}$, $C_{10}H_8^{+}$, $C_6H_6^{+}$, and $C_{10}H_8^{+}$, were created by an electron impact source (see Chapter 3).

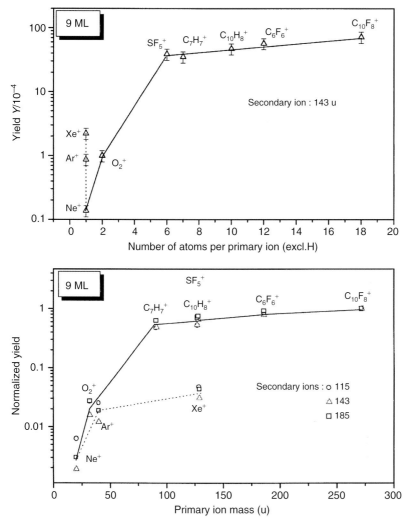

Figure 4.8   Yields and normalized yields of characteristic PMA fragment ions emitted from a multilayer sample under atomic (Ne^+, Ar^+, Xe^+) and molecular (O_2^+, SF_5^+, $C_7H_7^+$, $C_{10}H_8^+$, $C_6F_6^+$, $C_{10}F_8^+$) primary ion bombardment. (a) Yields for fragment at $m/z = 143$ are plotted as a function of the number of constituents per primary ion. (b) Normalized yields for fragments at $m/z = 115$, 143, and 185 are plotted as a function of primary ion mass. Reprinted from Stapel et al.,[14] with permission from Elsevier.

In Figure 4.8, an increase in secondary ion yields occurs within the range of one to about six atoms (SF_5^+). However, any additional yield increases from 6 to 18 atoms are minimal. Another important observation from this study was the fact that chemical effects when using fluorinated beams, such as SF_5^+, were not observed, as there was no difference in the plots when changing from fluorine-containing primary ions to hydrocarbon primary ions of comparable mass.

These saturation effects occur in samples containing less than 100 atoms. What happens then, when we switch to a massive cluster source, such as Ar_{400}^{4+} or $Ar_{(n>500)}^{m+}$? When directly comparing the sputtering yields of Ar_{1000} and C_{60} from recent round-robin studies[97,98] (see Chapter 5, Table 5.1), one observes that the range of sputtering yields is similar for both. So if there is no difference in the sputtering yields observed when employing these larger cluster sizes, why then is it beneficial to use them? As discussed in Chapter 2, and as will be discussed in greater detail in Chapter 5, having a large cluster size, allows for the ability to sputter below the so-called "sputter threshold," which has been recently defined as $\sim$10 eV/nucleon.[97,99–102] Below the sputter threshold, the fragmentation of a material is significantly suppressed (see Chapter 2, Fig. 2.17), allowing for extremely low damage sputtering, similar to what is observed more typically in matrix-assisted laser desorption/ionization (MALDI) experiments. Thus, molecular analysis of smaller-sized protein samples without a matrix becomes feasible. An example of protein analysis with Ar clusters is demonstrated in Figure 4.9, which shows the high mass range spectrum of insulin acquired using Ar_{1400} clusters.[100–103] As can be seen, the sodium-cationized molecular ion peaks are readily observed by SIMS without the use of a matrix.

4.3.4 Beam Energy

It has long been understood that increasing the primary ion beam energy will increase the sputter yield of a material. Furthermore, it is well known that increasing the beam energy results in a corresponding increase in the damage imparted to the sample; thus increasing the fragmentation in the corresponding mass spectrum. The same also holds true for cluster sources. However, for cluster ion beams, it is the energy per constituent atom that is the most critical factor in determining the sputter and damage characteristics of a particular beam. Therefore, while beam energy plays an important role in and of itself, it is inherently linked to the number of atoms in a cluster. For example, a 10 keV Ar_{2000} ion beam will likely impart similar damage as a 20 keV Ar_{1000} ion beam.

In order to minimize fragmentation and optimize molecular yields, it is beneficial to maintain the energy per unit atom as low as possible while still

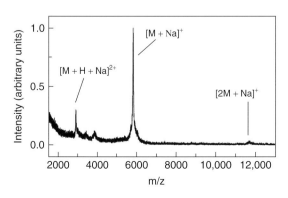

Figure 4.9 Secondary ion mass spectrum of a thin film ($\sim$40 nm) of human insulin on Si. Spectrum acquired using 5 keV Ar_{1400}. Recreated from Mochiji et al.,[101] with permission from Wiley.

maintaining sputtering of the sample. This is more readily achievable by using massive clusters that are capable of sputtering at much lower electronvolts per atom than other sources. Figure 4.10 shows representative mass spectra of arginine, acquired using an Ar gas cluster source under varying conditions.[104] In Figure 4.10a, the spectrum was acquired with a 10 keV Ar^+ atomic ion beam, which corresponds to 10,000 eV/atom. As can be seen, there is a significant amount of fragmentation observed in the mass spectrum. In Figure 4.10b, a 10 keV Ar_{300} cluster is used, which corresponds to energies of 33.3 eV/atom. In this case, the spectrum shows some evidence of decreased fragmentation. However, it is not very significant. Figure 4.10c, acquired with 10 keV Ar_{1500}, at 6.67 eV/atom, shows the most significant improvement, with a dramatic decrease in fragment ion intensities, and corresponding increases in the molecular signal. Note that this spectrum was acquired below the sputter threshold as defined previously (10 eV/atom). At these low energies, it is likely that only sputtering of the organic constituents occurs, and not the substrate constituents. Finally, the spectrum in Figure 4.10d represents that acquired with 20 keV Ar_{1500} (13.3 eV/atom),

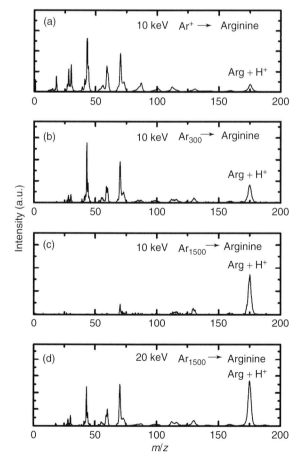

Figure 4.10 Examples of the mass spectra of positively charged secondary ions for an arginine target bombarded with (a) 10 keV Ar^+, (b) 10 keV Ar_{300}, (c) 10 keV Ar_{1500}, and (d) 20 keV Ar_{1500} ions. Reprinted from Ninomiya et al.,[104] with permission from Elsevier.

and appears to have an intermittent amount of fragmentation between that observed in Figure 4.10b,c, as is expected based on the electronvolts per atom values.

4.3.5   Sample Temperature

Recent experiments have indicated that there is a significant benefit regarding the application of low temperatures during organic depth profiling.[46,98,105−112] However, there has not been much published regarding changes in mass spectral data that occur with temperature in the static regime. Some preliminary studies have indicated that there are differences in the mass spectral data of polymers acquired at different temperatures.[105,106] In one study, for example, the authors observed changes in all the static SIMS spectra of a series of polymeric biomaterials, such as polyurethanes and poly(ethylene-*co*-vinyl acetate). More specifically, evidence of polycyclic aromatic hydrocarbon peaks was found to increase with decreasing temperatures. This same phenomenon was also observed for analysis of PMMA at varying temperatures.[105]

More importantly, significant differences were observed in mass spectra acquired from polyurethane copolymers at variable temperatures, whereby certain components of the copolymer are preferably observed at low temperatures. Figure 4.11 shows mass spectra acquired from a polyurethane copolymer, which contains methylene diisocyanate (MDI) and poly(ethylene glycol adipate) (PEGA) components.[106] The spectra shown are acquired at 25 $^\circ$C (Fig. 4.11a) and −100 $^\circ$C (Fig. 4.11b). The star labeled peaks are consistent with the structure of MDI component of the polymer, while the peaks labeled with a solid circle are characteristic of PEGA soft segment. Note that at room temperature, the intensities of the PEGA are enhanced relative to the MDI peaks, while the opposite is true at low temperatures. It is still unclear as to why this is the case, but it appears that sputter properties of the material are significantly changed at low temperatures. It is likely that the sputter yield of the PEGA component decreases, while that of the MDI is constant. It is also possible that the ion formation mechanisms change with temperature.

Low temperatures can also cause any water molecules present in the vacuum chamber to condense onto a surface.[113,114] A hydrated surface can serve as a source of protons for ionization, and help to prevent damage accumulation processes, as has been recently demonstrated in recent studies by Piwowar et al.[113,114]

4.3.6   Matrix-Enhanced and Metal-Assisted Cluster SIMS

One of the most important considerations during any SIMS experiment is sample preparation. Before the advent of cluster sources, for example, organic samples were usually prepared as submonolayer films on Ag substrates in order to increase secondary ion yields through Ag-cationization of the molecules.[115−118] This process, while useful for analysis of bulk molecular weight distributions, however, cannot be used to characterize the surface molecular weights in thick films, which may be different than the bulk. Therefore, instead of preparing a submonolayer

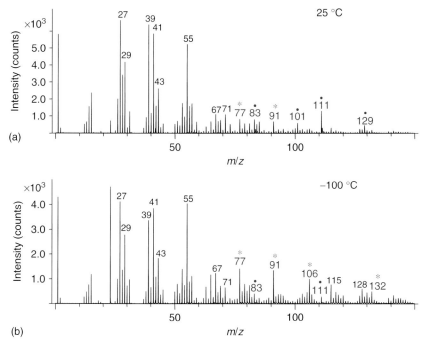

Figure 4.11   Positive ion mass spectra of a polyurethane copolymer with methylene diisociyanate (MDI) hard segments and poly(ethylene glycol adipate) (PEGA) soft segments measured at (a) 25 °C and (b) −100 °C. Peaks labeled with (•) are characteristic of the MDI-containing hard segment, while peaks labeled with (*) are characteristic of the PEGA soft segment of the polymer. Data acquired using 10 keV Ar$^+$. Recreated from Mahoney et al.,[106] with permission from Elsevier.

film on metal substrates, scientists have been moving toward the inverse situation; that is to say, the evaporation of small amounts of a metal onto a material's surface before analysis. In *metal-assisted* SIMS experiments or "MetA-SIMS,"[119−125] the sample is coated with a thin layer of a metal, such as Au or Ag in a controlled manner, such that significant enhancements in molecular signal are observed, along with the formation of additional adduct ions (e.g., M + Ag$^+$).[116,119] This enhances the sensitivity of the method, in particular for the higher mass region of the mass spectrum. The first observation of MetA-SIMS enhancement was shown by Bletsos and coworkers, who deposited a Ni grid onto poly(dimethyl siloxane) (PDMS) and PTFE samples in order to alleviate charging effects.[126] When using the Ni grid however, the authors were also able to observe a Ni-cationized molecular weight distribution on the surface of the PDMS. The cationized molecular weight distribution measured at the surface of the thick PDMS film was in the same general mass range as that observed from the corresponding submonolayer film.

Later, controlled deposition of Au and Ag overlayers was optimized for characterization of molecular weight distributions at polymer surfaces by Delcorte and coworkers.[119−125] An example is given in Figure 4.12a, b,[122] which shows

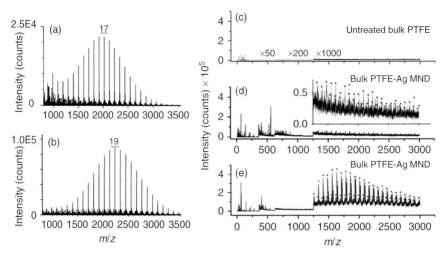

Figure 4.12   Molecular weight distributions measured by Meta-SIMS (a, b),[122] and metal nanoparticle SIMS (c–e),[127] both using a 15 keV Ga$^+$ primary ion beam. (a) Polystyrene (PS) film (>10 nm) with Au metal evaporated overlayer (20 nmol/cm^2); (b) PS submonolayer film prepared on a Au substrate; (c) polytetrafluoroethylene (PTFE) bulk; (d) PTFE with Au nanoparticle deposition; and (e) PTFE with Ag nanoparticle deposition. Recreated from Delcorte et al.[122] and Marcus and Winograd,[127] with permission from ACS publications.

representative positive secondary ion mass spectra acquired from a narrow molecular weight PS standard (M_n = 2180 g/mol) in the high mass range. Here submonolayer preparation and metal evaporation processes are directly compared. Figure 4.12a shows the Au-cationized oligomeric distribution obtained from a Au-metallized thick film (>10 nm) of the PS with a gold surface coverage of ~20 nmol/cm^2 Au. Figure 4.12b shows the Au-cationized PS oligomeric distribution obtained from the corresponding submonolayer film deposited on an Au substrate. Other than a slight shift in the molecular weight distribution in Figure 4.12a toward lower masses, and an increase in the fragmentation products, the molecular weight distributions shown are very similar.

A related method is also illustrated in Figure 4.12c–e.[127] In this example, metal nanoparticles were deposited from solution onto the surface of a thick PTFE film.[127] The mass spectrum taken from the surface of the film before deposition of the nanoparticles is shown in Figure 4.12c. No high mass range peaks were detected. After deposition of Au (Fig. 4.12d) and Ag (Fig. 4.12e) nanoparticles, however, a high mass oligomeric molecular weight distribution is observed. One can see from these results that metallization of a surface appears to be quite effective.

The mechanism of ionization enhancement in nanoparticle systems has been described recently and was also discussed in some detail in Chapter 2 (Fig. 2.21).[128] Briefly, when an atomic ion impinges upon a gold or silver nanoparticle, it creates a collision cascade in the nanoparticle. The resulting energy of the bombarded nanoparticle is then transferred directly into the polymer surface, creating a dense collision cascade in the polymer surface region and enhancing the sputtering yield of the polymer.

Other methods of enhancing ionization in SIMS experiments have been studied. For example, matrix-enhanced secondary ion mass spectrometry (ME-SIMS), whereby MALDI matrices are utilized for SIMS applications, show great promise for characterization of polymeric and protein samples.[129−134] Furthermore, with application of glycerol drops to a surface, the signal can be significantly enhanced, a process often referred to as *liquid SIMS* or *fast atom bombardment* (FAB).[135−138]

Although MetA-SIMS and ME-SIMS methods are quite promising, there is a lot of work put into sample preparation and optimization procedures. Cluster SIMS offers a unique matrix-free method of analysis of surfaces. However, the question is, can we benefit from the application of matrices and/or ionization enhancers in conjunction with cluster ion beams? So far, research shows that there is a much-reduced effect of metal deposition on the secondary ion yield enhancements when cluster ion beams are employed.[78,120,133,139,140]

Matrix-enhanced cluster SIMS, on the other hand, does show significant advantages whether an atomic ion beam or a cluster ion beam is used. Locklear et al.,[132] for example, utilized C_{60}^+ to investigate the emission of gramicidin S molecular ions embedded in a matrix of sinapinic acid. There was an observed increase in ion yield by up to eight times, simply by controlling the ratio of gramicidin S to sinapinic acid.

Application of a glycerol drop to the surface of a material has also proven to be beneficial when employing cluster ion beams.[135] For example, integrated secondary ion signals of selected organic compounds have been characterized with C_{60}^+ where the signal intensities were determined to increase as much as 100-fold when a drop of glycerol was deposited onto the sample surface.[135]

4.3.7 Matrix Effects

The application of matrices in SIMS can help to enhance ionization processes, making the method more sensitive. However, this same feature can also be quite problematic during SIMS experimentation, as any change in a sample's matrix composition can result in significant enhancement and/or suppression of certain ions. Jones et al.,[141] for example, compared Au_3^+ to C_{60}^+ for biological imaging applications, where the authors attempted to image the distribution of a drug, raclopride, in a rat brain section. Although the drug was detected with C_{60}^+, the distribution by SIMS did not corroborate with the known locations of the receptor sites in the brain. More specifically, the protonated molecular ion of the drug, $[M + H]^+$, was significantly enhanced in cholesterol-enriched regions of the brain, relative to phosphatidylcholine lipid-rich regions. This result was attributed to matrix effects, whereby ionization of the drug molecule was improved by the presence of the cholesterol (see Chapter 8 for more details).[141,142] Cholesterol can act as a source of protons, as it tends to form deprotonated molecular ions, $[M − H]^+$. Similarly, phosphatidylcholine removes a proton from the environment by forming protonated molecular ions, $[M + H]^+$, and therefore suppresses the formation of the protonated molecular ion in the drug.[141,142] Therefore, the imaged drug distribution was the opposite of the real distribution, due to matrix effects.

Matrix effects remain a significant problem in any surface mass-spectrometric-based method. One possible method of removing this effect is to use laser post-ionization, in which the sputtered neutrals are subsequently ionized with a laser beam. This separates the sputtering from the ionization process and allows for a more accurate representation of the sputtered material, which is >99% neutral material.[143–150] Application of laser desorption ionization has been fairly successful in minimizing matrix effects.[143–150] Matrix effects are particularly prominent in biological systems, and therefore a section of Chapter 8 will be devoted to this topic.

4.3.8 Other Important Factors

There are many other potential factors to consider when performing a surface analysis and/or imaging experiment with cluster ion beams.[1,35] These factors include the electron flood gun dose during the experiment, the ion beam angle, the ambient pressures and chemistries of the main chamber during the experiment, the primary ion beam dose, and many other parameters similar to that which should be considered during a dynamic SIMS experiment (see Chapter 5).

Changing the incident beam angle of the cluster ion beam with respect to the sample, for example, is known to affect the secondary ion and sputtering yields in organic materials.[100,151,152] Moreover, sputtering at glancing incidence angles with a cluster source, can result in decreased subsurface damage. The mass spectral data will be affected accordingly, exhibiting increased fragmentation in the mass spectra at more normal angles (perpendicular to the sample surface).

An exception exists where Ar gas clusters are concerned, which tend to exhibit increased fragmentation at more glancing angles.[100] Despite this increase in fragmentation, however, the damage cross sections associated with the Ar gas cluster decrease at more glancing angles, indicating that a decreased amount of damage accumulation and cross-linking is indeed occurring, similar to what is observed with other cluster ion beams. The increased fragmentation in the Ar cluster case has been attributed to changes in the mechanism of secondary ion emission and dissociation at variable angles when bombarded with massive gas cluster beams. As the cluster hits the surface at more normal angles, the emission of intact ions is directed toward the vacuum, whereas at more glancing angles, the emission of intact ions is directed backward into the sample.

4.4 DATA ANALYSIS METHODS

4.4.1 Principal Components Analysis

Multivariate statistical analysis methods, such as principal components analysis (PCA) and multivariate curve resolution (MCR), are commonly utilized for analysis of SIMS mass spectral datasets, in order to better assess correlations and anti-correlations among the mass spectral peaks.[105,153–170] This process allows for the isolation of mass spectral signatures that are correlated with different components in a sample, as well as clear differentiation between mass spectra acquired from different samples. Although there are many approaches for multivariate analysis of SIMS data, this chapter's discussion will be limited to PCA.

4.4.1.1 Basic Principles of PCA

The first step in the PCA process is to measure and tabulate the areas of the peaks of interest in the mass spectra in a manner demonstrated in Figure 4.13a (recreated from Belu et al.).[153] In this table, the rows (labeled as "samples") represent the various mass spectra acquired from varying samples or from different regions on the same sample. For example, rows labeled as "a" represent mass spectra acquired from sample "a," from differing regions, whereas the rows labeled as "b" represent mass spectra acquired from a different sample, "b," at different locations. The columns, or "variables," represent the peak areas in the corresponding mass spectra.

The variance–covariance matrix is calculated from the dataset in Figure 4.13a, which may or may not be preprocessed via mean centering, normalization, or other methods. This matrix is used to calculate the singular

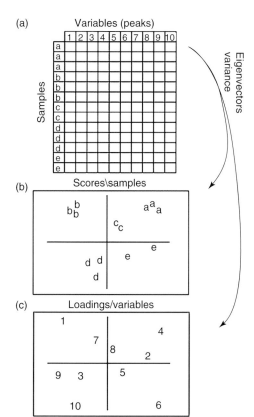

Figure 4.13 Schematic illustration of the principal component decomposition of a data matrix (a) into scores (b) and loadings (c), representing the samples (mass spectra of) and variables (peak intensities in the mass spectra), respectively. Reprinted from Belu et al.,[153] with permission from Elsevier.

value decomposition, thus reducing the dimensionality of the data to several key principal components (PCs), PC1, PC2, PC3, etc., where PC1 describes the greatest variance between the mass spectra, with each consecutively higher PC number describing decreasing amounts of variance in the data. From this process, a set of sample scores and peak loadings are calculated for each PC, which describe the correlations observed in the dataset in regard to that component.

Sample scores indicate how each "sample," or individual mass spectrum, is correlated with other mass spectra in the dataset. Similar scores will thus indicate mass spectral similarities. The scores plots in Figure 4.13b, for example, indicate that all samples labeled with an "a" have similar scores, and are therefore similar to each other. However, the samples labeled with "b" are located in a completely different quadrant on the scores plot, indicating that these mass spectra are different from those acquired from sample "a."

Peak loadings indicate the specific chemical differences observed in the mass spectral data at each PC. They show what peaks are correlated with the scores data, and also show how peaks within a mass spectrum correlate with each other. Therefore isolation of chemical signatures is possible using PCA. Figure 4.13c shows the values of the peak loadings calculated from the dataset. It is likely that peaks 1 and 7 are associated with one chemical species, say chemical species A, while peaks 3, 9, and 10 are characteristic of another species, chemical species B. Moreover, positive peak loadings are correlated with positive sample scores, and conversely with negative peak loadings. Hence, in Figure 4.13, sample "b" in the scores plot is highly correlated with chemical species A in the loadings plot, as they are both located in the upper left-hand quadrant in their respective plots. Similarly, sample "d" is highly correlated with chemical species B.

The PCA process is defined mathematically in Equation 4.2, where X is the mean-centered data matrix, P is the matrix of sample scores, and T is the matrix of peak loadings. The cross product of PT^T contains most of the original variance in X with the remaining variance (mostly noise) relegated to the residual matrix E.

$$X = PT^T + E \tag{4.1}$$

4.4.1.2   Examples of PCA in the Literature

PCA has been utilized heavily for proteomics applications of SIMS.[165,167–169,171–173] An example is shown in Figure 4.14, which shows the resultant scores (Fig. 4.14a) and loadings (Fig. 4.14b) plots from a series of mass spectra acquired from protein samples deposited on Si.[173] The scores plots in Figure 4.14a indicate that each of the different proteins analyzed have different compositions, even though they are mostly comprised of a series of similar amino acids. The ellipses drawn around each set of peaks represent the 90% confidence limit, indicating that the differences observed are statistically significant.

The corresponding loadings plot in Figure 4.14b indicates which peaks, or series of peaks, are correlated with each protein sample. In this particular case, iron is located on the far right-hand side of the plot, meaning that it is most highly

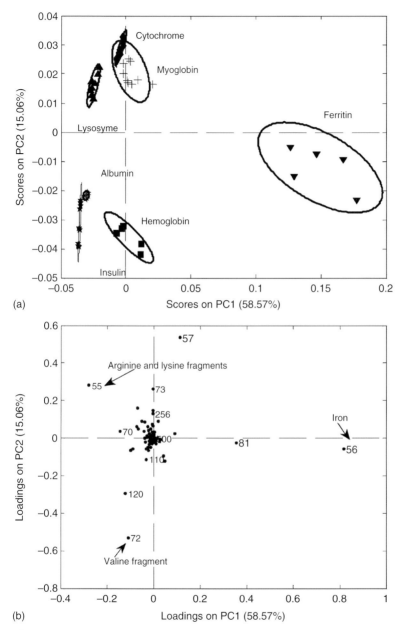

Figure 4.14 (a) Scores plot from PCA data reduction of ToF-SIMS positive ion spectra acquired from proteins spotted onto a silicon chip. Data points represent multiple regions of interest from a single spot. Ellipses represent 90% confidence intervals. (b) Loadings plot showing masses responsible for separation of protein spots. Data were acquired using 25 keV Ga[+]. Reprinted from Kulp et al.,[173] with permission from ACS publications.

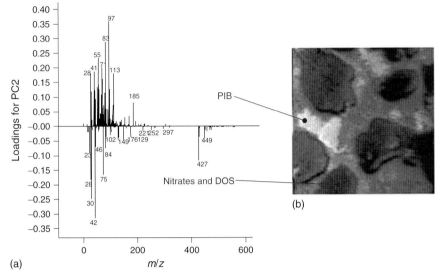

(a)

(b)

Figure 4.15 Example of image analysis by PCA in a poly(isobytylene) (PIB) containing plastic explosive (C-4). (a) Scores image (200 μm × 200 μm), where the bright regions represent positive scores and the dark regions represent negative scores. (b) The corresponding loadings plot indicates that the bright regions are correlated with PIB-rich domains, while the dark regions are correlated with the explosives (nitrates) and a di-isooctyl sebacate (DOS) binder. Data were acquired using 25 keV Bi_3^+. Reprinted from Mahoney et al.,[174] with permission from the ACS publications.

correlated with ferratin. The other peaks in the loadings plot represent various amino acid peaks, which are more prevalent in certain protein samples over others. For example, valine (m/z 72) and proline (m/z 120) are more prevalent in hemoglobin and albumin, relative to the other proteins.

PCA is also useful for isolation of mass spectral signatures from complex mixtures. An example is shown in Figure 4.15, which depicts the PCA image analysis from a solution-cast sample of a composition C-4 explosive, consisting mostly of cyclotrimethylenetrinitramine (RDX) explosive, poly(isobutylene) (PIB), plasticizers such as di-isooctyl sebacate (DOS) or dicapryl adipate (DCA), and fuel oil components.[174] The image shown in Figure 4.15b represents the PC2 scores image resultant from PCA analysis of the mass spectral dataset associated with the analyzed region. In this image, the bright regions represent areas with positive scores, while the dark regions represent areas with negative scores. The corresponding loadings plot (Fig. 4.15a) has positively loaded peaks that are highly characteristic of PIB, while the negatively loaded peaks are characteristic of a mixture of nitrates (from the RDX) and DOS. The white areas in the image are therefore highly correlated with PIB.

This example demonstrates quite nicely the utility of PCA for characterization of complex mixtures. The mass spectrum in this PIB-rich region is complex, containing several additional peaks characteristic of other components in the C-4.

These peaks can make it much more difficult to identify peaks specific to PIB. The loadings plot in the PCA however shows a mass spectrum that is close to identical to that of the pure PIB.

Although PCA is very useful as a tool to aid in mass spectral data analysis, it should *never be thought of as a substitute* for more detailed analysis of the mass spectral dataset by an expert. Failing to corroborate PCA results with actual mass spectral data can result in systematic errors and/or gross oversight of important features.

4.4.2   Gentle SIMS (G-SIMS)

Another analytical method worth mentioning here is a method developed by Ian Gilmore at the National Physical Laboratory (NPL) in the United Kingdom. This method, called *gentle secondary Ion mass spectrometry*, or *G-SIMS* has shown great promise as a tool for decreasing the influence of fragmentation in SIMS spectra in organic- and polymeric-type samples.[175−179] G-SIMS takes advantage of variable ion beam damage conditions to extrapolate information into low damage regimes.

The premise of G-SIMS is that during the ion bombardment process, a high density localized plasma is created, where the temperature of this plasma (T_p) varies with changing ion bombardment conditions (such as changing beam energies or using a different primary ion). Obviously, the variation in T_p will yield significant changes in the extent of fragmentation that occurs during the ion bombardment process. The G-SIMS process uses these differences to extrapolate into lower damage regions, thus yielding a mass spectrum with significantly reduced fragmentation. Nowadays, the G-SIMS process can be integrated into the software and hardware of certain instrument designs.[180]

An example representing the utility of G-SIMS for analysis of polymers is shown in Figure 4.16, which shows the representative positive ion SIMS spectrum (Fig. 4.16a) and the corresponding G-SIMS spectrum (Fig. 4.16b) acquired from a poly(glycolic acid) (PGA) sample cast on Si.[177] As can be seen, the G-SIMS spectrum has significant reductions in fragmentation allowing for increased visualization of higher mass fragmentation patterns (⋆, +). This method can be particularly useful for analysis of higher mass species, such as proteins and polymers.

Nowadays, with the advent of GCIB, one can perform a similar process simply by varying the size, and thus the energy, per constituent atom of the beam. As the size increases, fragmentation can be all but eliminated.

4.5   OTHER RELEVANT SURFACE MASS-SPECTROMETRY-BASED METHODS

Over the past decade, several new surface mass-spectrometry-based methods have been developed, which enable the rapid and direct analysis of samples at atmospheric pressures.[181,182] These methods, including atmospheric pressure matrix-assisted laser desorption ionization (AP-MALDI),[183] desorption electrospray ionization (DESI),[181] and plasma desorption ionization (PDI),[184] allow for

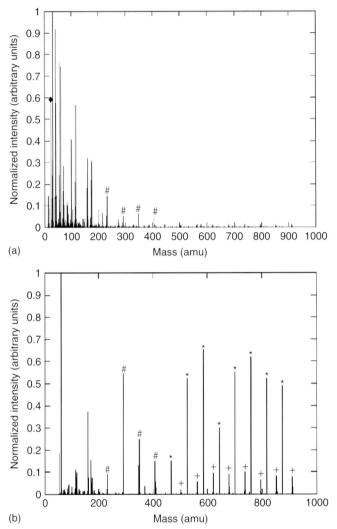

Figure 4.16   Representative positive ion SIMS (a) and G-SIMS (b) mass spectra from a sample of poly(glycolic acid) (PGA). Peak series of $[nM + H]^+$ (#), $[nM - OH]^+$ (+), $[nM - H_2O] + Na^+$ (⋆), and Na^+ (◇) are labeled accordingly. The SIMS spectrum (a) was acquired using 10 keV Cs^+, while the G-SIMS spectrum (b) was calculated from the mass spectra created with equal energy Cs^+ and Ar^+ primary ions. Reprinted from Ogaki et al.,[177] with permission from Elsevier.

the direct analysis of real-world samples, eliminating the need for a vacuum, and improving the overall versatility of surface mass spectrometry. This section of the chapter deals primarily with DESI- and PDI-based mass surface mass spectrometry. The pros and cons of each method are discussed, and compared with SIMS.

4.5.1   Desorption Electrospray Ionization (DESI)

DESI works by utilizing fast-moving charged solvent droplets to extract analytes from surfaces. Secondary microdroplets containing the analytes are then propelled (rebounded) toward a mass analyzer as is depicted in Figure 4.17a.[181,184,185]

DESI is turning out to be a robust method for characterization of organic materials. Its primary advantage is the ability to characterize samples at atmospheric pressure, therefore enabling the direct analysis of samples in real time. Furthermore, owing to its relatively soft ionization process, DESI surpasses SIMS in its ability to analyze higher mass species. DESI imaging is also possible,[181,186] although with significant reductions in the spatial resolution (values ranging from 180 to 220 μm,[181] although it has been reported to be as low as 40 μm under optimized conditions) as compared to what is typically observed with SIMS (50–200 nm).[187] Direct comparisons between DESI and SIMS are summarized in Table 4.1, created by Salter et al.[188]

DESI has shown great promise in particular, as a tool for characterization of polymers and biological samples.[189,190] An example of DESI for characterization of polymers is shown in Figure 4.18, which shows the DESI positive ion spectra acquired from a PEG sample of M_n = 3000 g/mol (M_n is the number average molecular weight of the PEG). Multiple molecular weight distributions are observed in the resulting mass spectrum. These distributions are associated with different charge state envelopes, a trait common to electrospray ionization (ESI) methods.[190] Note that the molecular weight distribution associated with charge state of +1 is not shown. This is because the mass spectrometer used in this study was only limited to a mass range of m/z 0 to m/z 2000. Despite this limitation, it is expected that the mass range of DESI will eventually rival that of MALDI.

TABLE 4.1   A Comparison of DESI and SIMS

	DESI	SIMS
Molecules detected	• Small molecules • Peptides and proteins • Inks and dyes • Molecules weakly bound to the substrate	• Small molecules • Inks and dyes • Molecules strongly and weakly bound to the substrate
Molecules not detected	• Molecules strongly bound to the substrate • Low solubility molecules	• Proteins and large peptides • Volatiles
Spectral information	Full MS/MS	MS with some structural information
Background signal	Significant background from source liquid	Background of fragmentation species and contamination
Sensitivity	Good (ng)	Excellent (fg or ag)
Repeatability	~15% over a day	Better than 2%
Spatial resolution	~150 μm	<100 nm

Recreated from Salter et al.,[188] with permission from Wiley.

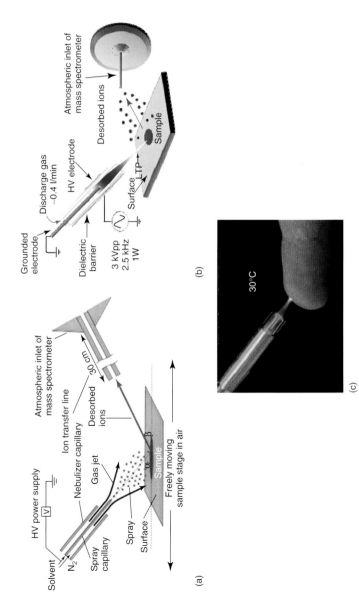

Figure 4.17 Schematic representations of DESI (a); low temperature plasma desorption ionization (LTPI) (b); and LTP interacting with a human finger (c). Reprinted from Takats et al.,[185] and Harper et al.,[184] with permission from the American Association for the Advancement of Science (AAAS) and ACS publications, respectively.

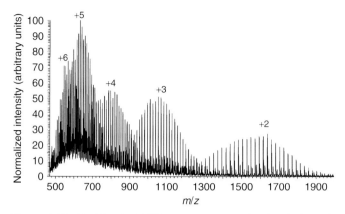

Figure 4.18   Positive ion DESI mass spectrum of poly(ethylene glycol) (PEG, MW = 3000 g/mol): 0.2 µg per sample spot, spray solvent of methanol/water, 1 : 1. Reprinted from Nefliu et al.,[190] with permission from RSC publishing.

DESI is also making major headway for biological imaging and cancer biomarker applications.[181,186,191] Although SIMS and MALDI are also being utilized for these applications,[4,173,192–194] the benefit of DESI is that the ionization occurs at atmospheric pressure. Furthermore, DESI is capable of detecting larger molecules, without the requirement of a matrix. This makes DESI a much more versatile technique for analysis of biological samples, and allows for direct analysis of tissue samples *in vitro*.

An example of DESI bioimaging for cancer applications is shown in Figure 4.19, which shows mass spectral and imaging results from a recent study by Dill and coworkers.[191] In this study, the authors show that they are able to differentiate between healthy tissue and cancerous tissue sections obtained from a dog bladder, using DESI.[191] The mass spectrum acquired from the tumor tissue is shown in Figure 4.19b, while the corresponding mass spectrum from the healthy tissue is shown in Figure 4.19a. Clear differences are observed in the spectra; these differences are mainly representative of changing lipid compositions.

The corresponding molecular images of the tumor tissue and healthy tissue are shown in Figure 4.19c. As can be seen, specific phospholipids are present only in the tumor-containing tissue and are not observed in the healthy tissue, illustrating that these lipids are a potential biomarker for cancer detection.

4.5.2   Plasma Desorption Ionization Methods

PDI techniques, such as direct analysis in real time (DART), atmospheric pressure glow discharge ionization (APGDI), and low temperature plasma ionization (LTPI) can also be operated under ambient conditions.[184] These methods use some form of plasma to desorb ions from a surface (see Figure 4.17 b,c). In general, methods employing plasmas do not extend to the analysis of larger biomolecules.[182] However, this approach is showing particular promise as a molecular depth profiling

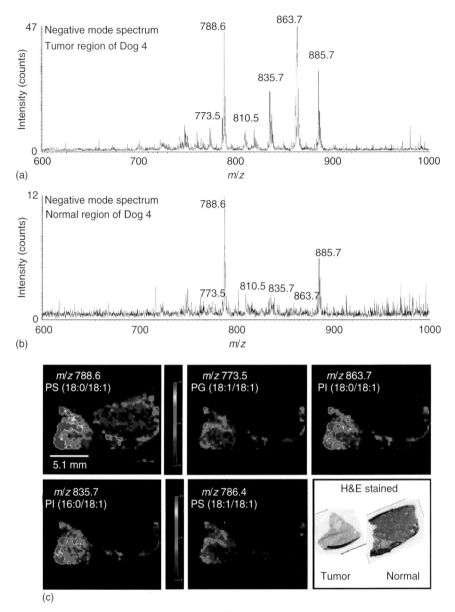

Figure 4.19 Typical negative ion mode DESI mass spectra of normal (a) and cancerous (b) dog bladder tissue in the mass range of m/z 600 to m/z 1000. (c) Corresponding secondary ion image maps in the tissues, where the tumor tissue is located on the left-hand side and the adjacent (mostly) healthy tissue is located on the right-hand side. Note that the adjacent healthy tissue does contain some tumor cells on the lower left edge of the specimen. Recreated from Dill et al.,[191] with permission from the ACS publications.

tool.[195] The results are promising, particularly for Type I polymers that tend to cross-link when irradiated. This will be discussed in greater detail in Chapter 5.

4.5.3 Electrospray Droplet Impact Source for SIMS

Electrospray droplet impact (EDI) has been utilized in conjunction with SIMS (EDI-SIMS) and other methods such as X-ray photoelectron spectroscopy (EDI-XPS) in order to obtain lower damage mass spectra. In this approach, charged droplets of 1 M acetic acid are accelerated into the vacuum system in a similar manner as any SIMS ion beam, resulting in a high momentum collision of the droplet with the sample surface. These collisions consequently result in the emission of secondary ions from the sample's surface.[196] This mechanism is significantly different from that employed with DESI, in which the electrospray droplets dissolve/ionize the samples under atmospheric pressure conditions.

EDI-SIMS has been particularly promising as a tool for characterization of polymers, as the fragmentation is significantly reduced when using EDI sources as compared to conventional ion beam sources. This makes EDI-SIMS a more viable approach for characterization of high molecular weight samples, typically observed in the realm of MALDI.[196,197] An example of EDI-SIMS for polymer characterization is depicted in Figure 4.20, which compares the EDI-SIMS mass spectra acquired from PS at two different molecular weights (m/z 2500 and m/z 4000) to MALDI mass spectra of the same polymers.[196] Both EDI-SIMS and MALDI yield similar molecular weight distributions that are consistent with the known distributions in the standard. However, while MALDI requires the use of

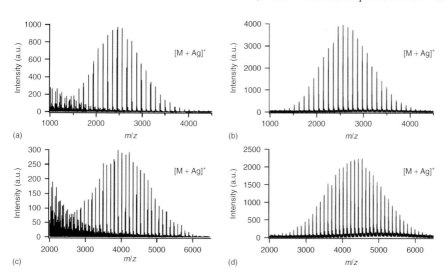

Figure 4.20 Positive ion mass spectra of polystyrene (PS) 2500 and 4000 g/mol using EDI-MS. (a) PS2500 obtained by EDI-SIMS; (b) PS2500 obtained by MALDI; (c) PS4000 obtained by EDI-SIMS; and (d) PS4000 obtained by MALDI. Silver tri-fluoroacetic acid (AgTFA) was added as the cationization salt for both EDI-SIMS and MALDI and dithranol was used as the MALDI matrix. Reprinted from Asakawa et al.,[196] with permission from Wiley.

a matrix, no matrix was required when employing EDI-SIMS spectrum. EDI has also proven to be capable of low damage etching in polymers, such as poly(vinyl chloride) (PVC), PS, and PMMA.[198−201] This will be discussed in greater detail in Chapter 5.

4.6 ADVANCED MASS SPECTROMETERS FOR SIMS

Nowadays, the mass spectrometry community is moving away from time-of-flight (ToF) mass analyzers and moving toward the utilization of advanced mass spectrometers, with tandem mass spectrometry (MS/MS) capabilities in conjunction with the use of state-of-the-art mass analyzers such as orbitraps and Fourier transform ion cyclotron resonance mass spectrometers (FT-ICR-MS). MS/MS, in particular, allows for increased specificity and enables more accurate identification and verification of the ions created during the sputtering process. This sort of instrument design in SIMS is ideal for identification of more complex samples, such as those encountered during the analysis of biological systems. New SIMS instruments have indeed been developed for this purpose and will be discussed in greater detail in Chapter 8.[202−204] However, these instruments do not employ FT-MS mass spectrometers. FT-ICR-MS[205] and orbitrap[206] systems have far superior mass resolutions (>1 million for FT-ICR-MS and >200,000 for orbitrap) and improved mass accuracies (∼0.65 ppm) as compared to magnetic sector and ToF mass analyzers, which can obtain mass resolutions of ∼15,000, at best. MALDI instrumentation is already using these advanced mass spectrometers for proteomics applications.[207,208]

With the advent of cluster sources, where analysis of soft samples has become a very real possibility, it is essential that we now develop these advanced mass spectrometers for SIMS. Recently a C_{60} SIMS-FT-ICR was developed by the Environmental Molecular Sciences Laboratory (EMSL) at the Pacific Northwest National Laboratory (PNNL).[209] This prototype instrument is depicted in Figure 4.21, which

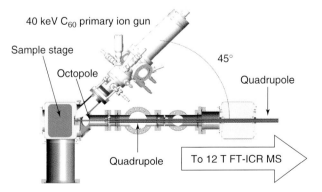

Figure 4.21   Schematic drawing of a C_{60} SIMS source that has been coupled to a 12 T FT-ICR MS. Reprinted from Smith et al.,[208] with permission from the ACS publications.

shows the front end of the instrument before entry into the FT-ICR. As can be seen with this design, MS/MS can be used in conjunction with the FT-ICR for increased chemical specificity. Moreover, the improvements in mass resolution are readily observed in the corresponding mass spectra. Figure 4.22 shows a narrow region of the mass spectrum obtained from a mouse brain sample and the corresponding molecular images. The peaks are not identified in the text. However, one can very clearly see the mass resolving power of the FT-ICR instrument, particularly for the peaks located at m/z 277.097 and m/z 277.101, which cannot be possibly be resolved using ToF-SIMS.

4.7 CONCLUSIONS

This chapter represents an overview of the current state of static SIMS using cluster primary ion beams and other emerging surface mass spectrometric–based techniques. Over the past few decades, the field of imaging mass spectrometry has rapidly expanded into the area of organic, polymeric, and biological analysis. SIMS, originally developed for characterization of inorganic materials and semiconductors, is now used on a regular basis for the surface and in-depth characterization of polymers and other soft samples. This is, in large part, a direct result of the advent of cluster SIMS. Cluster primary ion sources have allowed for significant increases in molecular sensitivities for static SIMS applications. This, in conjunction with the low damage sputtering process and decreased charging affects, has allowed for

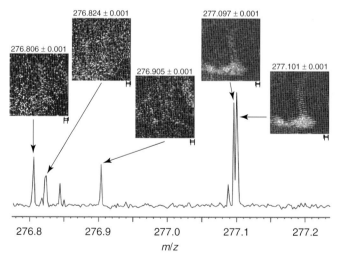

Figure 4.22 Zoom of the average mass spectrum from a C_{60} SIMS FTICR-MS imaging run of a mouse brain. This 0.4 m/z mass window shows no less than nine distinct peaks and illustrates the complexity of the mass spectra. Ion-selected images are shown with a m/z bin size of 0.002. Scale bar = 120 μm. Reprinted from Smith et al.,[208] with permission from the American Chemical Society.

SIMS to make a significant impact in the previously unattainable fields of biology and biomaterials characterization.

Similar to what is typically observed when depth profiling with cluster ion beams, considerations of optimum parameters, including sample temperature and ion beam conditions, are of utmost importance. As with all mass-spectrometric-based methods, data analysis remains a critical component to obtaining useful data. Statistical methods, such as PCA are rapidly becoming commonplace, expanding these methods further to include the analysis of complex mixtures and biological systems.

Although this is a time of expansion in the entire field of surface mass spectrometry, DESI and PDI methods are right at the beginning of their growth curve. It is expected therefore that significant developments will be observed in the coming years as these techniques mature.

The benefits of employing cluster ions for surface analysis of soft materials are undeniable. In combination with advanced FT-MS, it is expected that SIMS may dominate the field of surface mass spectrometry in the future, particularly because of its high spatial resolution molecular imaging capabilities. However, the most remarkable advances thus far have been made in the area of molecular depth profiling; a feat not possible using atomic primary ions. The next chapter will be devoted to this very important topic.

APPENDIX A: USEFUL LATERAL RESOLUTION

The spatial resolution in molecular samples is limited by the number of molecules that can be detected in a given area: the greater the molecular sputtering yield and ionization yield of a material, the greater the probability that molecules will be detected. Thus, even though the beam size can be very small, the probability of detecting a molecule from this small area is unlikely, particularly if the material is sensitive to radiation damage. The useful lateral resolution,[72] Δl is defined in Equation 4.A.1:

$$\Delta l = \left(\frac{N}{E}\right)^{1/2} = \left(\frac{N\sigma}{Y}\right)^{1/2} \qquad (4.A.1)$$

where N is the number of secondary ions detected from the area Δl^2, σ is the disappearance cross section of the material, Y is the secondary ion yield, and E is the secondary ion formation efficiency ($E = Y/\sigma$). As can be seen, the minimum useful lateral resolution that can be obtained with a reasonable amount of signal will decrease in cases where there is a decreased yield and/or an increased damage cross section.

REFERENCES

1. Mahoney, C. M. *Mass Spectrom. Rev.* **2010**, 29 (2), 247–293.
2. Fletcher, J. S. *Analyst* **2009**, 134 (11), 2204–2215.
3. Fletcher, J. S.; Vickerman, J. C., *Anal. Bioanal. Chem.* **2010**, 396 (1), 85–104.

4. Fletcher, J. S.; Lockyer, N. P.; Vickerman, J. C. *Mass Spectrom. Rev.* **2011**, 30 (1), 142–174.

5. Gronlund, F.; Moore, W. J. *J. Chem. Phys.* **1960**, 32 (5), 1540–1545.

6. Andersen, H. H.; Bay, H. L. *J. Appl. Phys.* **1974**, 45 (2), 953–954.

7. Andersen, H. H.; Bay, H. L. *J. Appl. Phys.* **1975**, 46 (6), 2416–2422.

8. Thompson, D. A.; Johar, S. S. *Appl. Phys. Lett.* **1979**, 34 (5), 342–345.

9. Wong, S. S.; Stoll, R.; Rollgen, F. W. *Z. Naturforsch.A J. Phy. Sci.* **1982**, 37 (7), 718–719.

10. Appelhans, A. D.; Delmore, J. E.; Dahl, D. A. *Anal. Chem.* **1987**, 59 (13), 1685–1691.

11. Appelhans, A. D.; Delmore, J. E. *Anal. Chem.* **1989**, 61 (10), 1087–1093.

12. Kotter, F.; Benninghoven, A. *Appl. Surf. Sci.* **1998**, 133 (1–2), 47–57.

13. Stapel, D.; Brox, O.; Benninghoven, A. *Appl. Surf. Sci.* **1999**, 140 (1–2), 156–167.

14. Stapel, D.; Thiemann, M.; Benninghoven, A. *Appl. Surf. Sci.* **2000**, 158, 362–374.

15. Stapel, D.; Benninghoven, A. *Appl. Surf. Sci.* **2001**, 174 (3–4), 261–270.

16. Davies, N.; Weibel, D. E.; Blenkinsopp, P.; Lockyer, N.; Hill, R.; Vickerman, J. C. *Appl. Surf. Sci.* **2003**, 203, 223–227.

17. Nagy, G.; Gelb, L. D.; Walker, A. V. *J. Am. Soc. Mass Spectrom.* **2005**, 16 (5), 733–742.

18. Nagy, G.; Walker, A. V. *Int. J. Mass Spectrom.* **2007**, 262 (1–2), 144–153.

19. Bryan, S. R.; Belu, A. M.; Hoshi, T.; Oiwa, R. *Appl. Surf. Sci.* **2004**, 231, 201–206.

20. Aimoto, K.; Aoyagi, S.; Kato, N.; Iida, N.; Yamamoto, A.; Kudo, M. *Appl. Surf. Sci.* **2006**, 252 (19), 6547–6549.

21. Weibel, D.; Wong, S.; Lockyer, N.; Blenkinsopp, P.; Hill, R.; Vickerman, J. C. *Anal. Chem.* **2003**, 75 (7), 1754–1764.

22. Zhu, Z.; Kelley, M. J. *Appl. Surf. Sci.* **2006**, 252 (19), 6619–6623.

23. Hill, R.; Blenkinsopp, P. W. M. *Appl. Surf. Sci.* **2004**, 231, 936–939.

24. Gillen, G.; Roberson, S. *Rapid Commun. Mass Spectrom.* **1998**, 12, 1303–1312.

25. Boschmans, B.; Van Royen, P.; Van Vaeck, L. *Rapid Commun. Mass Spectrom.* **2005**, 19 (18), 2517–2527.

26. Van Royen, P.; Taranu, A.; Van Vaeck, L. *Rapid Commun. Mass Spectrom.* **2005**, 19 (4), 552–560.

27. Braun, R. M.; Cheng, J.; Parsonage, E. E.; Moeller, J.; Winograd, N. *Anal. Chem.* **2006**, 78 (24), 8347–8353.

28. Touboul, D.; Kollmer, F.; Niehuis, E.; Brunelle, A.; Laprevote, O. *J. Am. Soc. Mass Spectrom.* **2005**, 16 (10), 1608–1618.

29. Licciardello, A.; Puglisi, O.; Calcagno, L.; Foti, G. *Nucl. Instrum. Meth. Phys. Res. B.* **1988**, 32 (1–4), 131–135.

30. Briggs, D.; Hearn, M. J. *Vacuum* **1986**, 36 (11–12), 1005–1010.

31. Hearn, M. J.; Briggs, D. *Surf. Interface Anal.* **1988**, 11 (4), 198–213.

32. Leggett, G. J.; Vickerman, J. C. *Appl. Surf. Sci.* **1992**, 55 (2–3), 105–115.

33. Benninghoven, A. *Phys. Status Solidi.* **1969**, 34, K169.

34. Benninghoven, A.; Loebach, E. *J. Radioanal. Chem.* **1972**, 12, 95–99.

35. Briggs, D.; Brown, A.; Vickerman, J. C. *Handbook of Static SIMS*; John Wiley & Sons: Chichester, UK; **1989**.

36. Benninghoven, A.; Jaspers, D.; Sichtermann, W. *Appl. Phys. A.* **1976**, 11 (1), 35–39.

37. Piwowar, A. M.; Vickerman, J. C. *Surf. Interface Anal.* **2010**, 42 (8), 1387–1392.

38. Piwowar, A. M.; Lockyer, N.; Vickerman, J. C. *Appl. Surf. Sci.* **2008**, 255 (4), 912–915.

39. Hand, O. W.; Majumdar, T. K.; Graham Cooks, R. *Int. J. Mass Spectrom.* **1990**, 97 (1), 35–45.

40. Benguerba, M.; Brunelle, A.; Della-Negra, S.; Depauw, J.; Joret, H.; Le Beyec, Y.; Blain, M. G.; Schweikert, E. A.; Assayag, G. B.; Sudraud, P. *Nucl. Instrum. Meth. Phys. Res. B.* **1991**, 62 (1), 8–22.

41. Hirata, K.; Saitoh, Y.; Narumi, K.; Kobayashi, Y. *Appl. Phys. Lett.* **2002**, 81, 3669.

42. Hirata, K.; Saitoh, Y.; Chiba, A.; Narumi, K.; Kobayashi, Y.; Arakawa, K. *Appl. Phys. Lett.* **2003**, 83, 4872.

43. Xu, J.; Szakal, C. W.; Martin, S. E.; Peterson, B. R.; Wucher, A.; Winograd, N. *J. Am. Chem. Soc.* **2004**, 126 (12), 3902–3909.

44. Cheng, J.; Winograd, N. *Anal. Chem.* **2005**, 77 (11), 3651–3659.

45. Wagner, M. S.; Gillen, G. *Appl. Surf. Sci.* **2004**, 231–2, 169–173.

46. Mahoney, C. M.; Fahey, A. J.; Gillen, G.; Xu, C.; Batteas, J. D. *Anal. Chem.* **2007**, 79 (3), 837–845.

47. Mahoney, C. M.; Gillen, G.; Fahey, A. J. *Forensic Sci. Int.* **2006**, 158 (1), 39–51.

48. Gillen, G.; Mahoney, C.; Wight, S.; Lareau, R. *Rapid Commun. Mass Spectrom.* **2006**, 20 (12), 1949–1953.

49. Miyayama, T.; Sanada, N.; Bryan, S. R.; Hammond, J. S.; Suzuki, M. *Surf. Interface Anal.* **2010**, 42 (9), 1453–1457.

50. Diehnelt, C. W.; Van Stipdonk, M. J.; Schweikert, E. A. *Int. J. Mass Spectrom.* **2001**, 207 (1–2), 111–122.

51. Harris, R. D.; Baker, W. S.; Van Stipdonk, M. J.; Crooks, R. M.; Schweikert, E. A. *Rapid Commun. Mass Spectrom.* **1999**, 13 (14), 1374–1380.

52. Czerwinski, B.; Samson, R.; Garrison, B. J.; Winograd, N.; Postawa, Z. *Vacuum* **2006**, 81 (2), 167–173.

53. Postawa, Z.; Czerwinski, B.; Winograd, N.; Garrison, B. J. *J. Phys. Chem. B* **2005**, 109 (24), 11973–11979.

54. Brison, J.; Muramoto, S.; Castner, D. G. *J. Phys. Chem. C* **2010**, 114 (12), 5565–5573.

55. Muramoto, S.; Brison, J.; Castner, D. *Surf. Interface Anal.* **2011**, 43 (102), 58–61.

56. Sun, S.; Szakal, C.; Roll, T.; Mazarov, P.; Wucher, A.; Winograd, N. *Surf. Interface Anal.* **2004**, 36 (10), 1367–1372.

57. Gillen, G.; Walker, M.; Thompson, P.; Bennett, J. *J. Vac. Sci. Technol. B.* **2000**, 18, 503.

58. Gillen, G.; King, L.; Freibaum, B.; Lareau, R.; Bennett, J.; Chmara, F. *J. Vac. Sci. Technol. A.* **2001**, 19, 568.

59. Fujiwara, Y.; Kondou, K.; Nonaka, H.; Saito, N.; Itoh, H.; Fujimoto, T.; Kurokawa, A.; Ichimura, S.; Tomita, M. *Japanese J. Appl. Phys.* **2006**, 45, L987–L990.

60. Tomita, M.; Kinno, T.; Koike, M.; Tanaka, H.; Takeno, S.; Fujiwara, Y.; Kondou, K.; Teranishi, Y.; Nonaka, H.; Fujimoto, T. *Appl. Phys. Lett.* **2006**, 89, 053123.

61. Tomita, M.; Kinno, T.; Koike, M.; Tanaka, H.; Takeno, S.; Fujiwara, Y.; Kondou, K.; Teranishi, Y.; Nonaka, H.; Fujimoto, T. *Nucl. Instrum. Meth. Phys. Res. B.* **2007**, 258 (1), 242–245.

62. Fujiwara, Y.; Kondou, K.; Teranishi, Y.; Watanabe, K.; Nonaka, H.; Saito, N.; Itoh, H.; Fujimoto, T.; Kurokawa, A.; Ichimura, S. *Appl. Surf. Sci.* **2008**, 255 (4), 916–921.

63. Tomita, M.; Kinno, T.; Koike, M.; Tanaka, H.; Takeno, S.; Fujiwara, Y.; Kondou, K.; Teranishi, Y.; Nonaka, H.; Fujimoto, T. *High Depth Resolution SIMS Analysis Using Metal Cluster Complex Ion Bombardment*; IOP Publishing; Bristol, UK; **2008**; p 012001.

64. Henkel, T.; Rost, D.; Lyon, I. C. *Rapid Commun. Mass Spectrom.* **2009**, 23 (21), 3355–3360.

65. Gillen, G.; Batteas, J.; Michaels, C. A.; Chi, P.; Small, J.; Windsor, E.; Fahey, A.; Verkouteren, J.; Kim, K. J. *Appl. Surf. Sci.* **2006**, 252 (19), 6521–6525.

66. Fletcher, J. S.; Rabbani, S.; Henderson, A.; Blenkinsopp, P.; Thompson, S. P.; Lockyer, N. P.; Vickerman, J. C. *Anal. Chem.* **2008**, 80 (23), 9058–9064.

67. Touboul, D.; Halgand, F.; Brunelle, A.; Kersting, R.; Tallarek, E.; Hagenhoff, B.; Laprevote, O. *Anal. Chem.* **2004**, 76 (6), 1550–1559.

68. Nygren, H.; Borner, K.; Hagenhoff, B.; Malmberg, P.; Mansson, J. E. *Biochim. Biophys. Acta* **2005**, 1737 (2–3), 102–110.

69. Nygren, H.; Hagenhoff, B.; Malmberg, P.; Nilsson, M.; Richter, K. *Micros. Res. Tech.* **2007**, 70 (11), 969–974.

70. Nygren, H.; Malmberg, P. *Trends Biotechnol.* **2007**, 25 (11), 499–504.

71. Sjovall, P.; Lausmaa, J.; Johansson, B. *Anal. Chem.* **2004**, 76 (15), 4271–4278.

72. Kollmer, F. *Appl. Surf. Sci.* **2004**, 231, 153–158.

73. Seah, M. P. *Surf. Interface Anal.* **2007**, 39 (7), 634–643.

74. Rabbani, S.; Barber, A. M.; Fletcher, J. S.; Lockyer, N. P.; Vickerman, J. C. *Anal. Chem.* **2011**, 83 (10), 3793–3800.

75. Tempez, A.; Schultz, J. A.; Della Negra, S.; Depauw, J.; Jacquet, D.; Novikov, A.; Lebeyec, Y.; Pautrat, M.; Caroff, M.; Ugarov, M. *Rapid Commun. Mass Spectrom.* **2004**, 18 (4), 371–376.

76. Hager, G. J.; Guillermier, C.; Verkhoturov, S. V.; Schweikert, E. A. *Appl. Surf. Sci.* **2006**, 252 (19), 6558–6561.

77. Guillermier, C.; Negra, S. D.; Rickman, R. D.; Pinnick, V.; Schweikert, E. A. *Appl. Surf. Sci.* **2006**, 252 (19), 6529–6532.

78. Guillermier, C.; Pinnick, V.; Verkhoturov, S. V.; Schweikert, E. A. *Appl. Surf. Sci.* **2006**, 252 (19), 6644–6647.

79. Verkhoturov, S. V.; Rickman, R. D.; Guillermier, C.; Hager, G. J.; Locklear, J. E.; Schweikert, E. A. *Appl. Surf. Sci.* **2006**, 252 (19), 6490–6493.

80. Van Stipdonk, M. J.; Harris, R. D.; Schweikert, E. A. *Rapid Commun. Mass Spectrom.* **1996**, 10 (15), 1987–1991.

81. Pinnick, V.; Verkhoturov, S. V.; Kaledin, L.; Schweikert, E. A. *Surf. Interface Anal.* **2011**, 43 (1–2), 551–554.

82. Fujiwara, Y.; Kondou, K.; Watanabe, K.; Nonaka, H.; Saito, N.; Fujimoto, T.; Kurokawa, A.; Ichimura, S.; Tomita, M. *Japanese J. Appl. Phys.* **2007**, 46 (11), 7599.

83. Blain, M. G.; Della-Negra, S.; Joret, H.; Le Beyec, Y.; Schweikert, E. A. *Phys. Rev. Lett.* **1989**, 63 (15), 1625–1628.

84. Reuter, W. *Anal. Chem.* **1987**, 59, 2081–2087.

85. Reuter, W.; Clabes, J. G. *Anal. Chem.* **1988**, 60, 1405–1408.

86. Yamazaki, H.; Mitani, Y. *Nucl. Instrum. Meth. Phys. Res. B.* **1997**, 124, 91–94.

87. Delmore, J. E.; Appelhans, A. D.; Peterson, E. S. *Int. J. Mass Spectrom.* **1995**, 146, 15–20.

88. Groenewold, G. S.; Cowan, R. L.; Ingram, J. C.; Appelhans, A. D.; Delmore, J. E.; Olson, J. E. *Surf. Interface Anal.* **1996**, 24 (12), 794–802.

89. Groenewold, G. S.; Delmore, J. E.; Olson, J. E.; Appelhans, A. D.; Ingram, J. C.; Dahl, D. A. *Int. J. Mass Spectrom.* **1997**, 163 (3), 185–195.

90. Groenewold, G. S.; Gianotto, A. K.; Olson, J. E.; Appelhans, A. D.; Ingram, J. C.; Delmore, J. E.; Shaw, A. D. *Int. J. Mass Spectrom.* **1998**, 174 (1–3), 129–142.

91. Groenewold, G. S.; Appelhans, A. D.; Gresham, G. L.; Olson, J. E.; Jeffery, M.; Wright, J. B. *Anal. Chem.* **1999**, 71 (13), 2318–2323.

92. Groenewold, G. S.; Appelhans, A. D.; Gresham, G. L.; Olson, J. E.; Jeffery, M.; Weibel, M. *J. Am. Soc. Mass Spectrom.* **2000**, 11 (1), 69–77.

93. Ingram, J. C.; Bauer, W. F.; Lehman, R. M.; O'Connell, S. P.; Shaw, A. D. *J. Microbiol. Meth.* **2003**, 53 (3), 295–307.

94. Boussofiane-Baudin, K.; Bolbach, G.; Brunelle, A.; Della-Negra, S.; Hakansson, P.; Le Beyec, Y. *Nucl. Instrum. Meth. Phys. Res. B.* **1994**, 88 (1–2), 160–163.

95. Simko, S. J.; Bryan, S. R.; Griffis, D. P.; Murray, R. W.; Linton, R. W. *Anal. Chem.* **1985**, 57 (7), 1198–1202.

96. Webb, R. P.; Kerford, M.; Kappes, M.; Brauchle, G. *Nucl. Instrum. Meth. Phys. Res. B.* **1997**, 122 (3), 318–321.

97. Lee, J. L. S.; Ninomiya, S.; Matsuo, J.; Gilmore, I. S.; Seah, M. P.; Shard, A. G. *Anal. Chem.* **2009**, 82 (1), 98–105.

98. Shard, A. G.; Ray, S.; Seah, M. P.; Yang, L. *Surf. Interface Anal.* **2011**, 43 (9), 1240–1250.

99. Delcorte, A.; Garrison, B. J.; Hamraoui, K. *Surf. Interface Anal.* **2011**, 43 (1–2), 16–19.

100. Oshima, S.; Kashihara, I.; Moritani, K.; Inui, N.; Mochiji, K. *Rapid Commun. Mass Spectrom.* **2011**, 25 (8), 1070–1074.

101. Mochiji, K.; Hashinokuchi, M.; Moritani, K.; Toyoda, N. *Rapid Commun. Mass Spectrom.* **2009**, 23 (5), 648–652.

102. Moritani, K.; Hashinokuchi, M.; Nakagawa, J.; Kashiwagi, T.; Toyoda, N.; Mochiji, K. *Appl. Surf. Sci.* **2008**, 255 (4), 948–950.

103. Moritani, K.; Houzumi, S.; Takeshima, K.; Toyoda, N.; Mochiji, K. *J. Phys. Chem. C.* **2008**, 112 (30), 11357–11362.

104. Ninomiya, S.; Nakata, Y.; Ichiki, K.; Seki, T.; Aoki, T.; Matsuo, J. *Nucl. Instrum. Meth. Phys. Res. B.* **2007**, 256 (1), 493–496.

105. Mahoney, C. M.; Fahey, A. J.; Gillen, G.; Xu, C.; Batteas, J. D. *Appl. Surf. Sci.* **2006**, 252 (19), 6502–6505.

106. Mahoney, C. M.; Patwardhan, D. V.; McDermott, M. K. *Appl. Surf. Sci.* **2006**, 252 (19), 6554–6557.

107. Mahoney, C. M.; Fahey, A. J.; Gillen, G. *Anal. Chem.* **2007**, 79 (3), 828–836.

108. Mahoney, C. M.; Fahey, A. J.; Belu, A. M. *Anal. Chem.* **2008**, 80 (3), 624–632.
109. Moellers, R.; Tuccitto, N.; Torrisi, V.; Niehuis, E.; Licciardello, A. *Appl. Surf. Sci.* **2006**, 252 (19), 6509–6512.
110. Zheng, L.; Wucher, A.; Winograd, N. *Anal. Chem.* **2008**, 80 (19), 7363–7371.
111. Sjovall, P.; Rading, D.; Ray, S.; Yang, L.; Shard, A. G. *J. Phys. Chem. B.* **2010**, 114 (2), 769–774.
112. Shard, A. G.; Goster, I. S.; Gilmore, I. S.; Lee, L. S.; Ray, S.; Yang, L. *Surf. Interface Anal.* **2011**, 43 (1–2), 510–513.
113. Piwowar, A.; Fletcher, J.; Lockyer, N.; Vickerman, J. *Surf. Interface Anal.* **2011**, 43 (1–2), 207–210.
114. Piwowar, A. M.; Fletcher, J. S.; Kordys, J.; Lockyer, N. P.; Winograd, N.; Vickerman, J. C. *Anal. Chem.* **2010**, 82 (19), 8291–8299.
115. Grade, H.; Winograd, N.; Cooks, R. G. *J. Am. Chem. Soc.* **1977**, 99 (23), 7725–7726.
116. Grade, H.; Cooks, R. G. *J. Am. Chem. Soc.* **1978**, 100 (18), 5615–5621.
117. Bletsos, I. V.; Hercules, D. M.; VanLeyen, D.; Benninghoven, A. *Macromolecules* **1987**, 20 (2), 407–413.
118. Bletsos, I. V.; Hercules, D. M.; Greifendorf, D.; Benninghoven, A. *Anal. Chem.* **1985**, 57 (12), 2384–2388.
119. Linton, R. W.; Mawn, M. P.; Belu, A. M.; DeSimone, J. M.; Hunt Jr, M. O.; Menceloglu, Y. Z.; Cramer, H. G.; Benninghoven, A. *Surf. Interface Anal.* **1993**, 20 (12), 991–999.
120. Adriaensen, L.; Vangaever, F.; Gijbels, R. *Anal. Chem.* **2004**, 76 (22), 6777–6785.
121. Ruch, D.; Muller, J. F.; Migeon, H. N.; Boes, C.; Zimmer, R. *J. Mass Spectrom.* **2003**, 38 (1), 50–57.
122. Delcorte, A.; Medard, N.; Bertrand, P. *Anal. Chem.* **2002**, 74 (19), 4955–4968.
123. Nittler, L.; Delcorte, A.; Bertrand, P.; Migeon, H. N. *Surf. Interface Anal.* **2011**, 43 (1–2), 103–106.
124. Delcorte, A.; Bour, J.; Aubriet, F.; Muller, J. F.; Bertrand, P. *Anal. Chem.* **2003**, 75 (24), 6875–6885.
125. Delcorte, A.; Bertrand, P. *Appl. Surf. Sci.* **2004**, 231, 250–255.
126. Bletsos, I. V.; Hercules, D. M.; Magill, J. H.; VanLeyen, D.; Niehuis, E.; Benninghoven, A. *Anal. Chem.* **1988**, 60 (9), 938–944.
127. Marcus, A.; Winograd, N. *Anal. Chem.* **2006**, 78 (1), 141–148.
128. Restrepo, O. A.; Delcorte, A. *Surf. Interface Anal.* **2011**, 43 (1–2), 70–73.
129. Nicola, A. J.; Muddiman, D. C.; Hercules, D. M. *J. Am. Soc. Mass Spectrom.* **1996**, 7 (5), 467–472.
130. Wu, K. J.; Odom, R. W. *Anal. Chem.* **1996**, 68 (5), 873–882.
131. Jones, E. A.; Lockyer, N. P.; Kordys, J.; Vickerman, J. C. *J. Am. Soc. Mass Spectrom.* **2007**, 18 (8), 1559–1567.
132. Locklear, J. E.; Guillermier, C.; Verkhoturov, S. V.; Schweikert, E. A. *Appl. Surf. Sci.* **2006**, 252 (19), 6624–6627.
133. McDonnell, L. A.; Heeren, R.; de Lange, R. P. J.; de Fletcher, I. W. *J. Am. Soc. Mass Spectrom.* **2006**, 17 (9), 1195–1202.
134. Wittmaack, K.; Szymczak, W.; Hoheisel, G.; Tuszynski, W. *J. Am. Soc. Mass Spectrom.* **2000**, 11 (6), 553–563.
135. Brewer, T. M.; Szakal, C.; Gillen, G. *Rapid Commun. Mass Spectrom.* **2010**, 24 (5), 593–598.
136. Barber, M.; Bordoli, R. S.; Sedgwick, R. D.; Tyler, A. N. *Nature* **1981**, 293 (5830), 270–275.
137. Rinehart, K. L. *Science* **1982**, 218 (4569), 254.
138. Ligon Jr, W. V.; Dorn, S. B. *Int. J. Mass Spectrom.* **1987**, 78, 99–113.
139. Delcorte, A.; Poleunis, C.; Bertrand, P. *Appl. Surf. Sci.* **2006**, 252 (19), 6494–6497.
140. Delcorte, A.; Yunus, S.; Wehbe, N.; Nieuwjaer, N.; Poleunis, C.; Felten, A.; Houssiau, L.; Pireaux, J. J.; Bertrand, P. *Anal. Chem.* **2007**, 79 (10), 3673–3689.
141. Jones, E. A.; Lockyer, N. P.; Vickerman, J. C. *Int. J. Mass Spectrom.* **2007**, 260 (2–3), 146–157.
142. Jones, E. A.; Fletcher, J. S.; Thompson, C. E.; Jackson, D. A.; Lockyer, N. P.; Vickerman, J. C. *Appl. Surf. Sci.* **2006**, 252 (19), 6844–6854.
143. Coon, S. R.; Calaway, W. F.; Burnett, J. W.; Pellin, M. J.; Gruen, D. M.; Spiegel, D. R.; White, J. M. *Surf. Sci.* **1991**, 259 (3), 275–287.
144. Mathieu, H. J.; Leonard, D. *High Temp. Mater. Process.* **1998**, 17 (1–2), 29–44.

145. Becker, C. H.; Gillen, K. T. *Anal. Chem.* **1984**, 56 (9), 1671–1674.
146. Terhorst, M.; Kampwerth, G.; Niehuis, E.; Benninghoven, A. *J. Vac. Sci. Technol. A.* **1992**, 10 (5), 3210–3215.
147. Tehorst, M.; Mollers, R.; Niehuis, E.; Benninghoven, A. *Surf. Interface Anal.* **1992**, 18 (12), 824–826.
148. Tanaka, K.; Waki, H.; Ido, Y.; Akita, S.; Yoshida, Y.; Yoshida, T.; Matsuo, T. *Rapid Commun. Mass Spectrom.* **1988**, 2 (8), 151–153.
149. Wucher, A.; Sun, S.; Szakal, C.; Winograd, N. *Anal. Chem.* **2004**, 76 (24), 7234–7242.
150. Willingham, D.; Brenes, D. A.; Wucher, A.; Winograd, N. *J Phys. Chem. C. Nanomater. Interfaces* **2010**, 114 (12), 5391–5399.
151. Kozole, J.; Wucher, A.; Winograd, N. *Anal. Chem.* **2008**, 80 (14), 5293–5301.
152. Miyayama, T.; Sanada, N.; Iida, S.; Hammond, J. S.; Suzuki, M. *Appl. Surf. Sci.* **2008**, 255 (4), 951–953.
153. Belu, A. M.; Graham, D. J.; Castner, D. G. *Biomaterials* **2003**, 24 (21), 3635–3653.
154. Kono, T.; Iwase, E.; Kanamori, Y. *Appl. Surf. Sci.* **2008**, 255 (4), 997–1000.
155. Klerk, L. A.; Broersen, A.; Fletcher, I. W.; van Liere, R.; Heeren, R. *Int. J. Mass Spectrom.* **2007**, 260 (2–3), 222–236.
156. Ito, H.; Kono, T. *Appl. Surf. Sci.* **2008**, 255 (4), 1044–1047.
157. Choi, C.; Jung, D.; Moon, D. W.; Lee, T. G. *Surf. Interface Anal.* **2011**, 43 (1–2), 331–335.
158. Mahoney, C. M.; Kushmerick, J. G.; Steffens, K. L. *J. Phys. Chem. C.* **2010**, 114 (34), 14510–14519.
159. Eynde, X. V.; Bertrand, P. *Surf. Interface Anal.* **1998**, 26 (8), 579–589.
160. Coullerez, G.; Lundmark, S.; Malmström, E., Hult, A.; Mathieu, H. *J. Surf. Interface Anal.* **2003**, 35 (8), 693–708.
161. Jasieniak, M.; Suzuki, S.; Monteiro, M.; Wentrup-Byrne, E.; Griesser, H. J.; Grondahl, L. *Langmuir* **2008**, 25 (2), 1011–1019.
162. Baytekin, H. T.; Wirth, T.; Gross, T.; Sahre, M.; Unger, W. E. S.; Theisen, J.; Schmidt, M. *Surf. Interface Anal.* **2010**, 42 (8), 1417–1431.
163. Lau, Y. T. R.; Weng, L. T.; Ng, K. M.; Chan, C. M. *Surf. Interface Anal.* **2010**, 42 (8), 1445–1451.
164. Chilkoti, A.; Ratner, B. D.; Briggs, D. *Anal. Chem.* **1993**, 65 (13), 1736–1745.
165. Wagner, M. S.; Shen, M.; Horbett, T. A.; Castner, D. G. *J. Biomed. Mater. Res. A.* **2003**, 64A (1), 1–11.
166. Wagner, M. S.; Shen, M.; Horbett, T. A.; Castner, D. G. *Appl. Surf. Sci.* **2003**, 203, 704–709.
167. Wagner, M. S.; Graham, D. J.; Ratner, B. D.; Castner, D. G. *Surf. Sci.* **2004**, 570 (1–2), 78–97.
168. Wagner, M. S.; Castner, D. G. *Appl. Surf. Sci.* **2004**, 231–2, 366–376.
169. Wagner, M. S.; Castner, D. G. *Langmuir* **2001**, 17 (15), 4649–4660.
170. *NIST/SEMATECH e-Handbook of Statistics.* http://www.itl.nist.gov/div898/handbook/ (accessed Nov 24, 2012).
171. Wagner, M. S.; Horbett, T. A.; Castner, D. G. *Langmuir* **2003**, 19 (5), 1708–1715.
172. Wagner, M. S.; Castner, D. G. *Appl. Surf. Sci.* **2003**, 203, 698–703.
173. Kulp, K. S.; Berman, E. S. F.; Knize, M. G.; Shattuck, D. L.; Nelson, E. J.; Wu, L.; Montgomery, J. L.; Felton, J. S.; Wu, K. J. *Anal. Chem.* **2006**, 78 (11), 3651–3658.
174. Mahoney, C. M.; Fahey, A. J.; Steffens, K. L.; Benner Jr, B. A.; Lareau, R. T. *Anal. Chem.* **2010**, 82 (17), 7237–7248.
175. Ogaki, R.; Shard, A. G.; Li, S. M.; Vert, M.; Luk, S.; Alexander, M. R.; Gilmore, I. S.; Davies, M. C. *Surf. Interface Anal.* **2008**, 40 (8), 1168–1175.
176. Ogaki, R.; Green, F. M.; Gilmore, I. S.; Shard, A. G.; Luk, S.; Alexander, M. R.; Davies, M. C. *Surf. Interface Anal.* **2007**, 39 (11), 852–859.
177. Ogaki, R.; Green, F.; Li, S.; Vert, M.; Alexander, M. R.; Gilmore, I. S.; Davies, M. C. *Appl. Surf. Sci.* **2006**, 252 (19), 6797–6800.
178. Hawtin, P. N.; Abel, M. L.; Watts, J. F.; Powell, J. *Appl. Surf. Sci.* **2006**, 252 (19), 6676–6678.
179. Gilmore, I. S.; Seah, M. P. *Appl. Surf. Sci.* **2000**, 161 (3–4), 465–480.
180. Green, F. M.; Kollmer, F.; Niehuis, E.; Gilmore, I. S.; Seah, M. P. *Rapid Commun. Mass Spectrom.* **2008**, 22 (16), 2602–2608.
181. Ifa, D. R.; Wu, C.; Ouyang, Z.; Cooks, R. G. *The Analyst* **2010**, 135 (4), 669–681.

182. Weston, D. J. *The Analyst* **2010**, 135 (4), 661–668.
183. Laiko, V. V.; Baldwin, M. A.; Burlingame, A. L. *Anal. Chem.* **2000**, 72 (4), 652–657.
184. Harper, J. D.; Charipar, N. A.; Mulligan, C. C.; Zhang, X.; Cooks, R. G.; Ouyang, Z. *Anal. Chem.* **2008**, 80 (23), 9097–9104.
185. Takats, Z.; Wiseman, J. M.; Gologan, B.; Cooks, R. G. *Science* **2004**, 306 (5695), 471.
186. Eberlin, L. S.; Ifa, D. R.; Wu, C.; Cooks, R. G. *Angew. Chem. Int. Ed.* **2010**, 49 (5), 873–876.
187. Kertesz, V.; Van Berkel, G. J. *Rapid Commun. Mass Spectrom.* **2008**, 22 (17), 2639–2644.
188. Salter, T. L.; Green, F. M.; Gilmore, I. S.; Seah, M. P.; Stokes, P. *Surf. Interface Anal.* **2011**, 43 (1–2), 294–297.
189. Jackson, A. T.; Williams, J. P.; Scrivens, J. H. *Rapid Commun. Mass Spectrom.* **2006**, 20 (18), 2717–2727.
190. Nefliu, M.; Venter, A. *Chem. Commun.* **2006**, (8), 888–890.
191. Dill, A. L.; Ifa, D. R.; Manicke, N. E.; Costa, A. B.; Ramos-Vara, J. A.; Knapp, D. W.; Cooks, R. G. *Anal. Chem.* **2009**, 81 (21), 8758–8764.
192. McDonnell, L. A.; Heeren, R. M. A. *Mass Spectrom. Rev.* **2007**, 26 (4), 606–643.
193. Crecelius, A. C.; Cornett, D. S.; Caprioli, R. M.; Williams, B.; Dawant, B. M.; Bodenheimer, B., *J. Am. Soc. Mass Spectrom.* **2005**, 16 (7), 1093–1099.
194. Chaurand, P.; Schwartz, S. A.; Caprioli, R. M. *Anal. Chem.* **2004**, 76 (5), 86–93.
195. Tuccitto, N.; Lobo, L.; Tempez, A.; Delfanti, I.; Chapon, P.; Canulescu, S.; Bordel, N.; Michler, J.; Licciardello, A. *Rapid Commun. Mass Spectrom.* **2009**, 23 (5), 549–556.
196. Asakawa, D.; Chen, L. C.; Hiraoka, K. *J. Mass Spectrom.* **2009**, 44 (6), 945–951.
197. Hiraoka, K.; Mori, K.; Asakawa, D. *J. Mass Spectrom.* **2006**, 41 (7), 894–902.
198. Hiraoka, K.; Iijima, Y.; Sakai, Y.. *Surf. Interface Anal.* **2011**, 43 (1–2), 236–240.
199. Sakai, Y.; Iijima, Y.; Takaishi, R.; Asakawa, D.; Hiraoka, K. *J. Vac. Sci. Technol. A.* **2009**, 27, 743.
200. Hiraoka, K.; Takaishi, R.; Asakawa, D.; Sakai, Y.; Iijima, Y. *J. Vac. Sci. Technol. A.* **2009**, 27, 748.
201. Sakai, Y.; Iijima, Y.; Asakawa, D.; Hiraoka, K. *Surf. Interface Anal.* **2010**, 42 (6–7), 658–661.
202. Carado, A.; Passarelli, M. K.; Kozole, J.; Wingate, J. E.; Winograd, N.; Loboda, A. V. *Anal. Chem.* **2008**, 80 (21), 7921–7929.
203. Fletcher, J. S. *Analyst.* **2009**, 134 (11), 2204–2215.
204. Hill, R.; Blenkinsopp, P.; Thompson, S.; Vickerman, J.; Fletcher, J. S., *Surf. Interface Anal.* **2011**, 43 (1–2), 506–509.
205. Marshall, A. G.; Hendrickson, C. L.; Jackson, G. S. *Encyclopedia of Analytical Chemistry*; John Wiley & Sons: Hoboken, NJ; **2000**.
206. Hu, Q.; Noll, R. J.; Li, H.; Makarov, A.; Hardman, M.; Graham Cooks, R. *J. Mass Spectrom.* **2005**, 40 (4), 430–443.
207. Cornett, D. S.; Frappier, S. L.; Caprioli, R. M. *Anal. Chem.* **2008**, 80 (14), 5648–5653.
208. Taban, I. M.; Altelaar, A. F.; van der Burgt, Y. E. M.; McDonnell, L. A.; Heeren, R.; Fuchser, J.; Baykut, G. *J. Am. Soc. Mass Spectrom.* **2007**, 18 (1), 145–151.
209. Smith, D. F.; Robinson, E. W.; Tolmachev, A. V.; Heeren, R.; Pasa-Tolic, L. *Anal. Chem.* **2011**, 83, 9552–9556.

CHAPTER 5

MOLECULAR DEPTH PROFILING WITH CLUSTER ION BEAMS*

Christine M. Mahoney and Andreas Wucher

5.1 INTRODUCTION

In Chapter 1, the concept of molecular depth profiling with cluster ion beams was introduced, citing several examples of successful cluster secondary ion mass spectrometry (SIMS) sputter depth profile analyses in soft samples. It was demonstrated that whereas the more conventional atomic primary ion beams can cause extensive subsurface damage in molecular samples, cluster ion beams create damage that is restricted to the near-surface region, preserving the underlying chemical structure, and therefore enabling molecular depth profiling. The polyatomicity of the incoming ion also causes a collisional spike, which results in nonlinear enhancements in the sputtering yields. Thus, one achieves an equilibrium state whereby any damage created by the incoming cluster ion is efficiently removed and does not accumulate as is the case with atomic ion beam sources.[1-4]

---

*This document was prepared as an account of work sponsored by an agency of the US Government. Neither the US Government nor any agency thereof, nor Battelle Memorial Institute, nor any of their employees, makes any warranty, express or implied, or assumes any legal liability or responsibility for the accuracy, completeness, or usefulness of any information, apparatus, product, or process disclosed, or represents that its use would not infringe privately owned rights. Reference herein to any specific commercial product, process, or service by trade name, trademark, manufacturer, or otherwise does not necessarily constitute or imply its endorsement, recommendation, or favoring by the US Government or any agency thereof, or Battelle Memorial Institute. The views and opinions of authors expressed herein do not necessarily state or reflect those of the US Government or any agency thereof.

Cluster Secondary Ion Mass Spectrometry: Principles and Applications, First Edition.
Edited by Christine M. Mahoney.
© 2013 John Wiley & Sons, Inc. Published 2013 by John Wiley & Sons, Inc.

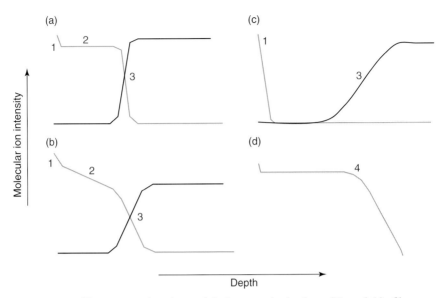

Figure 5.1 Different scenarios observed during organic depth profiling of thin films (a–c) and bulk samples (d): (a) results obtained under optimum conditions: the ideal depth profile shape, (b,c) unoptimized conditions exhibiting damage accumulation effects where damage in (c) is greater than (b). Three regions are observed in all depth profiles: (1) An initial drop or increase in the signal intensity is observed in the surface transient region of the profile; (2) a steady-state region, or pseudo-steady-state region in the case of (b), and (3) an interfacial region defined by a decrease in molecular signal, with commensurate increase in substrate signal. In bulk materials (d), there is usually a loss of signal intensity after a certain critical fluence (4). This critical fluence varies with the ion source used and other important parameters as discussed in this chapter. Reproduced from Mahoney,[5] with permission from Wiley.

While this physical picture of the sputtering process in molecular samples can explain many features of molecular depth profiles with cluster ion beams, there are also complex chemistries involved; chemistries that do not play a role during the more conventional sputtering of inorganic samples.[5,6] In this chapter, we discuss these chemistries, in addition to the important physical sputtering aspects, both of which play a critical role in low damage sputter depth profiling with cluster SIMS.

In general, attempts to depth profile through organic materials will result in one of the scenarios depicted in Figure 5.1. Under optimal sputtering conditions, as will be defined later in this chapter, the scenario shown in Figure 5.1a is observed. In region 1 of the depth profile, the molecular signal intensity will typically undergo a rapid exponential decay.* This surface transient region is

---

*In some cases, there is an increase in intensity in this region, particularly if larger cluster sources, such as $C_{60}{}^{n+}$ or $Ar_{n>500}{}^{+}$ sources are used, as they create a large number of protons during the bombardment process.[39-41] This can result in an enhancement in the ionization efficiency of molecular species.

resultant from the initial damage saturation processes, and its shape can be described by the erosion model which will be introduced in Section 5.4.[2−4] Next, a constant signal intensity, or steady-state signal region is observed (region 2), where the damage accumulation processes come into equilibrium with sputter removal processes. Finally, an interfacial region is observed (region 3), where the signal intensity drops with commensurate increase in the substrate signal, as would be expected when one completely erodes through an organic overlayer.

Under non-ideal sputtering conditions, the signal degrades immediately, displaying a depth profile shape similar to the one shown in Figure 5.1c. The substrate may eventually be reached, but the interface will typically be broadened and significantly diminished in its intensity. All of this is consistent with increased beam-induced structural damage. Some samples will exhibit an intermediate behavior as depicted in Figure 5.1b, whereby one can obtain some information as a function of depth, but the signal steadily declines with depth.

Depth profiling in bulk films can represent a particular challenge. In most cases, even in the materials that are amenable to being depth profiled, there will be signal degradation under higher primary ion fluences, typically in a similar manner to that depicted in Figure 5.1d.

What are the "optimized" conditions under which one can obtain the ideal depth profile shape displayed in Figure 5.1a as opposed to that displayed Figure 5.1b or Figure 5.1c? What causes the eventual loss of signal shown in Figure 5.1d? These questions and several others will be addressed here, in order to help define optimum experimental parameters for successful organic depth profiling.

We will begin this chapter with a brief timeline, describing the development of cluster SIMS as a molecular depth profiling tool. A historical review of events is provided, from the viewpoint of the authors.

We will then move on to examine the important aspects of molecular depth profiling in heterogeneous systems, where the concepts of qualitative and quantitative depth profiling with cluster ion beams is discussed, and examples of rudimentary 3D data reconstruction are given.

Next, in-depth discussions ensue, regarding molecular depth profile shapes and their meaning. Basic erosion dynamics theory is introduced, citing examples and fits to experimental data. Depth profile shapes are also discussed in terms of basic ion irradiation chemistry of organic and polymeric materials. It is demonstrated that free radical chemistry and cross-linking mechanisms play a vital role in the damage accumulation processes occurring during ion beam irradiation with cluster sources.

Finally, we present an overview of the important experimental parameters that should be considered during any depth profiling experiment where soft materials are being analyzed, from small molecules, to large complex biological molecules. Optimum experimental conditions are described, including the effects of sample temperature, beam energy, beam angle, and the type of cluster ion source used, all while remaining cognizant of the physical and chemical aspects of sputtering. State-of-the-art cluster sources will be introduced, and a brief discussion of related mass spectrometric based methods that could play a more significant role in the future will be touched upon. Overall, this chapter is intended to serve as a fundamental review of organic depth profiling as well as a guide for scientists interested in learning about the practical aspects of molecular depth profiling.

5.2 HISTORICAL PERSPECTIVES

The process of "molecular depth profiling" using SIMS was first demonstrated by Gillen et al.,[2] using atomic beams. In this study, simple molecular structures such as quaternary ammonium salts and amino acids were bombarded by an Ar^+ ion beam. It was observed that intact parent molecular ion signals characteristic of these particular organic species remained constant with dose, even at high primary ion doses. The authors defined favorable conditions for molecular depth profiling, whereby subsurface molecules must avoid being fragmented by the previous ion collisions necessary to expose them. This condition was said to be met when the sputtering yield is high and the ion penetration depth is low, such that a fresh supply of undamaged subsurface molecules is available every time an ion strikes the surface. A model was developed to describe the surface transient region during depth profiling of organics, which is the basis for the erosion dynamics model that will be introduced in Section 5.4.

While employing atomic ion beams for molecular depth profiling does work for some few molecular species, most molecules will exhibit extensive beam-induced damage effects when bombarded by high energy atomic ions, as was demonstrated in Chapter 1 (Fig. 1.6a). Looking at the required conditions for molecular depth profiling as defined above, one can easily predict that employing molecular beams instead of atomic beams will be favorable for low damage sputtering.

The first demonstration of molecular depth profiling with *cluster* ions was published by Cornett et al.,[7] where the authors utilized massive glycerol clusters to characterize a series of biologically relevant molecules. The primary results of this work are depicted in Figure 5.2, which shows secondary ion intensities from a protein sample (pentagetide), plotted as a function of increasing sputter time for both glycerol cluster ions and Xe^+ atomic ions. The investigators were able to demonstrate that when glycerol cluster ions were employed instead of atomic ions, the signal remained constant with increasing sputtering time. In contrast, when the experiments were performed with Xe^+ ions, there was rapid signal decay with sputtering time.

This very exciting result led to the expansion of cluster SIMS for molecular depth profiling applications in amino acid thin films, acetylcholine films, and even in some polymers.[1] In the latter example, Gillen and Roberson[1] showed that if cluster sources were employed, one could retain molecular information as a function of depth from macromolecular species such as polymers. Although it is not possible to obtain molecular information as a function of depth from high molecular weight polymers, fragment molecule intensities characteristic of the chemical structure can be observed, even under prolonged polyatomic ion bombardment. As an

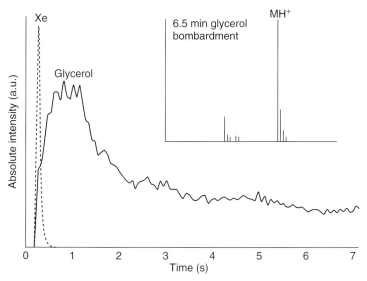

Figure 5.2   Plot of protonated molecular ion intensity versus exposure time for a dry sample of 2 nmol pentagetide using a 15 kV massive glycerol cluster ion beam (solid line) and a 6 kV xenon beam (dashed line). Data are plotted on the same intensity scale. Inset is the massive glycerol cluster ion mass spectrum acquired after 6.5 min of exposure. Reproduced from Cornett et al.,[7] with permission from Wiley.

example from this work, Figure 5.3 shows characteristic fragment ion depth profiles of poly(methyl methacrylate) (PMMA) samples using an SF_5^+ cluster source as compared to an Ar^+ atomic ion source. As can be seen, the fragment ion signal characteristic of PMMA at m/z 69 rapidly decays with increasing sputtering time when the Ar^+ source is employed (Fig. 5.3a). Furthermore, the Si substrate was not reached during the timing of the depth profile experiment. However, when the SF_5^+ primary ion source is employed under similar conditions (Fig. 5.3b), the signal remains relatively constant in comparison to what was observed with Ar^+. Moreover, the authors were able to demonstrate that with the cluster source they could etch through the entire film, as indicated by the rise in the Si intensity from the substrate with commensurate decay in molecular signals associated with the PMMA.

Although this result was very exciting, it was generally accepted by the scientific community at the time, that the utility of cluster SIMS for depth profiling in polymeric materials was severely limited. Brox et al.,[8] for example, compared depth profiles of several different polymers using Ar^+, SF_5^+, Cs^+, and Xe^+ in order to further explore the capabilities and limitations of cluster ion beams for polymer depth profiling.[8] While Ar^+ atomic ion bombardment resulted in rapid signal decay in all cases, Cs^+, Xe^+, and SF_5^+ were indeed able to profile through certain polymers without complete loss of characteristic signal. The best depth profile results were obtained with SF_5^+, which exhibited increased constancy in signals as a function of depth in thin films of poly(propylene glycol) (PPG), poly(ethylene glycol) (PEG), and PMMA.

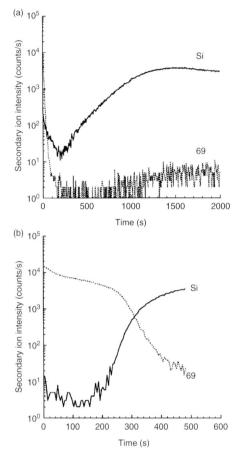

Figure 5.3 Comparison of depth profiles obtained from a 50 nm thick, spun-cast film of PMMA using (a) Ar^+ and (b) SF_5^+ primary ion bombardment. Both the characteristic PMMA fragment at m/z 69 ($C_4H_5O^+$) and the Si substrate signal at m/z 28 (Si^+) are shown. Reproduced from Gillen and Roberson,[1] with permission from Wiley.

While certain polymer thin films (<100 nm) did maintain characteristic molecular signals as a function of increasing primary ion fluence, there were still many more polymers that did not, including polycarbonate (PC), polystyrene (PS), polyethylene (PE), and polyisoprene (PIP). Moreover, all bulk polymers examined, except for polytetrafluoroethylene (PTFE), showed an immediate and rapid decrease in characteristic secondary ion emission for all applied primary ions. The authors concluded that polymeric depth profiling was more of an exception than a rule.[8]

However, starting with the discovery of certain well-behaved polymers,[9,10] and followed by the determination and optimization of important experimental parameters,[5,11] including temperature,[5,11−17] beam chemistry, size and type,[5,11,18−21] beam angle,[5,11,22−24] beam energy,[5,11,25−28] and sample rotation,[11,17] it is now understood that this is not the case, and that even polymers that were originally thought of as "impossible to depth profile" (e.g., PS),[5] can be depth profiled using cluster beams, thus opening the door for 3D molecular characterization in more complex systems, such as biological tissue and complex organic devices.

5.3 DEPTH PROFILING IN HETEROGENEOUS SYSTEMS

5.3.1 Introduction

Thus far, we have discussed molecular depth profiling in one-component systems, consisting of single layers of pristine organic films, with no additional components, additives, or layers. In order for cluster SIMS depth profiling to be useful for real-world applications, it is necessary to develop metrology and methodology for the surface and in-depth characterization of more complex samples, ranging from thick multicomponent polymer coatings and films, to complex biological systems. Thus, we introduce the concept of molecular depth profiling in multicomponent systems in Figure 5.4 and Figure 5.5, which show some simple examples from the literature, of molecular depth profiling in two-component model systems using cluster ion beams.

Figure 5.4a shows a representative depth profile in a model film (200 nm) used for studies of drug-eluting cardiac stents.[18,29−31] In this study, the authors employed a 5 keV SF_5^+ source for sputtering in conjunction with a Bi_3^+ analysis source for image analysis, and the sample was maintained at $-75°C$ during the depth profile experiment for optimum results. As can be seen from the figure, the drug segregates to the surface and interface regions. A diffusion profile is readily observed, with four distinctive regions: (i) a surface-enriched drug region, followed by (ii) a drug depletion zone, where the drug signal decays to a minimum value with a commensurate increase in polymer signal, (iii) a constant bulk composition region, and (iv) an interfacial region, which often exhibits characteristics similar to those observed at the surface. This type of behavior is repeatedly observed in multicomponent systems, in which preferential migration of one component occurs and has been described in detail in the literature.[32] Figure 5.4b shows a z-scale image overlay reconstructed from the data in Figure 5.4a, where the distribution of the drug is readily visualized. In this image, the drug-rich regions are colored green, the polymer-rich regions are colored red, the regions containing both drug and polymer are yellow; the blue color represents the signal from the Si substrate.

This example shows the utility of SIMS not only for studying the distribution of drugs in polymer coatings and films but also for studying the thermodynamic driving forces inherent in these systems. Not all systems, for example, will exhibit a drug-enriched surface region and will, in fact, have a drug-depleted region at the surface.[10] Furthermore, these systems may behave differently when different solvents and/or sample preparation methods are employed (e.g., spin cast, vs spray cast).

SIMS is also extremely useful for the study of highly ordered and/or multilayer molecular films.[11,17,20,27,28,33,34] Figure 5.5 shows an excellent example of this, displaying molecular information as a function of depth for perfectly alternating Langmuir−Blodgett (LB) layers of arachidic acid (AA) and dimyristoyl phosphatidate (DMPA).[32] The profile shown in Figure 5.5a was acquired using a 40 keV C_{60}^+ source, operated at glancing incidence angles of $73°$ with respect to the surface normal, and the sample was maintained at liquid nitrogen temperature for the duration of the experiment. A schematic view of the known 3D molecular

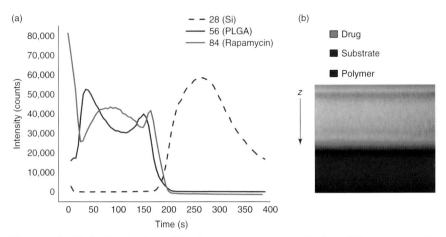

Figure 5.4 Molecular depth profiling in two-component model drug delivery system for study of drug-eluting cardiac stents: (a) depth profile of thin film (~200 nm) of rapamycin (50%) in a poly(lactic acid) (PLA) matrix; (b) z-scale image profile overlay of film. In both (a) and (b), m/z 56 (PLA), m/z 84 (rapamycin), and m/z 28 (Si substrate) are shown as green, red, and blue, respectively. Reproduced from Mahoney et al.,[31] with permission from the *Journal of Surface Analysis*.

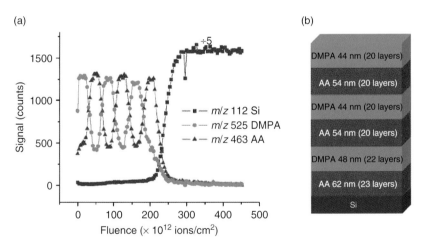

Figure 5.5 Molecular depth profiling in perfectly alternating organic layers, prepared via Langmuir–Blodgett (LB) method; (a) molecular ion intensities of arachidic acid (AA) at m/z 525, dimyristoyl phosphatidate (DMPA) at m/z 463, and Si (m/z 112, Si$_4^+$), monitored as a function of ion fluence; (b) visual representation of multilayer structures profiled in (a).[33] Reproduced from Zheng et al.,[33] with permission from ACS publications.

structure of the film is depicted in Figure 5.5b. The depth profile in Figure 5.5a is consistent with this structure.

Note that in both these examples, low temperatures were employed during the depth profile experimentation for optimal results, and in the case of the LB layers, glancing incidence angles were employed as well. The reasons behind this will be discussed in greater detail later on in Section 5.6.

5.3.2 Quantitative Depth Profiling

Quantitative depth profiling is also possible using cluster ion beams, although there are very few examples of this found in the literature.[32,35] Figure 5.6 illustrates the process used by Mahoney et al. to obtain quantitative depth profiles from a series of poly(lactic acid) (PLA)/Pluronic® P104 blends (~1 μm thick, cast on Si) which have potential as protein drug delivery vehicles.[32,36] Pluronic P104 is a triblock copolymer, consisting of poly(ethylene oxide) (PEO) and poly(propylene oxide) (PPO) components. The amphiphilic nature of the P104 polymer in combination with its increased biocompatibility means that it is well suited to protect protein components via micelle encapsulation, and thus increase their stability in hydrophobic polymer matrices, such as PLA.

Figure 5.6a shows a typical depth profile acquired from a PLA/P104 blend system having a P104 mass fraction (ω_{P104}) of 10%, where intensities characteristic of P104 (m/z 59; solid gray line), PLA (m/z 56, solid black line), and Si (m/z 28, dashed line) are displayed as a function of SF_5^+ sputter time (Bi_3^+ primary ions used for analysis). A very similar diffusion profile shape is observed in the P104/PLA blend system, as was discussed previously with the model drug delivery system in Figure 5.4a. In other words, the P104 is segregating to the surface and interfacial regions, creating a diffusion profile shape whereby P104 depletion zones are observed immediately below the P104 surface-enriched region. As these films are thicker than those described in Figure 5.4a (>1 μm as opposed to 200 nm), the constant bulk composition region is more readily observed.

These data can be plotted as a ratio for quantitative purposes. For example, in Figure 5.6b, the ratio of P104 (m/z 59) to PLA (m/z 56) signals as a function of sputtering time is shown, giving a qualitative assessment of the P104 content relative to PLA over the course of the depth profile. It may also be noted that the data shown in Figure 5.6b have been converted to a depth scale using measured sputter yield volumes.

It can be expected that the intensity ratio observed in the steady-state region of the profile will be associated with the known bulk content in the sample (in this case, 10%). Therefore, one can create calibration curves by analyzing a series of polymer blends with varying bulk compositions and plotting the bulk intensity ratios in the steady-state region of the depth profile as a function of known sample composition. The resulting calibration curves acquired for this series of samples are depicted in Figure 5.6c and d, with corresponding curve fit data indicated in the inset. The resulting equations with fitted parameters were utilized to transform the intensity ratio profiles into quantitative depth profiles as shown in Figure 5.6e. This concentration depth profile was calculated by converting the profile in (b)

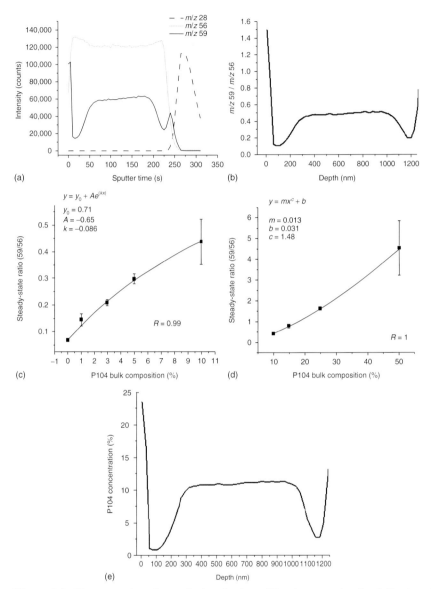

Figure 5.6   Demonstration of quantitative depth profiling in a sample of poly(lactic acid) (PLA) containing $\omega_{P104} = 10\%$ (see text). (a) signal characteristic of P104 (m/z 59), PLA (m/z 56), and Si (m/z 28) plotted as a function of increasing sputter time with a 5 keV SF_5^+ polyatomic primary ion source; (b) ratio of intensities characteristic of P104 (m/z 59) to PLA (m/z 56) plotted on a depth scale; (c) average steady-state intensity ratios plotted as a function of P104 mass fraction for a series of samples with known bulk compositions (ranging from 0% to 10%); (d) average steady-state intensity ratios plotted as a function of P104 mass fraction for a series of samples with known bulk compositions (ranging from 10% to 50%); and (e) P104 concentration depth profile representing the mass fraction of P104 as a function of depth. Values plotted in (c) and (d) are the averaged values from at least four measurements with the error bars representing the corresponding standard deviations. Reproduced from Mahoney and Weidner,[170] with permission from Wiley.

into concentration values using the calibration curves shown in (c) and (d), where the curve in (c) was employed for ratios less than 0.5, and the curve in (d) was employed for values greater than 0.5. The concentration depth profile shown is surprisingly consistent with known bulk compositions and surface compositions as measured by X-ray photoelectron spectroscopy (XPS).[32]

One of the major shortcomings of this method is that quantitative analysis in the surface region can be difficult due to surface transient effects; these effects were minimized in the example shown in Figure 5.6. In the case where there are significant surface transient effects, one should first experiment with peak selections, preferentially selecting peaks that exhibit decreased transient effects in the pure organic film, and if necessary, employ correction factors based on extrapolation techniques. Regardless of this, it is expected that the largest source of error will reside at the surface and therefore the concentrations obtained from this region should be verified by a secondary method, such as XPS. Laser post-ionization may be one method of alleviating this surface transient effect, as will be shown later in Section 5.4.

5.3.3   Reconstruction of 3D Images

Each depth profile shown in Figure 5.6 has a series of images associated with it, which can be reconstructed to obtain 3D molecular information. The depth profile shown in Figure 5.6, for example, was acquired by alternating sputtering with an SF_5^+ source, with an analysis cycle using Bi_3^+, where the Bi_3^+ analysis source was focused and rastered over a defined area (in this case, 300 μm × 300 μm). Each pixel in the image therefore contains a localized mass spectrum that can be used to reconstruct molecular images as a function of depth. These images can be stacked and visualized in three dimensions. This will be discussed in greater detail in the following chapter. However, a brief example is given here for the PLA/P104 system.

Figure 5.7 shows the 3D volumetric representations of P104 in a PLA matrix. In this example, ω_{P104} is 15%, where Figure 5.7a and b also contain bovine serum albumin (BSA) at a mass fraction, ω_{BSA}, of 8%. The BSA was added via an oil-and-water emulsion method described in greater detail in Reference 36.

Figure 5.7c and d show the 3D distribution of Pluronic in the control sample containing PLA and P104 only. Figure 5.7c shows a side view of the image (surface facing toward the bottom of the image) and Figure 5.7d shows a frontal view. P104 crystalline domains are clearly observed and are particularly noticeable in the subsurface region where the P104 depletion layer is located.

The corresponding images of the BSA-containing system in Figure 5.7a and b show a very different 3D structure. In both cases, the 2D diffusion profile in the z-direction is clearly observed, whereby the pluronic component is enriched at the surface and depleted in the subsurface region. However, spherical domains are observed in the depletion region in the BSA-containing sample, as opposed to the crystalline P104 domains that were observed in the control. These spherical domains are consistent with the formation of BSA-containing micelles.

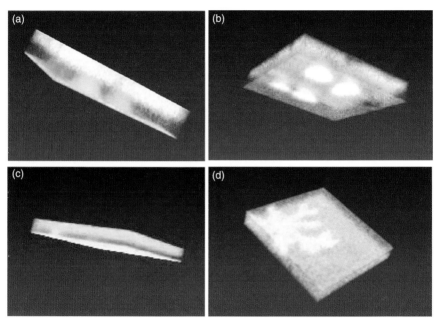

Figure 5.7   3D volumetric representations of Pluronic (P104) in a model protein delivery device (200 μm$\times$ 200 μm $\times$ 0.9 μm). Data were acquired using a Bi_3^+ cluster source for imaging and an 8 keV SF_5^+ polyatomic source for sputtering. All images contain a ω_{P104} of 15% in PLA. (a + b) also contains bovine serum albumin (BSA) (ω_{BSA} of 8%); (a) side view and (b) frontal view. (c + d) control sample containing no BSA; (c) side view and (d) frontal view. Data shown represent all signals acquired prior to the polymer/Si interface region. Therefore, the approximate thickness of the displayed area is approximately 900 nm in both cases. The visual differences in scaling in the z-direction are resultant from differing sputter rates (sputter rates were slower in the BSA-containing system). Reproduced from Mahoney and Weidner,[170] with permission from Wiley.

Although SIMS 3D imaging does show promise for characterization of organic and polymeric materials, there are certain points that need to be considered when performing any sputter depth profile experiment, particularly for multicomponent systems, which can exhibit differential sputtering rates and different damage accumulation rates. This can be corrected for, in part, by measuring the topography [e.g, atomic force microscopy (AFM)] before and after the experiment, and correcting the z-scaling accordingly. Wucher et al.[37] define these and other important protocols for obtaining accurate 3D images in soft materials, all of which will be discussed in greater detail in the following chapter.

5.3.4   Matrix Effects in Heterogeneous Systems

Even with the advent of cluster ion beam sources for SIMS, matrix effects remain an elusive problem in the field and can be particularly problematic in complex biological systems, which consist of complex matrices.[38] However, it can cause serious problems for quantitative depth profiling as well. For example, in most depth profiles, there is a surface transient effect, where an initial increase or decrease in signal intensity is observed. This change may or may not reflect actual compositional changes in the sample.[39-41]

The system introduced in Figure 5.7 also shows an example of matrix effects in SIMS depth profiling. Note that in the figure, only the P104 signal is discussed and there is no mention of where the protein is located in this system. This is simply because there was no evidence of any protein in the mass spectral dataset. Any peaks characteristic of the protein were completely masked by the polymer signal. In some cases, the negative ion signal did confirm the location of the BSA (at the site of the P104 micelles), but the signals were extremely low and not consistent from sample to sample.

To better understand and confirm where the protein in this system was located, another sample was prepared with a similar composition as that shown in Figure 5.7a and b, where 1% trifluoroacetic acid (TFA) was added to the protein-containing water phase, prior to emulsification with the PLA/P104 solution. It was hoped that TFA would serve as a viable source of protons for the amino acids in the protein and therefore help with its ionization.

The secondary ion images acquired during the depth profiling in the TFA-containing sample are compared with that of the sample prepared without TFA in Figure 5.8. Secondary ion image overlays of fragment ions characteristic of P104 (m/z 59), PLA (m/z 99), and BSA ($CH_2NH_2^+$, m/z 30) averaged from the depletion region of the depth profile, are displayed in Figure 5.8, where Figure 5.8a shows the sample prepared without TFA, and Figure 5.8c shows a sample of similar composition, but where the protein solution was prepared in 1% TFA. The green regions in the image are characteristic of the P104, the blue regions are characteristic of PLA, and the yellow regions contain both P104 and BSA signals. As can be seen, there are no yellow regions in the sample prepared in water. However, the sample prepared in 1% TFA does show signal characteristic of the protein, and further investigations with principal components analysis (PCA) show that the mass spectral data associated with these yellow regions are consistent with the protein composition of BSA. It can also be confirmed that the yellow regions are colocated at the site of the P104 micelles and not with the P104 crystalline regions surrounding the micelles.

The images in Figure 5.8b and d show the secondary ion yields of m/z 30 (characteristic of the BSA component) averaged over the entire depth profile for the sample prepared without (Fig. 5.8b) and with (Fig. 5.8d) TFA. The differences in ionization are clearly evident, whereby no discernable protein signal is observed in the original sample prepared in H_2O.

It is clear from these results that a better understanding of matrix effects is required for further advancement of the field, because with these effects looming, it can be almost impossible to obtain quantitative and sometimes even qualitative information. That being stated, these effects, when they do occur, are usually very obvious, particularly to the more experienced analyst. Many studies are currently underway to address these issues with hopes of finding means to minimize their effects in SIMS experiments, some of which will be discussed in Chapter 8. It

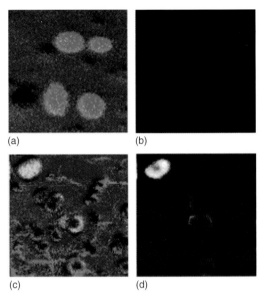

(a) (b)

(c) (d)

Figure 5.8 Illustration of matrix effects for the PLA/BSA/P104 system introduced in Figure 5.7 (15% P104, 8% BSA in PLA). (a,c) Image overlays reconstructed from the depletion region of the depth profile, where the green and blue regions of the images represent signals characteristic of P104 and PLA respectively. The yellow regions represent a mixture protein signal at m/z 30 and P104 signal. (a) Same data as Figure 5.7 (a and b). (c) Sample prepared with a similar composition as (a), but with 1 % trifluoroacetic acid (TFA) added to the protein solution prior to emulsion. (b,d) signal characteristic of BSA (m/z 30) averaged over the entire depth profile: (b) No TFA added; (d) 1% TFA added to protein solution before emulsion.

should be noted here, however, that these same matrix effects are observed in all surface mass spectrometry experiments, and are not exclusive to SIMS.

5.4 EROSION DYNAMICS MODEL OF MOLECULAR SPUTTER DEPTH PROFILING

Ever since the discovery that cluster ion beams enable molecular sputter depth profiling, a key question has been why the rapid destruction of the molecular integrity of the sample—which had been always observed when atomic ion beams were used for sample erosion—is reduced under cluster bombardment to such an extent that successful depth profiling becomes possible. In principle, ion bombardment–induced erosion, which forms the basis of sputter depth profiling, is an energetic process involving violent atomic collisions, which will inevitably lead to the breaking of bonds and, hence, will alter the chemical state of the surface. This is particularly important in the case of molecular samples, where chemical damage will be produced upon every single ion impact and accumulate to such an extent that the molecular information is destroyed during erosion of the first nanometer of the sample. Before the advent of cluster ion beams, it was therefore standard knowledge that molecular SIMS had to be done in a static mode (see Chapter 4), where no point at the surface was influenced by more than one single impact. This, of course, precludes sputter erosion and, hence, depth profiling.

In principle, the situation is not much different under cluster bombardment. Molecular dynamics (MD) simulations clearly show that the kinetic energy of the projectile is deposited in a region of several nanometers diameter and depth, where bonds are broken and the chemical integrity of the material is strongly altered (see Chapter 2). However, there is an important difference with respect to atomic ion bombardment: the simulations also show that the cluster deposits its energy in a shallower depth region than an isoenergetic atomic ion, thereby strongly increasing the energy density within that region. This leads to an important modification of the sputter erosion process, which changes from a linear collision cascade to a nonlinear collisional spike (see Chapter 2). As a consequence, a cluster impact can produce sputter yields that are orders of magnitude higher than those achievable under atomic ion bombardment. In the following section, we show that these enhanced sputter yields are the key to successful molecular depth profiling. For that purpose, we present a simple model describing the dynamics of surface erosion and chemical damage under high fluence ion bombardment in terms of simple equations, which are based on ideas of Gillen et al.,[2] and utilize a concept originally developed to describe preferential sputtering phenomena.[42,43] As is shown, these models can help the experimentalist to perform extrapolations into surface regions (thus minimizing surface transient effects), and define damage parameters such as the altered layer thickness (the thickness of the damaged region created by the impinging ion beam).

In a first step, we follow the fate of parent molecules that most directly represent the chemical state of the solid and give rise to the so-called "molecular ion" signal in SIMS. In a second step, we then concentrate on the dynamics of fragmentation products that are often used as "fingerprints" in SIMS analysis.

5.4.1 Parent Molecule Dynamics

The general scenario to be described here is the sputter-induced surface erosion of a molecular solid (for instance, an organic film) by means of an impinging projectile ion beam. At the beginning of the erosion process, the solid is assumed to be homogeneous with a bulk density of n molecules per unit volume. The analyte molecule of interest is assumed to be present with a dimensionless "concentration" c_M ($0 \leq c_M \leq 1$), which describes the fraction of intact analyte molecules in the sample. Hence, the number density of analyte molecules is nc_M, where n is assumed to be constant and c_M is allowed to vary both as a function of position and time (or projectile ion fluence). At the beginning of the analysis, c_M is assumed to be equal to the bulk concentration c_M^{bulk} everywhere in the sample. In principle, the analyte may be either the matrix itself ($c_M^{bulk} = 1$), or some other molecule doped into the matrix at an arbitrary concentration $c_M^{bulk} < 1$.

Upon start of the ion bombardment, the concentration of analyte molecules at or closely below the surface will be altered due to a number of effects. First, *molecules will be removed from the surface because of sputtering.* Second, the particle dynamics initiated by the projectile impact will lead to a *relocation of molecules within the solid*, which can be formally described as a time- and position-dependent flux density $\vec{j}\,(\vec{r},t)$. Third, the *projectile impact will lead to chemical damage*, that is, fragmentation of molecules and other damage processes as are described later. Although the debris from such fragmentation processes may not be sputtered and therefore remain in the sample, we consider these molecules as lost for the analysis, as they cannot contribute to the measured molecular ion signal any more. In the following, we therefore interpret c_M as the concentration of *intact* analyte molecules within the sample. Hence, c_M will be diminished by an additional damage rate, $r_D(\vec{r})$, which will be assumed to be constant in time but certainly depends on the position $\vec{r}$ within the sample.

In general, the variation of c_M as a function of position and time can be described by the following continuity equation:

$$n \cdot \frac{dc_M(z)}{dt} = \frac{\partial j_z(z,t)}{\partial z} - r_D(z) \cdot n \cdot c_M(z) \tag{5.1}$$

which describes the concentration profile $c_M(z)$ with the (momentary) surface being located at $z = 0$. In order to proceed, expressions are needed for both $j_z(z,t)$ and $r_D(z)$. A rigorous evaluation of these quantities requires detailed knowledge about both the ion impact–induced relocation and fragmentation dynamics, which need to be microscopically modeled in order to allow a general calculation of the concentration profile $c_M(z,t)$ from Equation 5.1.

In order to treat the problem within a simple, easily tractable model, we therefore revert to approximate solutions. As the conceptually simplest possible approach, we approximate the microscopic function $c_M(z,t)$ by a step profile, with the concentration being altered to a constant, average "surface" value c_M within a layer of width d beneath the surface. This model is illustrated in Figure 5.9.

At depths $z > d$, the concentration is assumed to be the unaltered bulk value c_M^{bulk}. In this approximation, the solution of Equation 5.1 reduces to the determination of $c_M(t)$ or, more accurately, $c_M(f)$, where f denotes the projectile ion fluence accumulated over the erosion time t.

The dynamics of c_M are determined by the flux of molecules into and out of the altered layer. It is seen in Figure 5.9 that three different flux contributions can be discerned. First, the flux of intact molecules leaving the surface by sputtering per unit surface area is determined by the projectile flux j_p and the partial sputter yield Y_M (i.e., the average number of analyte molecules sputtered per projectile impact) as $j_{sput} = j_p \cdot Y_M$. Under steady-state sputtering conditions, Y_M can be approximated as

$$Y_M \approx c_M \cdot Y_{tot} \tag{5.2}$$

where Y_{tot} denotes the total sputter yield, that is, the total number of molecules (or molecule equivalents) removed per projectile impact. Hence, we obtain

$$j_{sput} \approx j_p \cdot Y_{tot} \cdot c_M \tag{5.3}$$

Second, there is a virtual "supply" flux of intact molecules from the bulk into the altered layer, which is caused by the fact that the interface is receding with the

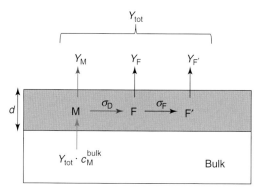

Figure 5.9 Schematic view of particle fluxes in a sputter depth profiling experiment. The arrows describe the direction of the flux. Y_M, Y_F, and $Y_{F'}$ denote the partial sputtering yields (see text) of parent molecules (Y_M) and fragment ions (Y_F and $Y_{F'}$). The corresponding surface concentrations of intact molecules and fragments are defined as c_M^s and c_F^s and/or $c_{F'}^s$, respectively.

erosion rate $v_{erosion} = \frac{j_p \cdot Y_{tot}}{n}$. The respective flux density is defined as

$$j_{supply} = v_{erosion} \cdot n \cdot c_M^{bulk} = j_p \cdot Y_{tot} \cdot c_M^{bulk} \qquad (5.4)$$

Third, a virtual flux of intact molecules is removed from the altered layer by fragmentation. In order to describe this contribution, we introduce the damage cross section σ_D by assuming that a single projectile impact destroys all molecules within an area σ_D around the impact point across the entire altered layer. The resulting fragmentation flux density is given by:[3]

$$j_{frag} = j_p \cdot n \cdot d \cdot \sigma_D \cdot c_M \qquad (5.5)$$

From a simple flux balance, we find

$$nd\frac{dc_M}{dt} = j_{supply} - j_{sput} - j_{frag} \qquad (5.6)$$

Inserting Equations 5.3 through 5.5, this yields

$$\frac{dc_M}{dt} = j_p \left\{ \frac{Y_{tot}}{nd} \left(c_M^{bulk} - c_M \right) - \sigma_D c_M \right\} \qquad (5.7)$$

or, with $df = j_p dt$,

$$\frac{dc_M}{df} = \frac{Y_{tot}}{nd} \left(c_M^{bulk} - c_M \right) - \sigma_D c_M \qquad (5.8)$$

Within the above approximations, Equation 5.8 describes the variation of the average surface concentration c_M of intact analyte molecules within the altered layer as a function of the accumulated projectile ion fluence, f. It can be used to compare predictions of the model to experimental data of molecular SIMS depth profiles.[3,4,22,28,33] In this form, the model contains three free parameters, namely, the total sputter yield Y_{tot}, the damage cross section σ_D and the width d of the altered layer.

5.4.2 Constant Erosion Rate

If all three parameters are assumed to remain constant throughout the acquisition of a depth profile, the solution of Equation 5.8 is given by

$$c_M(f) = c_M^{bulk} \left\{ \frac{Y_{tot}}{Y_{tot} + nd\sigma_D} + \left(1 - \frac{Y_{tot}}{Y_{tot} + nd\sigma_D}\right) \exp\left[-\left(\frac{Y_{tot}}{nd} + \sigma_D\right)f\right]\right\}$$
(5.9)

Examples of the fluence dependence of $c_M(f)$ defined by Equation 5.9 are plotted in Figure 5.10. In all cases, the surface concentration exhibits an exponential decay from the initial bulk value c_M^{bulk} toward a steady-state value c_M^{ss} defined as

$$c_M^{ss} = c_M^{bulk} \frac{Y_{tot}}{Y_{tot} + nd\sigma_D}$$
(5.10)

After c_M^{ss} is reached, the concentration remains constant throughout the rest of the depth profile, until finally the entire sample is removed.

Equation 5.10 can be rationalized as follows. While Y_{tot} characterizes the sputter-induced removal of material from the surface, the quantity $nd\sigma_D$ describes the number of damaged molecules per projectile impact. If the latter is large compared to the sputter yield (i.e., $Y_{tot} \ll nd\,\sigma_D$), much more material is damaged than is removed in each projectile impact. As a consequence, chemical damage accumulates with increasing projectile fluence and c_M^{ss} approaches zero. If, on the other hand, $Y_{tot} \gg nd\,\sigma_D$, the debris produced by chemical damage is efficiently removed in each impact, thereby exposing fresh, undamaged molecules from the bulk, which can then be analyzed in subsequent impact events. In this limit, no damage accumulation occurs and c_M^{ss} therefore approaches the bulk concentration c_M^{bulk}. Intermediate cases can be characterized by the "cleanup efficiency," ε, of the projectile to remove the fragmentation debris produced by its own impact, which can assume values between zero and infinity.

$$\varepsilon = \frac{Y_{tot}}{nd\sigma_D}$$
(5.11)

With this definition, Equation 5.9 and Equation 5.10 can be rewritten as

$$c_M(f) = c_M^{bulk} \left\{ \frac{1}{1 + \varepsilon} + \left(1 - \frac{\varepsilon}{1 + \varepsilon}\right) \exp[-(1 + \varepsilon)\sigma_D f]\right\}$$
(5.12)

Useful molecular depth profiling requires ε to approach values of 0.1 and above; otherwise, the situation demonstrated in Figure 5.10a is observed, where the signal rapidly decays to zero. In other words, the "yield volume"[44] Y_{tot}/n of

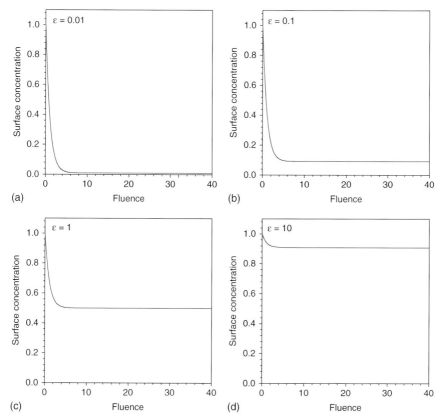

Figure 5.10 Surface concentration of intact analyte molecules versus accumulated projectile fluence for constant total sputter yield (Eq. 5.9). The fluence is given in (dimensionless) units of the inverse disappearance cross section σ (Eq. 5.13), and the concentration is plotted relative to the bulk concentration, c_M^{bulk}. The parameter ε denotes different values of the cleanup efficiency (Eq. 5.11). Figure reprinted from Wucher[3], with permission from Wiley.

sample which is removed per projectile impact must become comparable to the damaged volume, $\sigma_D \cdot d$.

As seen from Equations 5.9 and 5.12, the initial exponential decay is characterized by a "disappearance cross section" σ, which can be measured experimentally.

$$\sigma = \frac{Y_{tot}}{nd} + \sigma_D = (1 + \varepsilon)\sigma_D \qquad (5.13)$$

It is important to note that σ contains contributions from sputtering as well as from damage and is only equal to the damage cross section σ_D for $\varepsilon \ll 0.1$. As this has been the case for practically all earlier experiments performed with atomic projectiles, both quantities have often been confused in the literature.[45] If cluster projectiles are used, however, the cleanup efficiency may become quite large and the observed disappearance cross section σ may be largely determined by sputter removal rather than by damage accumulation.

5.4.3   Fluence-Dependent Erosion Rate

The solution depicted in Figure 5.10 is only valid if the parameters Y_{tot}, σ_D and d remain constant throughout the entire depth profile. It is easy to imagine that this may be the case for σ_D and d, as both the damage cross section and depth should not strongly depend on the analyte concentration or chemical damage (some exceptions do exist, and will be discussed in section 5.5). It has been demonstrated, however, that the sputter yield may well depend on the projectile ion fluence, particularly if more damaging projectiles are used. In this case, Equation 5.8 must be modified to

$$\frac{dc_M}{df} = \frac{Y_{tot}(f)}{nd}(c_M^{bulk} - c_M) - \sigma_D c_M \qquad (5.14)$$

where the solution of this equation critically depends on the function $Y_{tot}(f)$ and can in general only be derived numerically. For the prominent example of C_{60} cluster projectiles, a gradual decrease of the erosion rate with increasing fluence has been observed for some target materials and bombarding conditions.[25,44,46] Some theories will be addressed regarding this effect in the following section. For now, its influence on a sputter depth profile is explored.

Assuming, as a conceptually simple example, an exponential yield decay with a cross section a according to[28]

$$Y_{tot}(f) = Y_0 \cdot \exp[-af] + Y_\infty \qquad (5.15)$$

one obtains depth profiles similar to those displayed in Figure 5.11. For the sake of universality, the parameters in the figure have been scaled in the following way. First, the (initial) cleanup efficiency ε (Eq. 5.11) is used to combine the parameters $Y_{tot}(0), n, d$, and σ_D. As an example, the data of Figure 5.11 were calculated for a value of $\varepsilon = 0.5$. Then, the yield decay cross section, a, is given in units of the disappearance cross section, σ, and the projectile ion fluence is scaled in units of $1/\sigma$. The fluence-dependent total sputter yield, the surface concentration of intact molecules, and the corresponding signal intensities are plotted relative to their initial values at zero fluence. If the resulting dimensionless parameters $\tilde{c}_M = c_M/c_M^{bulk}$ and $\tilde{f} = f\sigma$ are inserted into Equation 5.14, we obtain

$$\frac{d\tilde{c}_M}{d\tilde{f}} = \frac{1}{\varepsilon + 1}[\varepsilon\{(1 - \tilde{y}_\infty)\exp(-\tilde{a}\tilde{f}) + \tilde{y}_\infty\}(1 - \tilde{c}_M) - \tilde{c}_M] \qquad (5.16)$$

for the yield dependence of Equation 5.15. For a given value of ε, the resulting curves only depend on the ratios $\tilde{a} = a/\sigma$ and $\tilde{y}_\infty = Y_{tot}(\infty)/Y_{tot}(0)$.

From the data depicted in Figure 5.11, limiting cases can be discussed. First, if $a \geq \sigma$, the surface concentration c_M simply decays exponentially to essentially zero. In this case, the behavior is qualitatively similar to that observed for low cleanup efficiency (cf. Figure 5.10), and no steady state is observed. In the other limit $a \ll \sigma$, the solution of Equation 5.14 is easy to guess. In this case, the surface concentration rapidly decays exponentially (with cross section σ)

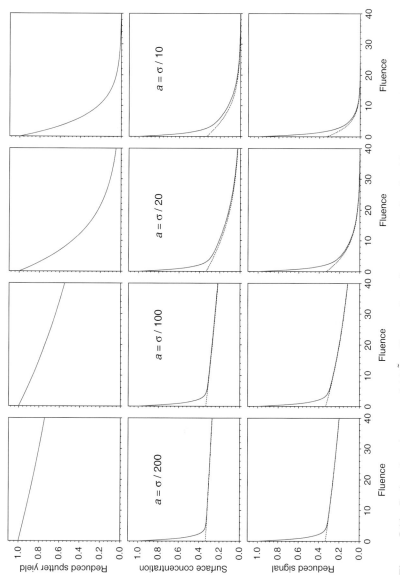

Figure 5.11 Reduced total sputter yield $\tilde{Y}_{tot}$ (first row), surface concentration $\tilde{c}_s$ of intact molecules (second row), and molecular secondary ion signal $\tilde{S}$ (third row) as a function of reduced projectile ion fluence $\tilde{f}$ for an exponentially decreasing total sputter yield. Red line: numerical solution of Equation 5.16 for $\varepsilon = 0.5$; blue line: quasi-steady-state approximation according to Equation 5.17 and Equation 5.18. The parameter of the different columns is the yield decay cross section a in Equation 5.15. Figure reprinted from Wucher,[3] with permission from Wiley.

toward a "quasi-steady-state" value, which then gradually decreases as the sputter yield decays. Even though the concentration strictly speaking never reaches a steady state, we deliberately use the term *quasi-steady state* here, as in this regime the surface concentration is described by the steady-state solution (Eq. 5.10), just inserting $Y_{tot}(f)$ instead of Y_{tot}. This is visible in Figure 5.11, where the numerical solution of Equation 5.16 and the quasi-steady state approximation are displayed as red and blue lines, respectively.

Owing to the changing value of the cleanup efficiency, the fluence dependence of the surface concentration does not necessarily map that of the total sputter yield. This would only be true for $\varepsilon \ll 1$, which is not generally the case under cluster ion bombardment. If the start-up value of ε is comparable or larger than unity, the surface concentration will first exhibit a weaker decrease and only at larger fluence follow the declining sputter yield.

5.4.4 Using Mass Spectrometric Signal Decay to Measure Damage Parameters

In a mass spectrometric depth profiling technique such as SIMS (or secondary neutral mass spectrometry—SNMS), the signal intensity of sputtered molecular analyte ions (or neutrals) is followed as a function of the projectile ion fluence. In principle, the measured signal, S, is proportional to the respective partial sputter yield Y_M, which is again approximated by Equation 5.2. If the total sputter yield is constant, the measured signal directly tracks the surface concentration described by Equation 5.9 and Equation 5.10, provided the detection probability of the analyte molecule remains constant throughout the depth profile. Under these conditions, the ratio between steady-state signal (S_{ss}) and initial signal (S_0) is given as[4]

$$\frac{S_{ss}}{S_0} = \frac{Y_{tot}}{Y_{tot} + nd\sigma_D} = \frac{\varepsilon}{1+\varepsilon} \tag{5.17}$$

Measuring this ratio therefore allows one to directly determine the cleanup efficiency, ε. In addition, the disappearance cross section, σ, can be determined from the initial exponential signal decay into the (quasi-)steady state. In connection with an independent measurement of the total sputter yield (for instance from a measured projectile fluence necessary to remove an overlayer of known thickness), it is therefore possible to use Equations 5.13 and 5.17 to determine both the damage cross section, σ_D, and the altered layer thickness, d, entirely from measured depth profile data. An analysis of this kind has been performed for different organic overlayers on silicon substrates under bombardment with C_{60} projectiles[3,4,22,33].

Typical values of the resulting parameters for C_{60} are $\sigma_D \sim 10$ nm^2 and $d \sim 20$ nm, both depending on the kinetic energy as well as the impact angle of the projectiles. With a typical molecule density of $n \sim 2$ nm^{-3}, one can estimate the quantity $nd\sigma_D$ to be of the order of several hundred damaged molecules per projectile impact. If the total sputter yield is small compared to this value, the cleanup efficiency is negligible and molecular depth profiling is not expected to work. As it appears, this condition is practically always fulfilled for atomic projectiles, which can only produce sputter yields of the order of 10 molecules per impact regardless of the projectile energy[47]. Useful molecular sputter depth profiling therefore appears to hinge on the use of cluster projectiles, which can produce sputter yields of hundreds of molecules per impact and, hence, generate cleanup efficiencies of the order of unity and above.

For varying sputter yield, both c_M (concentration of intact molecules at the *surface*) and Y_{tot} influence the dependence of the measured signal on the projectile ion fluence. For a sputter yield decay described by Equation 5.15, the resulting signal variation is depicted in the third row of Figure 5.11. If the yield decay cross section becomes comparable to the disappearance cross section, we obtain a rapidly decreasing signal that goes to zero within a fluence interval of about $(a + \sigma)^{-1}$ (e.g., a scenario similar to the depth profile on the far right of the figure is observed). In the limit of slowly changing sputter yield, on the other hand, the signal exhibits a rapid initial decay within a fluence interval of about $3\sigma^{-1}$ and then decreases approximately linearly with increasing fluence (e.g., a scenario similar to the depth profile toward the left-hand side of the figure is observed). In this regime, the signal variation should be described by

$$\frac{S_{ss}(f)}{S_0} \cong \frac{[Y_{tot}(f)]^2 / Y_{tot}(0)}{Y_{tot}(f) + nd\sigma_D} \tag{5.18}$$

Extrapolation of this behavior toward zero fluence yields the same value as would be observed for constant sputter yield. Note that this extrapolated value should be inserted into Equation 5.17 to calculate the (initial) cleanup efficiency, damage cross section, and altered layer thickness from a measured depth profile.

An example for such an analysis is depicted in Figure 5.12. The displayed experimental data were taken from a SIMS depth profile of a 350 nm cholesterol film on Si, which was measured using normally incident 40 keV C_{60}^+ projectile

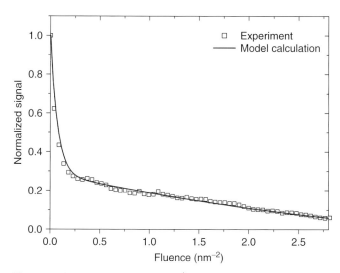

Figure 5.12 Measured $[M - OH]^+$ molecular ion signal as a function of projectile fluence for a 350 nm cholesterol film on Si analyzed by 40 keV C_{60} projectile ions. The solid line denotes the signal variation predicted by Equation 5.14 using $Y_{tot}(0) = 421$, $\sigma_D = 10 \text{ nm}^2$, $d = 60$ nm, and assuming a linear yield variation with $a = 0.21 \text{ nm}^2$. Figure reprinted from Wucher,[3] with permission from Wiley.

ions.[22] An evaluation of the parameters as described above yields an initial sputter yield of 421 molecules/impact, a damage cross section of 12 nm^2, an altered layer thickness of 63 nm, and a yield decay cross section of 0.21 nm^2.[22] The solid line depicts the prediction of the model using a full solution of Equation 5.14 for a sputter yield decay according to Equation 5.15. The calculation was performed using $Y_{tot} = 421$, $\sigma_D = 10$ nm^2, $d = 60$ nm, $n = 1.8$ nm^{-3} and $a = 0.21$ nm^2. It is seen that the simple model presented here almost perfectly describes the measured signal throughout the removal of the entire film.

In a similar way, we can analyze depth profile data of thick Irganox1010 films as published by Shard et al.[28] As an example, their molecular [M–H]$^-$ secondary ion signal measured under 20 keV C$_{60}$$^+$ bombardment is replotted versus ion fluence in Figure 5.13. From their analysis, the authors report values of $Y_{tot}(0)/n = 174$ nm^3, $Y_{tot}(\infty)/n = 24$ nm^3, $\varepsilon = 0.33$, and $a = 0.32$ nm^2. The solid line in Figure 5.13 was calculated using these values along with $\sigma_D = 15$ nm^2 and $d = 50$ nm in connection with a slightly increased $a = 0.48$ nm^2. Again, it is seen that the model describes the measured data very well. The increase of a by a factor 3/2 is due to the fact that we assume the molecular ion signal to be proportional to the total sputter yield in Equation 5.2, while a quadratic dependence is assumed in the original reference.[28] The authors justify this assumption empirically from an analysis of static SIMS data,[48,49] which shows the exponent in Equation 5.18 to change from 2 to 3 in the limit of low sputter yields.[49] For an exponent of 3, one indeed obtains a similarly good fit as displayed in Figure 5.13

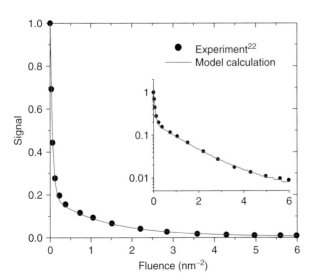

Figure 5.13 Measured [M$_{1010}$ – H]$^-$ molecular ion signal as a function of projectile fluence for a thick Irganox1010 film analyzed by 20 keV C$_{60}$ projectile ions. The solid line denotes the signal variation predicted by Equation 5.14 using $Y_{tot}(0)/n = 174$ nm^3, $\sigma_D = 15$ nm^2, $d = 50$ nm, and assuming an exponential yield decay with $a = 0.48$ nm^2 and $\tilde{y}_\infty = 0.14$. The experimental data were reproduced from Reference 28. Figure reprinted from Wucher,[3] with permission from Wiley.

for $a = 0.32\,nm^2$, which appears to approximate the measured yield decay slightly better (cf. Fig. 3 of Reference 28).

5.4.5 Surface Transients

It should be noted at this point that the signal behavior predicted by erosion dynamics only holds if the detection probability of a sputtered molecule remains constant throughout the acquisition of a depth profile. This assumption, however, is not obvious in SIMS, as the ionization of a sputtered particle during the emission process may change as a function of surface state. In fact, surface transients of the molecular secondary ion signal are frequently observed during the initial stages of a sputter depth profile, where the measured signal variation deviates from the prediction displayed in Figure 5.11. As an example, molecular ion signals observed during C_{60} depth profiles of trehalose films doped with peptide molecules exhibit a rapid rise at low fluence, which is then followed by the exponential decay predicted by erosion dynamics. Interestingly, the magnitude of this effect is different for matrix and dopant molecules, indicating that it is not produced by erosion dynamics. In principle, such transients may arise from fluence-dependent variations of the secondary ion formation probability. In order to investigate the role of this effect, Willingham et al.[50] have compared depth profiles of organic films by monitoring sputtered neutral molecules and their secondary ion counterparts. For the example of a guanine film on Si, their data are displayed in Figure 5.14. It is obvious that the surface transient observed for the molecular $[M + H]^+$ secondary ion signal (m/z 152, Fig. 5.14a) is caused by variations of the ionization probability. In contrast to the secondary ions, the signal of post-ionized neutral parent molecules (m/z 151, Fig. 5.14b) nicely follows the erosion dynamics prediction, with the steady-state level indicating a cleanup efficiency of $\varepsilon \approx 0.5$. Unfortunately, this is the only dataset available, and thus general conclusions cannot be made at this point. It appears feasible, however, that the surface transients observed under rare gas cluster bombardment[51]—which, depending on the projectile cluster size, may extend to such large fluence that they completely mask the erosion dynamics–induced initial exponential signal decay—may also be caused by similar ionization effects.

5.4.6 Fragment Dynamics

So far, the discussion has been limited to the dynamics of intact parent molecules. The treatment becomes more complicated if the variation of characteristic fragment signals is to be described, such as is often the case with larger molecules or polymers. In principle, one must distinguish between "static" and "dynamic" fragmentation. In the first case, fragmentation and emission occurs during the same projectile impact, and the fragment signal therefore reflects the concentration of *intact* parent molecules rather than that of fragments in the altered layer. In the second case, fragments may *accumulate* as a function of projectile fluence. In order to describe the dynamics of the resulting fragment concentration, an additional "population" term must be included into Equation 5.8, which describes the fragment production rate.

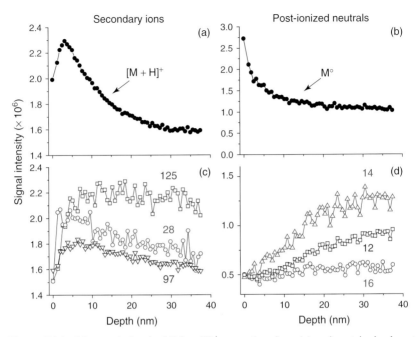

Figure 5.14 Measured signal of $[M + H]^+$ secondary ions (a) and post-ionized neutral M^0 molecules (b) versus projectile ion fluence in the beginning stage of a 40 keV C_{60}^+ depth profile of a guanine film on silver. Also shown are the signals of fragment secondary ions at m/z 125, 97, and 28, respectively (c), and those of neutral C, N, and O atoms (d). Data reproduced from Willingham et al.,[50] with permission from ACS publications.

Following the same concepts as outlined above, the concentration c_F of a fragment F within the altered layer should vary with projectile fluence as

$$\frac{dc_F}{df} = \sigma_{pop} \cdot c_P - \underbrace{\left\{ \frac{Y_{tot}}{nd} + \sigma_F \right\}}_{\sigma_{dep}} \cdot c_F \qquad (5.19)$$

where c_P denotes the concentration of the precursor (parent molecule or a larger fragment) from which the fragment F is derived, σ_{pop} denotes the corresponding "population" cross section for production of F, and σ_F is the cross section for further fragmentation of F to smaller fragments F' (Fig. 5.9). In a fragmentation chain, this leads to a system of coupled differential equations, which must be solved from top to bottom, starting at the largest fragment with the original molecule as its parent. In analogy to the disappearance cross section for parent molecules, the removal of F from the altered layer by sputtering and fragmentation can be described by a "depletion" cross section, σ_{dep}, as indicated in Equation 5.19. The equation is then easily solved yielding

$$c_F(f) = \sigma_{pop} \cdot \int_0^f c_P(f') \exp(\sigma_{dep} \cdot f') df' \cdot \exp(-\sigma_{dep} \cdot f) \qquad (5.20)$$

For a further evaluation, the variation of c_P is explicitly needed. Assuming, for simplicity, that F derives directly from the parent molecule M as indicated in Figure 5.9, we insert $c_P = c_M$, decaying with disappearance cross section σ according to Equation 5.14. Then, solution of Equation 5.20 yields

$$c_F(f) = c_M^{bulk} \cdot \left\{ \frac{\varepsilon}{1+\varepsilon} \cdot \frac{\sigma_{pop}}{\sigma_{dep}} \left[1 - e^{-\sigma_{dep} \cdot f} \right] + \frac{1}{1+\varepsilon} \cdot \frac{\sigma_{pop}}{\sigma_{dep} - \sigma} \left[e^{-\sigma \cdot f} - e^{-\sigma_{dep} \cdot f} \right] \right\}$$

(5.21)

If F is the only fragmentation channel for M, we must set $\sigma_{pop} = \sigma_D$. The solution of Equation 5.20 obtained under these conditions is depicted in Figure 5.15 for three different values of the cleanup efficiency, ε. In the limit $\varepsilon \ll 1$, the steady-state parent molecule concentration is small, yielding an approximate solution:

$$c_F(f) \cong c_M^{bulk} \frac{\sigma_D}{\sigma_{dep} + \sigma_D} \{ exp(-\sigma_{dep} \cdot f) - exp[-\sigma_D \cdot f] \}$$

(5.22)

Note that c_F starts at zero at the beginning of the depth profile and goes through a maximum at

$$f_{max} = \frac{1}{\sigma_{pop} - \sigma_{dep}} \ln \frac{\sigma_{pop}}{\sigma_{dep}}$$

(5.23)

Using $\sigma_{pop} = \sigma_D$, and $\sigma_{dep} = \sigma_F + \varepsilon \cdot \sigma_D$, Equation 5.23 allows one to determine the ratio σ_D/σ_F from the position of the observed fragment signal maximum. In the limit of large fluence, the fragment concentration approaches a steady-state value

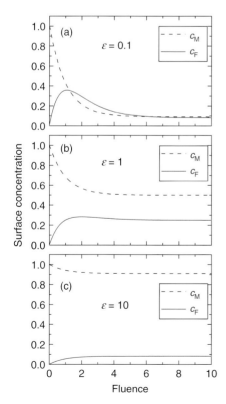

Figure 5.15   Surface concentration of fragments and parent molecules versus projectile ion fluence calculated from Equation 5.14 and Equation 5.20 for $\sigma_F = \sigma_D$ and three different values of the cleanup efficiency ε as defined by Equation 5.11.

given by

$$c_F^{ss} = \frac{\sigma_{pop}}{\sigma_{dep}} \cdot c_M^{ss} = \frac{\sigma_D}{\sigma_F + \varepsilon\sigma_D} \cdot \frac{\varepsilon}{1+\varepsilon} \cdot c_M^{bulk} \quad (5.24)$$

From Equation 5.22 through Equation 5.24, limiting cases can be discussed. If, on one hand, fragmentation of F is slow compared to that of the parent molecule M, one would observe a complementary behavior of fragment and parent, with the fragment signal rising exactly in correspondence with the parent decay, reaching a steady state just as the parent does, with a steady-state fragment concentration of $c_F^{ss} \approx c_M^{bulk}/(1+\varepsilon)$. This behavior is observed in Figure 5.16a, which shows the variation of c_F and c_M for a cross section ratio, $\sigma_D/\sigma_F = 10$. If, on the other hand, depletion occurs much faster than production, the fragment concentration will exhibit a fast rise and then essentially track the parent concentration multiplied by σ_D/σ_F, a behavior which is indicated in Figure 5.15c for $\sigma_D/\sigma_F = 0.1$. For the intermediate case where $\sigma_D = \sigma_F$ (Fig. 5.16b), we find a fragment signal maximum at $f_{max} \cong 1/\sigma_D$ and $c_F^{ss} \cong c_M^{ss}/(1+\varepsilon)$.

An example for these different dependencies that were observed experimentally is displayed in Figure 5.14 for the guanine film discussed earlier.[50] In Figure 5.14c, the signals of three fragment secondary ions at m/z 125, 97, and 28 are followed, which appear to be derived from the original guanine parent molecules. It is seen that the $[M + H - HCN]^+$ fragment at m/z 125 exhibits the behavior expected under slow refragmentation conditions. In fact, from the signal

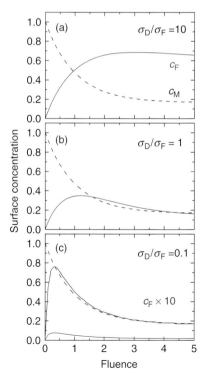

Figure 5.16 Surface concentration of fragments and parent molecules versus projectile ion fluence calculated from Equation 5.14 and Equation 5.20 for $\varepsilon = 0.2$ and three different values of the cross section ratio σ_D/σ_F.

maximum observed at about a depth of about 20 nm, one finds a cross section ratio $\sigma_D/\sigma_F \approx 37$. The $[HCN + H]^+$ fragment at m/z 28, on the other hand, manifests an example of the opposite behavior, with the position of the signal maximum indicating a cross section ratio $\sigma_D/\sigma_F \approx 0.13$. The $[M + H - HNCOC]^+$ fragment at m/z 97 constitutes an intermediate case with a cross section ratio of $\sigma_D/\sigma_F \approx 1.3$. In the neutral channel (Figure 5.14d), the most prominent fragments are C, N, and O atoms, which represent the final products at the end of multistep fragmentation chains. These fragments are expected to populate significantly slower than the decay of the parent molecule, a behavior that is indeed observed.

5.4.7 Conclusions

The model presented here demonstrates that the essential features of a molecular sputter depth profile can be understood in terms of simple erosion dynamics considerations. It is shown that this also holds true under conditions where the total sputter yield and, hence, the erosion rate is not constant. For the case of a slowly varying sputter yield, a quasi-steady state approximation can be used to describe the parent molecule signal variation after an initial exponential decay. In this regime, the concentration of intact molecules at the surface is determined by a balance between ion-induced damage on one hand and sputter removal of debris on the other hand. If the sputter yield is sufficiently large, this leads to an efficient cleanup in the course of a single projectile impact, thereby exposing fresh, undamaged molecules for analysis in further impacts. It is seen that this requires sputter yields of the order of hundreds of molecules per projectile impact, which are impossible to achieve with atomic projectile ions regardless of their kinetic energy or impact angle. If cluster projectiles are used, however, cleanup efficiencies of the order of unity can be reached, thus rendering molecular sputter depth profile analysis possible.

It should be kept in mind that the model presented here represents only a first approximative step toward an accurate description of molecular depth profiling. So far, no attempt has been made to predict the signal variation across the interface between, for example, an organic film and a metallic or semiconductor substrate. In this region, one has to deal with a rapidly varying matrix that may, among others, lead to a very rapid change of the erosion rate. The resulting erosion dynamics may become quite complicated and are outside the scope of the present description. Moreover, the notion of chemical damage as a purely destructive process removing molecules from the analysis is likely to be oversimplified. As seen above, experimental data reveal that different (characteristic) fragments of the same molecule may exhibit different damage cross sections. In addition, surface transients may overlay the erosion dynamics, leading in some cases to an initial *increase* of a molecular ion signal instead of an exponential decay. In principle, such transients may be caused by a multitude of different effects, ranging from ion bombardment–induced modifications of the surface chemistry to variations of the ionization probability of the ejected species, all of which are not included in the simple model presented here.

5.5 THE CHEMISTRY OF ATOMIC ION BEAM IRRADIATION IN ORGANIC MATERIALS

5.5.1 Introduction

At the beginning of this chapter (Fig. 5.1), a set of four scenarios was presented, describing the behavior of molecular ion signal intensities (or fragment ions in the case of polymer samples) as a function of depth. In the first scenario (Fig. 5.1a), the molecular ion intensity remained constant throughout the depth profile. We know now that this occurs when the beam-induced damage is low and the sputtering yield is high, such that the so-called cleanup efficiency, ε (Eq. 5.11), is large. In Figure 5.1b and c, the signal is not constant as function of dose or is lost completely. This scenario will occur when the beam-induced damage is high and/or the sputtering yields are low, such that ε is low. An important question to ask at this point is what exactly is meant by "beam-induced damage." From the physical models describing the erosion process introduced in Section 5.4, we know that in samples which are accumulating damage, the corresponding sputter rates are nonlinear. What is the cause of this nonlinearity and what can the experimentalist do to minimize this? This section purports to discuss the chemistry of ion beam irradiation and the corresponding damage mechanisms in organic materials in order to better understand, define, and improve upon molecular depth profile characteristics. Although most of these damage characteristics are absent with the recent advent of the gas cluster ion beam (GCIB) source, it is still important to understand the fundamentals of radiation chemistry that occur during ion beam sputtering with both atomic and polyatomic ion sources.

5.5.2 Understanding the Basics of Ion Irradiation Effects in Molecular Solids

When an ion beam irradiates a molecular sample, chemical changes occur, resulting from both primary interactions with the ion beam (i.e., created directly at the site of the ion beam crater or track) and secondary reactions that result from the diffusion of free radicals or ions created as a result of this interaction.[5,52–54] The end product of these interactions usually take the form of chemical cross-links. While the ion beam does indeed cause significant fragmentation of the molecules, these fragments will contain free radicals and other reactive species, which, if not removed during the sputtering process, will further react to form cross-links.

Radiation-induced cross-linking of molecular samples is a well-known phenomenon and radiation is often used by scientists and engineers to create novel polymeric materials, improve the mechanical strength of a material, or graft molecules onto surfaces.[52] For SIMS applications, however, this ion beam–induced cross-linking is oftentimes the root cause of sputter rate nonlinearities and eventual depth profile failure. Hence, in order to effectively depth profile through organic and polymeric samples, it is imperative to find ways to minimize the cross-linking that occurs during ion bombardment of molecular solids.

Ion beam irradiation of *polymers* is a special case, and there are often competing mechanisms of polymer scission and/or ion-induced depolymerization that can occur.[5] For this reason, polymers are usually labeled by radiation chemists as either Type I or Type II polymers, depending on how they respond to radiation.[5,53]

Type I polymers, for example, prefer cross-linking when irradiated by ion beams and/or other types of radiation. Type I polymers include polymers such as PE, PS, polypropylene (PP), and several other polymeric materials that contain a high concentration of aromatic components and/or have very little branching associated with them. In contrast, Type II polymers predominantly degrade under irradiation through a random main-chain scission process. Type II polymers include polymers such as poly(isobutylene) (PIB), PMMA, poly(α-methylstyrene) (PAMS), and other polymers with increased branching or "weakened" points located on the main-chain backbone, such as polyesters (e.g., PLA) and polyethers (e.g., PEO).[53,55]

Polymers with high concentrations of quaternary carbon atoms along the chain (e.g., CH_2-CR_2) will be more likely to undergo scission, while those with a structure of CH_2-CH_2 or CH_2-CHR tend to cross-link. This is because tetra-substituted carbons in the main chain can cause steric effects that can increase the strain in the polymer main-chain structure, making it easier to break. Other instances, such as changing tacticity (e.g., isotactic vs syndiotactic) can cause the steric repulsion to increase, and therefore can also effect the degradation processes in polymers.[6] Some of the most radiation-resistant (cross-linking) polymers contain a high concentration of aromatic substituents (e.g., PS), which are fairly resistant to radiation in general.[53] Any charges created are stabilized within the resonance of the aromatic ring structures and these rings are therefore often the site of the cross-linking.[5]

5.5.3 Ion Beam Irradiation and the Gel Point

Even in Type II samples, cross-linking will eventually predominate under higher primary ion fluences. A good example of this is PMMA, which behaves as a positive resist at low fluences (depolymerizes under irradiation), but behaves as a negative resist (cross-links under irradiation) at higher fluences.[56,57] Part of the reason for this is likely to be the approach of the gel-point of the polymer.

In most molecular samples undergoing ion bombardment, there is a critical fluence (I_{crit}) above which cross-linking rapidly accelerates and the corresponding solubility decreases rapidly to zero. This point is considered the gel point of the polymer and is defined as the point at which a 3D cross-linked network is formed. Theoretically, this will happen when there is one cross-link formed for every two molecules.[58] Once this occurs, the solubility will start to decay with increasing dose, eventually approaching zero as the number of cross-links per molecule increases to 1 and above. It is assumed that the corresponding SIMS spectra will have a similar drop in signal intensity at this point as it is well known that increased cross-linking is directly correlated to decreased secondary ion yields and sputtering yields in SIMS.[59,60]

Ion beam–induced gel formation can be monitored directly by measuring the insoluble fraction of a polymer, while simultaneously using gel permeation

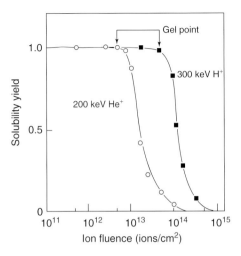

Figure 5.17 Solubility yield versus (mass fraction of soluble material) ion fluence for a polystyrene film bombarded with 200 keV He$^+$ and 300 keV H$^+$ ions. Reproduced from Calcagno,[55] with permission from Elsevier.

chromatography (GPC) to characterize the soluble fraction of the polymer.[54,56,58, 61–63] As an early example, Calcagno and coworkers studied the gel formation in several polymers using atomic ion beams.[55] Figure 5.17 shows the solubility yield of PS (in CHCl$_3$) as a function of increasing He$^+$ and H$^+$ ion doses operated at 200 and 300 keV beam energies, respectively. As can be seen in Figure 5.17, the PS sample is 100% soluble until primary ion doses of $\sim$7.5 $\times$ 10^{12} ions/cm^2 for 200 keV He$^+$ and 7.5 $\times$ 10^{13} ions/cm^2 for 300 keV H$^+$. At these critical fluencies (I_{crit}), the gel point of the polymer has been reached and the corresponding solubility of the polymer starts to decline rapidly with dose. At higher doses (>1 $\times$ 10^{15} ions/cm^2 for He$^+$, for example), the film is completely insoluble.

The gel point tends to be reached earlier for higher molecular weight samples and is highly dependent upon the number average molecular weight (M_n), in particular.[63] As there are, theoretically, a critical number of cross-links required per molecule that need to be attained before the gel point is reached (defined as 0.5 earlier, or 1 cross-link for every two molecules), this number is obviously reached earlier for the larger molecules. Similarly, in Figure 5.17, the transition observed from soluble to insoluble occurs very rapidly (under a very narrow fluence range). This was attributed to the polydispersity of the sample. The PS sample used in the study was a monodisperse sample, having a very narrow molecular weight distribution. Since the molecular weight of a polymer will affect I_{crit}, in more polydisperse samples, this transition is expected to be much broader.

Finally, as can be seen in Figure 5.17, the gel point is reached earlier for the 200 keV He$^+$ as compared to the 300 keV H$^+$. In this case, the change in I_{crit} is associated with both the increased mass of the He projectile relative to H and the decreased beam energy.[64] Calcagno and Foti show that it is the energy deposited per unit volume, or the local energy density, that is the most important factor in defining I_{crit}. The chemical yield of cross-links in PS, for example, increases with the local energy density. Ions with increased masses and decreased acceleration energies will keep the local energy density high, and therefore will increase the yield of chemical cross-links, thus diminishing the value of I_{crit}.

For cluster sources, approaching the thermal spike regime of sputtering, it is expected that different damage mechanisms will occur in terms of energy density. It is likely, for example, that the higher energy densities will increase the degradation rates in some polymers, and hence a corresponding increase in the sputtering yield will be observed in addition to or in lieu of an increase in the number of cross-links. Similar sputter yield enhancements will be observed in small molecular samples. Therefore, in the case of cluster SIMS, I_{crit} will likely be dependent upon the cross-link density remaining behind in the altered layer, d. This will be described in more detail in Section 5.6.

Figure 5.18 depicts GPC traces acquired from the soluble portions of the irradiated polymer in Figure 5.17 as a function of increasing H^+ ion dose, where Figure 5.18a and b show traces acquired before the gel point, I_{crit}, and Figure 5.18c and d show traces acquired from the soluble fraction of the polymer, after I_{crit}. Figure 5.18a shows the GPC trace acquired from the PS sample before and after irradiating the sample with 1.5×10^{12} H^+ ions/cm^2. The dashed line shows the initial trace acquired from the pristine PS sample before irradiation. The PS is a monodisperse sample with a number average molecular weight (M_n) of 9000. After irradiating the sample with H^+ ions, a shoulder appears in the trace, which is located at higher molecular weights. This is consistent with dimer formation resultant from chemical cross-links. Not surprisingly, PS dimer formation was also observed during extended SIMS analysis of PS.[65] The cross-linking intensifies in Figure 5.18b, after irradiation with 6.5×10^{12} ions/cm^2.

After the approach of the gel point, the distribution appears to shift a bit back toward lower molecular weights. This is because only the soluble fraction is represented in these traces. The higher molecular weight species are removed from the system via gel formation and are not analyzed in the GPC.

In addition, it may be noted that there is also a minor contribution from chain scission as is indicated in Figure 5.18d. This contribution is expected to be much higher while characterizing polymers such as PMMA or PLA, which prefer chain scission to cross-linking.

It should be mentioned that although the beam energies used in these series of experiments are much higher than what is typically used for SIMS applications, and therefore slightly different sputtering mechanisms apply, the overall result will be the same—the formation of a highly cross-linked material.[66] The only differences between higher beam energies described (100–500 keV) and beam energies employed for SIMS applications (1–40 keV) will be the rates of damage, the relative ratio of damage products, as well as the thickness of the damage layer affected.

Samples composed of smaller molecular species, as opposed to polymers, will also cross-link and eventually polymerize when irradiated with ion beams. Similarly, gel formation will occur if the material is not efficiently sputtered (e.g. using an atomic primary ion source, as opposed to a cluster source). Similar to polymers, smaller molecular species will behave differently under ion beam irradiation, depending on their chemical structure and size. Obviously, larger molecules,

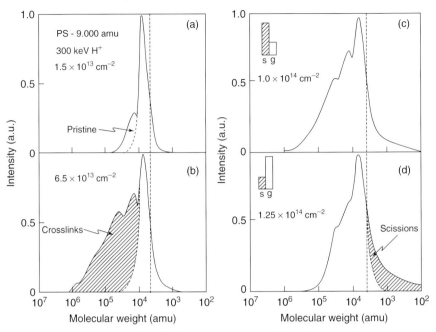

Figure 5.18 Molecular weight distributions as measured by gel permeation chromatography (GPC) from a polystyrene film before and after irradiation with 300 keV H$^+$ ions at varying doses: (a) 1.5×10^{13} ions/cm^2, (b) 6.5×10^{13} ions/cm^2, (c) 1.0×10^{14} ions/cm^2, and (d) 1.3×10^{14} ions/cm^2. The number average molecular weight (M_n) of the PS used in the experiment was 9000 g/mol. The original PS distribution (before irradiation) is indicated in (a) and (b) as a dashed line. In (a) and (b), the PS film was completely soluble, while in (c) and (d), the film was only partially soluble. The amount of soluble material relative to insoluble gel material are indicated in the histograms shown in the insets of (c) and (d). Vertical lines are indicated at the half max point of the original curve to help guide the eye in determining the extent of scission processes. Reproduced from Calcagno and Foti,[64] with permission from Elsevier.

which are more difficult to remove from the surface intact and require a lower cross-link density to form a gel, will tend to cross-link more readily then smaller molecules.

5.5.4 The Chemistry of Cluster Ion Beams

Thus far, we have been focusing primarily on the chemistry of atomic ion bombardment. But what about cluster beams? Do these same chemistries apply? Certainly cross-linking is delayed as compared to atomic ion beams because of the enormous increases in sputter yields that allow for simultaneous damage and sputter removal of the material. Furthermore, the damage processes are different and do not occur along a track, but rather are localized at the site of a sputtered crater bottom. However, the same chemistries do indeed apply and play a very

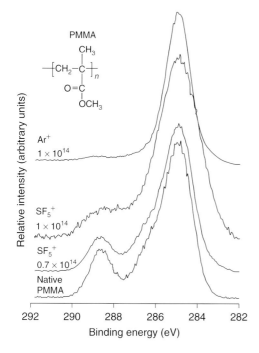

Figure 5.19 Comparison of the C1s core-level X-ray photoelectron (XPS) peak acquired form a film of PMMA bombarded by increasing fluencies of 3 keV SF_5^+ and 3 keV Ar^+. Reproduced from Fuoco et al.,[67] with permission from ACS publications.

important role, in particular, for the smaller polyatomic ions such as SF_5^+ in which the mechanisms of ion bombardment fall somewhere in between that of atomic (collision cascade regime) and larger cluster bombardment processes (thermal spike model).

Figure 5.19[67] shows the core-level XPS spectra of the C(1s) region acquired from a PMMA sample under different bombardment conditions. In the native PMMA surface, there are three major components, consistent with the chemical structure of the PMMA (see inset). The largest peak is the aliphatic carbon peak at a binding energy of 285.0 eV. There is also a methoxylic carbon component (C^*-O)), which is shifted to slightly higher binding energies (286.5 eV). Finally, the carboxylic carbon peak ($O-C^*=O$) occurs at 288.9 eV. As can be seen, the results are consistent with a significant loss of the carboxylic carbon moiety upon sputtering with Ar^+ as compared to SF_5^+, thus indicating decreased structural damage when employing cluster ion beams as compared to atomic ion beams. However, the peak decays further when increasing the SF_5^+ dose from 7.0×10^{13} to 1.0×10^{14} ions/cm^2. The damage processes are indeed still occurring, but at a much slower rate than is observed with the atomic ion beams. Over time, it is likely that the XPS spectra of the PMMA bombarded with SF_5^+ will look very similar to that obtained by Ar^+ under lower doses[5,26].

In the case of cluster bombardment, the damage processes and chemistries will occur in the crater bottom as opposed to along the site of the ion track. Therefore, it is assumed that the chemistry remaining in the altered layer (d) will play a vital role in the damage processes. In other words, this layer of thickness d will have a certain number of cross-links, degradation products, and free radicals resultant from the initial impact event, where the free radicals will react further to form additional cross-links and/or degradation products in this region. It is assumed that similar to what occurs under atomic bombardment along the track of the ion, this layer of thickness d will eventually form a critical number of cross-links such that the altered layer becomes a 3D cross-linked gel. After this gel is formed, the sputter yields will diminish dramatically, and it is expected that there will be a corresponding catastrophic loss of signal as described in Figure 5.1d; unless that is, the sputter source can also *efficiently remove this gel-like damage layer*, such as has been reported to be the case for the larger $Ar_{n>500}^+$ gas clusters.[68]

Figure 5.20 illustrates the process of gel formation during cluster bombardment of a 1 μm thick sample of PMMA cast onto a silicon substrate. In the figure, optical images are shown from a 300 μm × 300 μm crater (dark region in image) formed in a sample of PMMA under high SF_5^+ primary ion dose conditions ($\sim 3 \times 10^{16}$ ions/cm^2). Figure 5.20a shows the original image of the crater acquired shortly after the depth profile experiment. Figure 5.20b shows the same crater, after soaking in an acetone solution for 10 min. As can be seen, the unirradiated PMMA has been completely dissolved, while the ion-beam-modified PMMA has become insoluble and therefore remains after soaking. This is consistent with gel formation at the site of ion beam irradiation. Figure 5.20c shows the corresponding SIMS image of the crater after soaking, acquired for verification of compositions in the crater bottom shown in Figure 5.20b.

5.5.5 Chemical Structure Changes and Corresponding Changes in Depth Profile Shapes

We now have a general idea of how chemistry affects depth profiling of organic and polymeric materials. A detailed summary of the process is depicted in Figure 5.21, which shows a representative "molecular" depth profile of a polymer bulk film

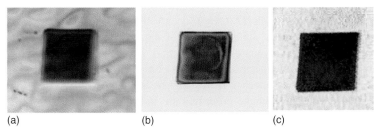

(a)   (b)   (c)

Figure 5.20 Optical image of crater bottom in PMMA thick film ($\sim$1 μm) acquired before and after soaking in acetone. The crater was formed using 5 keV SF_5^+ operated under high dose conditions (3×10^{16} ions/cm^2) with a raster size of 300 μm × 300 μm. (a) Before soaking, (b) after soaking for 10 min in acetone solution, and (c) SIMS positive ion image overlay of the sample after soaking with signal characteristic of the polymer (*m/z* 59) in blue and signal characteristic of the silicon substrate (*m/z* 28) in yellow.

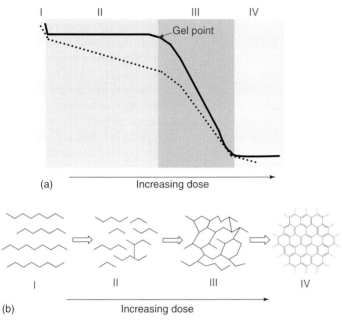

Figure 5.21   Illustration of damage processes in polymers during ion bombardment. (a) Representative SIMS depth profile of a bulk polymer, with four regions; (I) surface transient region representing the initial saturation of beam damage, (II) steady-state region where damage accumulation is balanced by sputter removal of damaged material, (III) gel point—where a three-dimensional cross-linked structure is formed, with a corresponding loss in SIMS signal, and (IV) formation of graphitic material or other C allotrope. (b) Corresponding polymeric structure with (a). Dashed line represents the case whereby there is an increased rate of cross-link formation or nonuniform sputtering.

(Fig. 5.21a) and the corresponding chemical structure changes that occur in each region of the profile (Fig. 5.21b).

The profile in Figure 5.21a shows the signal intensity as a function of increasing dose, where the signal is monitored beyond the point of failure (loss of molecular signal). The chemical structure of the polymer in the first two regions (I and II) reflects a transition from a pristine, undamaged polymer structure (Fig. 5.21b, structure I) to a polymer containing multiple chain scissions and some cross-links (Fig. 5.21b, structure II). This illustration also applies for smaller molecular species, with the exception that there would be fragment ion species in lieu of chain scission products, and fewer cross-linked molecules.

Because we are assuming the use of a cluster source, the chemistry would originally be located within the altered layer, d, but reactive species are also expected to diffuse into deeper layers of the polymer, depending on the conditions of ion bombardment. The number of cross-links relative to scissions will also vary depending on the experimental conditions.

The formation of cross-linked material creates a particular problem, as these larger species will be much more difficult to remove via sputtering. Therefore, over time, the number of cross-links concentrated at the surface is expected to accumulate. When enough cross-links are formed, the damaged surface layer becomes a 3D cross-linked gel (Fig. 5.21b, III). At this point, I_{crit}, the cross-linking rapidly accelerates, resulting in a more rapid decline in molecular signal as is shown in Figure 5.21a, region III.

Many "steady-state" regions are not flat as is indicated in Figure 5.21a, but rather will exhibit a constant decay with increasing dose as was the case described in Figure 5.11 and Figure 5.13. As discussed in Section 5.4, this region can be described by a steady decline in the total sputtering yield of the polymer or molecular species with dose (Eq. 5.15). The dashed line in Figure 5.21a demonstrates such a case. From a chemical perspective, this steady decay in sputtering yield results from increased rates of cross-linking (observed in certain materials over others for example) and/or nonuniform sputtering processes. Nonuniform sputtering could refer to the ion beam bombardment conditions, but will also likely be largely dependent upon the relaxation of the crater, as well as the spatial location of the chemistries within that crater. In other words, are the chemical cross-links uniformly distributed across the crater after an impact event? This is not likely to be true for samples under normal ion bombardment conditions.[†] In the case where there is nonuniform sputtering or increased sputter-induced topography during an experiment, one can imagine that there might be localized regions of highly cross-linked material with decreased sputter yields as compared to other regions in the sample, creating gel-like nucleation sites.

At higher doses, a so-called "graphitization" process begins (region IV—Fig. 5.21). This process is usually accompanied by an increased conductivity of the surface and the presence of graphitic-type peaks in the mass spectrum.[57,69–73] Figure 5.22 helps to visualize the graphitization process that occurs in a bulk PMMA film during SF_5^+ cluster ion bombardment, with the aid of principal components analysis (PCA), a multivariate statistical analysis approach often utilized to analyze complex mass spectral datasets (see Chapter 4 for introduction to PCA). The data were acquired from a thick film of PMMA (5 μm) sputtered with 5 keV SF_5^+ at cryogenic temperatures ($-100\ ^\circ C$) and using a separate Bi_3^+ primary beam for analysis. In Figure 5.22a, the molecular signal characteristic of PMMA (m/z 59 and m/z 69) is plotted as a function of depth profile cycle. As can be seen, the profile appears to be very similar in shape to that described in Figure 5.21a, showing a surface transient region, a steady-state region, and a gel-point region. Specific time points during the depth profile were selected for PCA analysis, which are also indicated in the figure. Figure 5.22b and c show the results from PCA analysis of the selected mass spectra, where Figure 5.22b shows the calculated scores as a function of depth profile cycle, and Figure 5.22c shows the corresponding loadings plots. Recall from Chapter 4, in which PCA was introduced, that each mass spectrum is given a score that indicates the correlation with a series of peak loadings. In other words, positive scores correlate with positive loadings and negative scores correlate with negative loadings. In this case, the positively loaded peaks in Figure 5.22c are characteristic of carbon-rich species and graphite, while the negatively loaded peaks are

---

[†]Normal conditions are defined as 45° incidence, no sample rotation, and at room temperature.

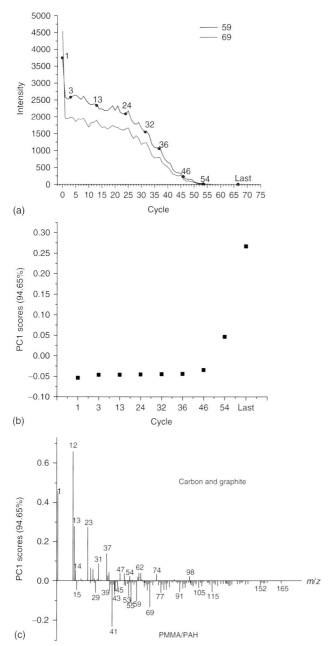

Figure 5.22 Principal components analysis (PCA) of a depth profile acquired from a thick PMMA film (1 μm) using a 5 keV SF_5^+ sputter source: (a) original depth profile, displaying the intensity of m/z 69 as a function of the number of depth profile cycles. PCA analysis of the depth profile is shown in (b) and (c), where (b) is the scores plot for PC1 (encompassing 94.7% of the variation in the data) and (c) is the loadings plot that corresponds to (b).

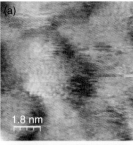

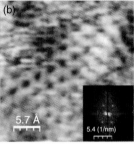

Figure 5.23 Scanning tunneling microscopy images of polyimide films after ion irradiation with 1 keV Ar^+ at 350 °C and under high primary ion dose conditions (1 × 10^{18} ions/cm^2). (a) 9.0 nm × 9.0 nm and (b) 2.9 nm × 2.9 nm. Inset represents the Fourier transform of the image in (b). Reproduced from Lazareva et al.,[69] with permission from the American Institute of Physics (AIP).

characteristic of PMMA and polycyclic aromatic hydrocarbons (PAH—consistent with slightly fragmented PMMA). Note also, that these peaks intensify rapidly only after complete loss of the molecular signal, consistent with the schematic of the depth profile process shown in Figure 5.21b.

Although the majority of studies describe this process as a graphitization process, this is not always the case. Other carbon allotropes, such as diamond and carbon black, have also been observed when sputtering with ion beams.[26,74–79] It should also be mentioned that there appears to be some controversy in the literature regarding the formation of graphite during ion beam irradiation.[76–79] The formation of graphite, if it does occur, will depend highly upon the ion bombardment conditions, and will not likely be continuous in nature. Beam energy, for example, has been shown to play a very critical role in the formation of graphite.[26,74,75]

Figure 5.23 shows some very nice scanning tunneling microscopy images acquired from an ion beam irradiated polyimide sample, under specified conditions, where the overlapping graphite regions are clearly observed.[69] These particular samples were irradiated under high temperature conditions. The graphitized regions were observed only in specified regions and were not continuous. It should be noted that direct comparisons of the graphitization process between *cluster* sources as compared to atomic sources have not yet been performed to the knowledge of the authors.

5.6 OPTIMIZATION OF EXPERIMENTAL PARAMETERS FOR ORGANIC DEPTH PROFILING

5.6.1 Introduction

This chapter cites several examples that employ "optimized" conditions for depth profiling, whereby the authors used special conditions, such as low temperatures and/or grazing ion beam incidence angles, to improve the depth profile character-istics. Here, we discuss these experimental conditions in greater detail, describing the important parameters that need be considered during a molecular depth pro-filing experiment (e.g., temperature, beam angle, and cluster size, among several others) in order to obtain optimal results. Furthermore, we discuss how changes in the experimental parameters directly affect the ion beam irradiation chemistries introduced in section 5.5, as well as the physical sputtering processes described in section 5.4, and in Chapter 2.

5.6.2   Organic Delta Layers for Optimization of Experimental Parameters

Before discussing optimal parameters for organic depth profiling, it is important to first introduce the concept of organic delta layers. Delta layers typically consist of multiple layers of two distinctive materials, whereby one of the layers is extremely thin (on the order of, or better than, the depth resolution of the instrument), such that it can be used to evaluate the performance of a particular instrument and/or the experimental conditions employed.

This type of reference material was originally utilized by the semiconductor industry for measuring and optimizing instrument parameters for elemental depth profiling (see Chapter 7 for more details),[80] and more recently has been developed for the organic depth profiling community,[28] where it has been utilized in a recent round-robin study on organic depth profiling conducted by the Versailles Project on Advanced Materials and Standards (VAMAS).[11,17,20,27,28,81] Much of the following section refers to the results from this very important series of studies by Shard et al., and therefore, a brief introduction of the sample structure and typical results observed during depth profiling of this sample is warranted.

The reference material, which is depicted in Figure 5.24a, consists of alter-nating layers of two UV stabilizers (Irganox 1010 and Irganox 3114), prepared by a thermal evaporation process, where the Irganox 3114 layer is extremely thin (~4 nm) relative to the thicker Irganox 1010 layers (50–100 nm, depending on the layer). A representative depth profile of this sample acquired using 20 keV C_{60}^{2+} ions and 14 kHz sample rotation[11] is shown in Figure 5.24b, where molecular signal characteristic of the Irganox 3114 (m/z 564) is plotted as a function of the erosion depth. In the ideal case, all four layers will be resolved with very narrow and constant depth resolutions. Obviously, the depth resolutions degrade slightly with depth in the example shown. However, all four peaks are indeed resolved under the conditions employed and are actually greatly improved as compared to what is observed without sample rotation.

Experimental conditions can be directly compared by plotting the full width at half maximum (FWHM) of the peaks shown in (b). As an example, Figure 5.24c plots the FWHM measured from each of the four peaks under different experi-mental conditions.[11] The curve on the top (○) shows the control values acquired using 20 keV C_{60}^{+} at 45° incidence and 25 °C. The depth resolution of the peaks (FWHM) clearly decays with depth. The corresponding sputter rates are also nonlin-ear with increasing depth.[28] The best results for this particular system are obtained when employing low temperatures (■) and sample rotation (□). This is followed

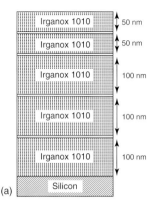

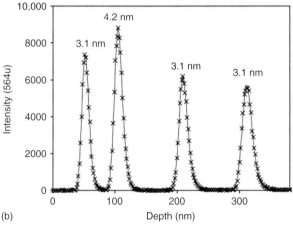

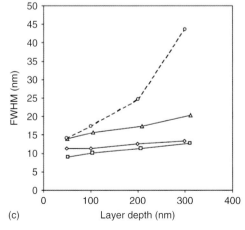

Figure 5.24 Depth profiling in Irganox multilayer reference standard developed by the National Physical Laboratory (NPL) in the United Kingdom.[28] The reference material consists of alternating layers of Irganox 1010 and Irganox 3114 (3–4 nm) and is prepared on a Si substrate via an evaporation process. (a) Visual representation of standard, showing the thickness of the Irganox layers; (b) example of a depth profile acquired using 20 keV C_{60}^{2+} ions with 14 Hz sample rotation; the thickness of each Irganox 3114 layer is indicated; and (c) FWHM measured for each layer, using improved parameters such as 20 keV C_{60} at a 76° incidence angle (□), 10 keV C_{60} at −80 °C (◇), and 20 keV C_{60} at 14 Hz rotation (△), compared to mean 20 keV C_{60} values (○). Reproduced from Shard et al.,[11,85] with permission from Elsevier.

closely by the results obtained at glancing incidence angles ($\triangle$). The reasoning behind each of these results will now discussed in detail.

5.6.3   Sample Temperature

It has been observed that controlling the temperature of a sample during depth profiling analysis of soft materials with cluster ion beams is very important.[5,6,11–17,26,29,30,33,34,81–84] Most commonly observed is a drastic improvement in the quality of the depth profiles at cryogenic temperatures, whereby the resulting depth profiles exhibit increased signal constancy and longevity. Because most cross-linking and degradation mechanisms rely heavily on the formation of free radical intermediates, temperature can play a particularly important role in the degradation and cross-linking processes in organic and polymeric materials.[53] This is because free radicals will lose their reactivity at lower temperatures and have a much longer lifetime before secondary interactions occur (on the order of months in some cases).[53] Therefore, at low temperatures (-100 °C range), cross-linking is suppressed during the time frame of a depth profile experiment and it will typically take a lot longer before the critical fluence is reached (I_{crit}) for gel formation. When the sample is later "unfrozen" again, for example, by increasing the temperature or by addition of a solvent, the radical is again free to react and will typically do so, causing deferred cross-linking and/or degradation of the ion-bombarded region. Although decreased cross-linking is the most likely reason for the improvement observed at low temperatures, particularly for larger molecules and polymer samples, there is also evidence of decreased beam-induced mixing for smaller molecular species when employing cryogenic temperatures.[5,81] Overall, the decreased mobility of the bombarded species at low temperatures proves beneficial for molecular depth profiling.

Increasing temperature above room temperature during sputtering will also affect the chemistries occurring during ion bombardment. For example, increasing the temperature (particularly above the glass transition temperature or melting point of a sample), will allow for increased mobility of the molecules and free radicals contained in these samples, and therefore cross-linking mechanisms can be sped up. For some polymers, ion-induced depolymerization mechanisms will start playing a role at high temperatures, causing a decrease in the molecular weight of the sample with corresponding increases in sputter rates. In some cases, increasing the temperature of the sample during depth profiling can even result in successful depth profile analysis of materials that do not typically perform well when irradiated with cluster beams at room temperature.[16]

The effect of temperature on cluster SIMS depth profiling of soft materials was first studied by Mahoney and coworkers, and later published in a series of works describing the effects of temperature on depth profiling in various polymeric materials.[12–15] The effect of temperature on polymer depth profiles can be quite impressive. Figure 5.25 shows data from the first of these studies, where the authors employed varying temperatures during $SF_5{}^+$ depth profiling in order to elucidate its effect on damage characteristics in the depth profile shape and topography formation in PMMA thin films.[12]

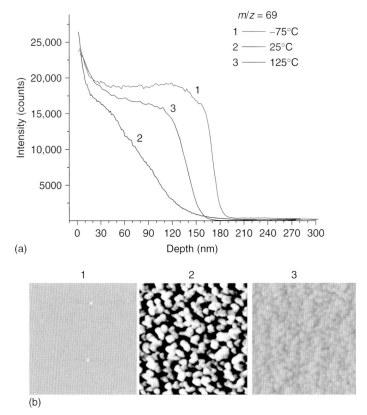

Figure 5.25 Effect of temperature on depth profiling and topography formation in PMMA under 5 keV polyatomic ion bombardment (SF_5^+). (a) Positive secondary ion intensities of m/z 69 plotted as a function of depth for a PMMA film ($\sim$160 nm) on Si, where the sample is maintained at varying temperatures during the experiment: (1) -75 °C, (2) 25 °C, and (3) 125 °C. All profiles shown utilized 10 keV Ar^+ for analysis. (b) Atomic force microscopy (AFM) topography images (1 μm × 1 μm) of crater bottoms formed at different temperatures: (1) -75 °C, $R_{rms} = 0.27$ nm; (2) 25 °C, $R_{rms} = 11.37$; and (3) 125 °C, $R_{rms} = 1.00$ nm. Note: R_{rms} for PMMA control acquired before sputtering and Si substrate alone is 0.85 nm and 0.20 nm, respectively. Reproduced from Mahoney et al.,[12] with permission from Elsevier.

Figure 5.25a shows an overlay of three representative depth profiles obtained from PMMA thin films (160 nm) cast on Si, where the sample was maintained at -75, 25, and 125 °C during the course of the depth profile experiment. Also shown in Figure 5.25b is the resultant topography formed at the bottom of the crater after the depth profile experiment. As can be seen, the highest quality depth profiles were obtained at -75 °C, as evidenced by the increased signal constancy and decreased apparent interface widths observed.

At room temperature, significant signal degradation was apparent prior to the interface location. It should be noted that this extensive signal degradation observed in the PMMA film is not typically observed with thinner films (<80 nm),[1] and is not observed at all for PMMA thin films characterized with C_{60}^+.[16] This is because the beam-induced damage accumulation is much greater for SF_5^+ than for C_{60}^+, and beyond 50 nm or so, the polymer changes from a degrading polymer to a cross-linking polymer (I_{crit} for gel formation is approached much more rapidly with SF_5^+ than with C_{60}^+).

The corresponding AFM topography maps (Fig. 5.25b) show a significant reduction in the corresponding sputter-induced topography formation at low temperatures. The root mean square roughness values (R_{rms}), as determined by AFM, are comparable to those of bare Si for the crater formed at $-75\,^\circ C$ ($R_{rms} = 0.3$ nm), whereas the R_{rms} value for the crater formed at room temperature is much greater ($R_{rms} = 11.4$ nm). Further investigations into the topography formation and accumulation processes over the entire profile have been described[15], where it was shown that the topography formed by the ion beam at low temperatures remains constant throughout the profile, while at 25 $^\circ C$, the topography steadily increases with dose.

The corresponding chemistry of the crater bottoms was also investigated using XPS, where it was shown that the material remaining in the crater formed at 25 $^\circ C$ contained significant amounts of a highly damaged C species, with little or no evidence of C=O functionalities that are consistent with the undamaged structure of PMMA. Conversely, the $-75\,^\circ C$ crater bottom contained very little carbon, and much more of silicon. We can therefore conclude that while at low temperatures the polymer is efficiently removed due to decreased cross-linking, at room temperature the bombardment of the polymer results in the rapid formation of a highly cross-linked gel, with significantly reduced sputtering yields.

Similar improvements in depth profiles at low temperatures have been observed in other organic systems such as Irganox[11,17,85] and AA[33,34,81,82] layers. Figure 5.26 shows depth profiling results obtained from the Irganox delta layer standard, acquired at variable temperatures, using 10 keV C_{60}^+ ions for depth profile analysis.[17] At 20 $^\circ C$, only three layers are resolved, with significant broadening and decreased intensities of the peaks with increasing dose. At $-40\,^\circ C$, there is some improvement, and the sputter yields increase slightly, as indicated by the decreased doses required to reach the first layer. Upon decreasing the temperature even further to $-80\,^\circ C$, however, all the layers are resolved, with improved depth resolutions and increased signal constancies. Recall that the FWHM measured for each peak in this system were plotted and compared in Figure 5.24. As can be seen, the best overall results were obtained when employing low temperatures during depth profiling ($\square$).

While diminishing the extent of cross-linking is ideal for any molecular depth profiling experiment, decreasing the temperature can also result in decreased sputtering yields, and in the case of polymers, there may also be a decreased rate of polymer degradation kinetics, which have proven to be beneficial for depth profiling.[5,15,16] For example, when employing low temperatures for depth profiling of PMMA with C_{60}^+, which works well at room temperature, the only real effect observed is a decrease in the sputtering rate.[16]

The depth profile acquired from PMMA at 125 $^\circ C$ with 5 keV SF_5^+ is shown in Figure 5.25a. As can be seen, there is also a considerable improvement

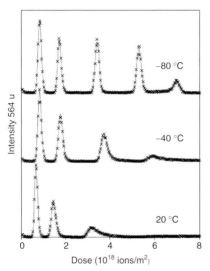

Figure 5.26 Organic depth profiles of Irganox multilayer samples introduced in Figure 5.24, showing the normalized intensity of m/z 564 as a function of C_{60}^+ dose. Profiles taken using 10 keV C_{60}^+, while maintaining the sample stage at varying temperatures ranging from -80 to 20 °C. Reproduced from Sjovall et al.,[17] with permission from ACS publications.

in the quality of the depth profile obtained at high temperatures as compared to room temperature conditions. Furthermore, there is a corresponding decrease in the sputter-induced topography formation, as is evident in Figure 5.25b, and a decrease in the C/Si ratio measured by XPS, indicating that a lower amount of damaged material remains in the crater bottom after sputtering.[12,15]

The improvement at high temperatures in this particular example is attributed to increased ion beam–induced depolymerization chemistries, more commonly referred to as *polymer unzipping*.[15,16,86] Polymer unzipping is a process whereby polymers degrade via a rapid main-chain scission process to form lower molecular weight products and monomeric species under high temperature conditions. PMMA, when heated to temperatures greater than $\sim$360 °C, will rapidly break down to lower molecular weight products and release large quantities of methyl methacrylate (MMA) monomeric species in the process.

When atomic ion beam irradiation is employed during the heating, the temperature at which polymer unzipping occurs is known to decrease.[86] For PMMA, for example, polymer unzipping occurs in the temperature range of $\sim$100–200 °C when being irradiated with 300 keV He$^+$.[86] With cluster sources, this temperature is expected to decrease even further, likely occurring at room temperature for C_{60}^+ and other larger cluster sources.[16] Depolymerization mechanisms can be readily detected through use of a residual gas analyzer (RGA). RGA was used during the sputtering of PMMA with SF$_5^+$, to detect the gaseous evolution of MMA monomers.[15] The results from this study confirmed that sputtering of a PMMA sample at increased temperatures does indeed result in a shift in the ion beam irradiation chemistry toward a depolymerization process, and that the start of these depolymerization mechanisms are directly correlated with increases in the sputtering yields.[15] Depolymerization mechanisms have also been found to play an important role during high temperature sputtering of PAMS, when employing 10 keV $C_{60}{}^+$ primary ion bombardment,[16] where it was shown that PAMS could only be depth profiled when the temperature was increased to above 160 °C.

Both SIMS and XPS have been used to characterize crater bottoms formed in PMMA at varying temperatures, where the results provide additional confirmation of the changing damage mechanisms occurring at high temperatures.[15] High resolution C1s spectra acquired from PMMA crater bottoms bombarded at −75, 25, and 125 °C are shown in Figure 5.27, where the dashed line represents the PMMA surface (no ion beam sputtering) and the solid line represents the sputtered crater bottom after 5 keV $SF_5{}^+$ bombardment, where the sputtering was stopped at the midpoint of the depth profile. As can be seen, the C=O moiety in the samples bombarded at −75 °C (Fig. 5.27b) and 25 °C (Fig. 5.27c) decreases slightly after sputtering. However, there is no discernable difference in the crater bottoms formed at −75 and 25 °C, a result that was initially quite surprising. On the other hand, the craters formed at 125 °C show very little difference in the C1s spectrum before and after sputtering, a result that is consistent with preservation of the monomeric structural integrity.[15]

Figure 5.28 shows the results from PCA of the positive ion SIMS spectra acquired before and after sputtering at variable temperatures. Figure 5.28a and b represent the scores and loadings plots from principal component 1 (80.5% of total variance). As can be seen, there is a large difference in the mass spectra acquired before and after sputtering. In general, after sputtering there are more peaks indicative of the formation of PAH and a decrease in the peaks associated with PMMA. In the negative spectra (not shown here), there is an increase in the C peak intensities after sputtering, and a corresponding decrease in the O-containing peak intensities. All of these results are indications of increased damage.[87]

Note that the smallest differences in the scores before and after sputtering are observed at 125 °C. In fact, where all other crater bottoms have negative scores, the craters formed at 125 °C have scores that are close to neutral, and, in actuality, are slightly positive. This indicates that there is less damage occurring to the monomeric structural unit, once again consistent with increased depolymerization mechanisms, whereby the material is only breaking up into smaller units while maintaining the integrity of its monomeric structure.

Similar to what is observed in the XPS spectra, there is no evidence of decreased damage observed at lower temperatures (−75 °C) as compared to room temperature; in fact, there is evidence to the contrary. As can be seen, the scores obtained from the sputtered crater bottoms actually decrease with temperature. This difference is also evident in the mass spectral data acquired from the surface at 25 °C as compared to −75 °C.[12] This effect is quite surprising, given the improvement in depth profiles obtained at these low temperatures.

Temperature effects on SIMS surface and depth profile analysis of several other polymeric materials with $SF_5{}^+$ were investigated by Mahoney et al. as well, including PLA, PS, polyurethanes (PU), poly(ethylene-*co*-vinyl acetate) (PEVA),

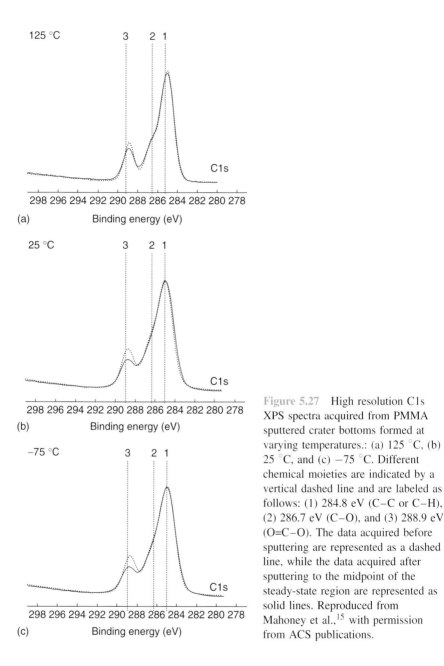

Figure 5.27 High resolution C1s XPS spectra acquired from PMMA sputtered crater bottoms formed at varying temperatures.: (a) 125 °C, (b) 25 °C, and (c) −75 °C. Different chemical moieties are indicated by a vertical dashed line and are labeled as follows: (1) 284.8 eV (C–C or C–H), (2) 286.7 eV (C–O), and (3) 288.9 eV (O=C–O). The data acquired before sputtering are represented as a dashed line, while the data acquired after sputtering to the midpoint of the steady-state region are represented as solid lines. Reproduced from Mahoney et al.,[15] with permission from ACS publications.

and poly(styrene-*co*-isobutylene).[12−14,30] Most polymers tended to have increased PAH peaks when characterized at low temperatures, similar to that observed with PMMA. These differences are observed even in the static SIMS spectra acquired

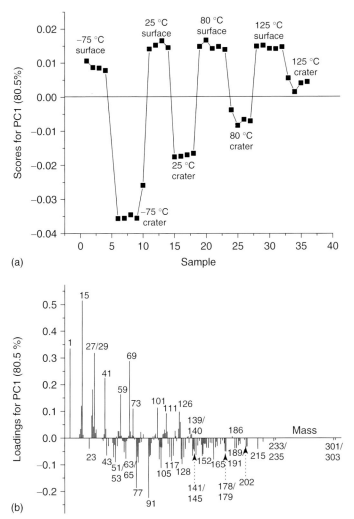

Figure 5.28 Principal components analysis (PCA) of positive ion mass spectral data acquired from sputtered PMMA crater bottoms at varying temperatures. (a) PC1 scores plots and (b) PC1 loadings plots. All data acquired using an 10 keV Ar$^+$ analysis source. Crater bottoms were formed using 5 keV SF$_5^+$. Each crater was prepared by sputtering to the midpoint of the steady-state region in a PMMA film. Reproduced from Mahoney et al.,[15] with permission from ACS publications.

(see chapter 4, Figure 4.11). It appears as though in all the analyses performed thus far, aromatic-type structures are preferred in the mass spectra acquired at low temperatures in polymer samples. At this juncture, we can only speculate as to why this occurs.

It may be that increases in PAH signal are only associated with a decrease in depolymerization/degradation mechanisms. It could also be that the decreased polymer chain mobilities are the result of increased PAH signal. For example, cyclization (formation of PAH) is known to occur more frequently in crystalline regions under irradiation as compared to noncrystalline regions.[53] Similarly, low temperature materials exhibit decreased chain mobility and therefore may show increased cyclization mechanisms as well.

Results from smaller molecules show contrary evidence, indicative of decreased damage accumulation in the form of fragmentation at lower temperatures.[83,84] Therefore, this effect of having increased peaks consistent with damage is likely observed solely in polymer materials, in which the damage accumulation and ion formation processes are much more complex. While molecular species are typically ionized as intact species, polymers are broken apart into fragment ions, which are produced by more than one bond scission and/or several rearrangements of the polymer backbone.[83,84] These processes will likely be heavily influenced by temperature changes. Furthermore, it was found that in some cases, the adsorption of water at the surface can occur at low temperatures. This will benefit molecular samples, by provision of protons for ionization; an affect that will obviously be less effective for large polymer molecules.

Finally, it should be mentioned that the temperature during the depth profile experiment will affect the crater relaxation processes, which in turn will directly affect the sputter-induced topography formation and the chemical uniformity of the sputtering. In studies by Papaleo et al.[88,89] for example, PMMA films were bombarded with 20 MeV Au^+ ions at low fluences and at variable temperatures from -196 to $150\ °C$. Scanning force microscopy (SFM) was then used to image the bombarded regions after cooling. The SFM images revealed nanometer-sized craters and/or raised regions (hillocks) around the point of each ion impact, where the size of the craters and hillocks was independent of the temperature up to 80 °C, but at higher temperatures, the crater dimensions increased steeply and no hillock was observed. When the sample was rapidly cooled, therefore quenching the relaxation of the crater, the hillocks disappeared closer to the glass transition temperature of the polymer (150 °C). The presence of the hillocks at room temperature, under slow cooling conditions, therefore indicates that the transient local heating induced by the incoming ion itself cannot be higher than 100 °C at the hillock position, during a time interval greater than several seconds. In other words, *if* after the bombardment of the surface (at 25 °C), the local temperature then increases above the glass transition, it remains at this temperature for a *shorter timescale than is required for large-scale movement of macromolecules;* otherwise, the hillocks would not be observed. This is true, given that the relaxation times for large-scale chain movements of PMMA after mechanical deformation is on the order of seconds to hours depending on the temperature, while the increased temperature at the site of the bombardment is increased for approximately 300 ps as predicted by thermal spike models.[88,89] These results are consistent with the topography observed during SF_5^+ bombardment, where the initial topography features in the PMMA were shown to be broader at higher temperatures.[15]

5.6.4 Understanding the Role of Beam Energy During Organic Depth Profiling

Beam energy plays a critical role in organic depth profiling experiments, as it affects both the physical and chemical sputtering processes that occur. Similar to what is observed when employing atomic primary ion species at variable energies, increasing cluster ion beam energies will typically result in corresponding increases in sputtering yields.[39,20,26,33,44,90−93] Figure 5.29a shows the sputtered depth into a leucine molecular film plotted as a function of ion fluence for a massive gas cluster ion beam, $Ar_{2000}{}^{+}$,[‡] operated at four different acceleration energies.[94] All four curves are linear, indicating that very little damage is occurring over the range of fluence used during the experiment and that the sputtering yield is constant with increasing fluence. Furthermore, the slopes increase with increasing beam energy, as a result of the increased sputter yield volumes.

The corresponding sputtering yields are plotted as a function of ion beam energy in Figure 5.29b. Once again, a linear dependence is observed. Also shown in the figure are the sputtering yields obtained for Au. Note the dramatic increase in the sputtering yields of organic leucine molecules as compared to inorganic Au atoms; a common occurrence when employing cluster sources (see Chapter 4).

In addition to the increased sputtering yields observed at higher beam energies, there is also an increase in the maximum achievable erosion depth obtained before signal decay is observed in SIMS as a result of the gel effect.[25,26,28] Figure 5.30 represents the first observation of this phenomenon, where the authors were attempting to better understand the limits of cluster SIMS depth profiling in bulk PMMA films and its relationship to beam energy.[25] The figure shows three representative depth profiles acquired from a bulk sample of PMMA, using $C_{60}{}^{n+}$ ions at beam energies of 10 keV (green), 20 keV (blue), and 40 keV (red). The signal monitored at m/z 185 is characteristic of two PMMA monomer units minus a methyl group $[2n − CH_3{}^{-}]$ and is normalized to the steady-state intensity just before signal decay occurs. The results indicate that signal loss occurs much later with higher energy beams. The same effects were also observed with $SF_5{}^{+}$, indicating that this is not merely a C-deposition effect.[26] These results can be explained in terms of sputter yield, whereby a greater removal of material is achieved before the critical fluence, I_{crit}, is reached. However, while the sputtering yield is certainly expected to play an important role here, this is not necessarily the only parameter, as will become clear in the ensuing discussions.

Although higher beam energies can be useful for the characterization of thicker layers, there is actually increased structural damage imparted to the sample when higher energy ion beams are employed. The topography formation increases with beam energy and a corresponding decay in the depth resolution

---

[‡]Any reference to the number of Ar atoms in a massive gas cluster is represented as an average value with standard deviations defined in the corresponding references cited. For example, $Ar_{2000}{}^{+}$ refers to an Ar gas cluster consisting of an average of 2000 Ar atoms per ion.

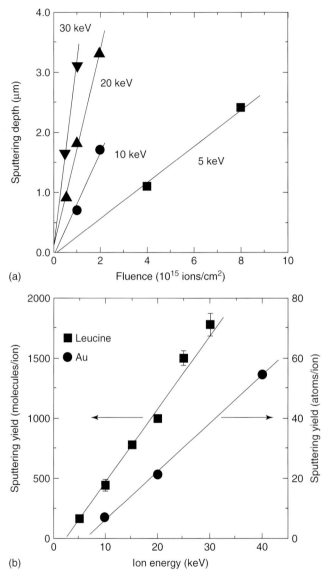

Figure 5.29   Effect of ion beam energy on the sputtering yield in thick (3 μm) leucine films evaporated on Si. Experiments performed using a GCIB source with an average Ar cluster size of 2000 (Ar_{2000}^{+}): (a) sputtered depth plotted as a function of beam fluence, at four different beam energies; (b) sputter yields as a function of beam energy; (■) leucine sputter yield, (●) Au sputter yield (shown for comparison purposes). Reproduced from Ichiki et al.,[94] with permission from Elsevier.

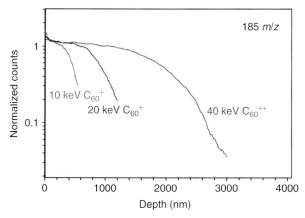

Figure 5.30 Negative ToF-SIMS depth profiles of bulk PMMA generated using 10 keV (green), 20 keV (blue), and 40 keV C_{60}^{n+} (red) primary ions for both sputtering and acquisition. All profiles show the normalized intensity of m/z 185 (characteristic of PMMA) as a function of depth. Reproduced from Fisher et al.,[25] with permission from Elsevier.

is observed.[39,11,23,25,28,33,81,85] Furthermore, increased fragmentation and beam damage effects (initial signal decay/damage cross sections) are observed in the mass spectral data at higher beam energies.[39,23,26,28,33,95−97] Finally, XPS results are also consistent with increased structural damage at higher beam energies.[25,26] Therefore, if surface or subsurface features are of interest, thin films are being analyzed, or if high depth resolution is a requirement, lower beam energies are always preferred.

Figure 5.31 illustrates the effect of beam energy on the depth resolution on the NPL reference sample introduced earlier.[85] The depth resolution very clearly decays with increasing beam energy. However, similar to the results obtained from the PMMA samples, the fourth layer in the standard is only resolved when employing higher ion beam energies. The depth resolution of the first layer was found to be linearly dependent on the cube root of the sputter yield volume that occurs at a particular energy, which provides the characteristic length scale of the sputter crater from an individual ion impact (d). This linear relationship would be expected if the increase in depth resolution is strictly a sputter-induced topography effect.[85]

As was mentioned earlier, there is a significant increase in beam-induced topography formation observed at higher beam energies. This was observed for the PMMA sample introduced earlier in Figure 5.30, and is also observed for the Irganox standard. Figure 5.32, shows the root mean square roughness values (R_{rms}) measured from the Irganox sample, as a function of ion beam fluence at four different beam energies. As can be seen, the beam-induced roughening is much greater for 30 keV C_{60} bombardment as compared to 5 keV bombardment. In all cases, the roughness increased with fluence.

To summarize, it appears as though increased beam energies result in significant increases in sputter yield volumes, which allows for much more material

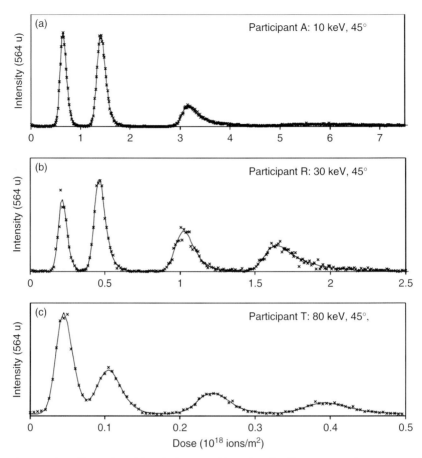

Figure 5.31 Examples of SIMS depth profiles using C_{60}^{n+} as a sputtering source, where the intensity of the secondary ion at m/z 564, characteristic of Irganox 3114 is shown as a function of C_{60}^{n+} dose. The kinetic energy of C_{60}^{n+} ions are (a) 10, (b) 30, and (c) 80 keV. Data are shown as crosses and solid lines connecting points from the fit using the Dowsett function (see Reference 85 for more information regarding fitting parameters). Reproduced from Shard et al.,[85] with permission from Wiley.

to be removed before the yield of chemical cross-links reaches its critical value for gel formation. On the other hand, there is significant evidence of increased ion beam-induced damage at higher primary ion beam energies. The material remaining in the crater bottom after sputtering at higher beam energies is highly fragmented and shows increased damage to the chemical structure. Furthermore, the depth resolution decreases with increasing beam energy, as a direct result of increased sputter-induced topography formation. Wucher et al.[39] also report significant increases in the damage cross sections when employing higher energy C_{60} primary ions in trehalose films. The authors determined that increased beam energies deposit their energy deeper into the sample, creating a thicker altered layer (d).

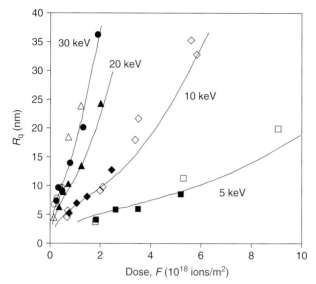

Figure 5.32 The development of root mean square roughness, R_q, with ion dose for four different C_{60}^{n+} beam energies. The filled symbols represent values calculated from SIMS data (see Reference 28 for more information): (■) 5 keV, (◆) 10 keV, (△) 20 keV, and (●) 30 keV. Open symbols represent roughness values measured using atomic force microscopy (AFM): (□) 5 kev, (■) 10 keV, (△) 20 keV, and (○) 30 keV. Solid curves are intended as guides to the eye. Reproduced from Shard et al.,[28] with permission from ACS publications.

It is very possible that although there is increased beam-induced damage per primary ion impact occurring at higher beam energies and an increased altered layer thickness, there is also a corresponding decrease in the cross-link density of the damaged layer. This is definitely consistent with Fishers' findings, showing evidence of significant degradation in the crater bottoms of polymers bombarded at higher beam energies.[25] It is expected that at lower beam energies, even though the altered layer thickness is diminished, there is also a corresponding increase in the cross-link density associated with this region, yielding more rapid gel formation, particularly in polymer samples and other larger molecular samples that are more prone to cross-linking.

5.6.5 Optimization of Incidence Angle

It has been demonstrated that glancing ion beam angles are the most ideal for organic and polymeric depth profiling with C_{60}^{n+} beams, where the resulting depth profiles acquired with both XPS and SIMS exhibit evidence of decreased beam-induced chemical damage effects at glancing ion beam incidence.[11,22,24,33,85,98,99] More specifically, the chemistry of the structure is preserved at glancing angles and the steady-state region of the profile exhibits an increased signal constancy, consistent with a more uniform sputtering yield.[11,22,24,33,85,98,99] This improvement has been attributed to the increased surface localization of the damage created by ion beams impinging at glancing incidence angles, combined with the direction of the energy deposition, which is directed away from the bulk. The net result is a decrease in the altered layer thickness, d.[22]

Molecular dynamics simulations studying the effect of beam angle on C_{60} bombardment of benzene molecules are consistent with this, showing that at normal incidence, all of the energy is deposited deeper into the substrate (within 2.5 nm), while at more grazing incidence angles, only 25% of the projectile energy is deposited into the sample, affecting only the first 1.5 nm.[100]

One of the first demonstrations of improved depth profiling at glancing incidence angles is depicted in Figure 5.33, which shows molecular depth profiling in a thin cholesterol film (~300 nm) using 40 keV C_{60} operated at two different beam angles.[22] The depth profile in Figure 5.33a was acquired at 25° with respect to the sample normal, while the depth profile in Figure 5.33b was acquired at a grazing incidence angle (73°). The change in angle results in significant changes to the depth profile characteristics, as is clearly observed in the Figure. More specifically, at grazing incidence angles, there is a significant improvement in the quality of the depth profile obtained in terms of the initial signal loss (ratio of signal in the steady-state region, region II, of the depth profile to the initial intensity observed in region I), signal constancy in the steady-state region, and interface widths.

Even though there are significant enhancements observed at glancing incidence angles with C_{60}, there are also decreased sputtering yields observed. Furthermore, an elongated beam shape is associated with odd angles and it is not yet confirmed as to whether this will work for all ion beams. In the example shown (Fig. 5.33), only two beam angles are represented. The original work also displayed results for other beam angles, specifically 5°, 25°, 40°, 57°, and 73° with respect to the surface normal. The sputtering yield was found to first increase, and then decrease with beam angle. The two angles shown in Figure 5.33 were selected because they were determined to have comparable sputter rates and damage cross sections, meaning that the differences observed are associated only with the altered layer thickness, d.

For depth profiling in materials that do not work under conventional beam conditions, Type I polymers for example, glancing beam angles offer a significant advantage. Figure 5.34 illustrates this by showing depth profile attempts in a thin PS film, a material that was previously thought to be impossible to depth profile due to its cross-linking nature. In Figure 5.34a, fragment molecules consistent with the PS structure at m/z 91 are monitored as a function of depth at 48° and 76° incidence angles. While the signal rapidly diminishes at 48°, the signal remains constant with depth in the case of the more grazing 76° angle. These differences were also observed in PC, which also has been shown to be a poor performer in terms of polymer depth profiling under normal conditions.[iv]

A direct comparison of the mass spectral data acquired before and after sputtering with 20 keV C_{60}^+ is shown in Figure 5.34b and c. The spectrum acquired from the pristine undamaged PS surface is shown in Figure 5.34b, where all of the characteristic PS peaks are observed. After sputtering with C_{60}^+ primary ions

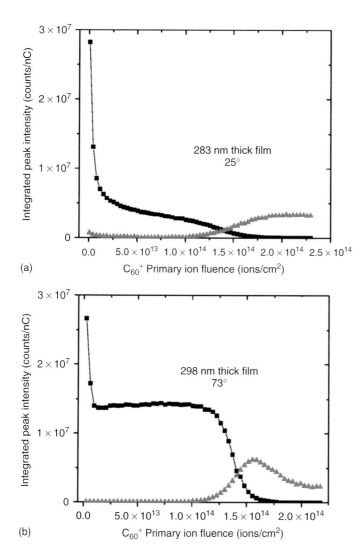

Figure 5.33 Integrated peak intensity of the cholesterol quasi-molecular ion at m/z 369 $[M - OH]^+$ (black squares) and Si at m/z 28 (green triangles) plotted as a function of 40 keV C_{60}^+ primary ion fluence; (a) acquired at $25°$ incidence angle with respect to the surface normal and (b) acquired at $73°$ incidence angle with respect to the surface normal. The Si intensity is multiplied by a factor of 10 in each plot. Reproduced from Kozole et al.,[22] with permission from ACS publications.

at $76°$ incidence angle, these characteristic peaks remain, indicating that there is no damage accumulating in the sample. In Figure 5.34d, however, all the mass spectral peaks characteristic of the PS structure are diminished and replaced by a series of C-rich peaks. This could be attributed to a graphitization process, although these peaks may also just be a result of C-deposition. The end result however is clear.

Grazing incidence angles are superior for depth profile analysis in cross-linking materials when employing C_{60}^+ beams.

This result also gives us insight into the mechanism of cross-linking during cluster ion bombardment of polymers. When a cluster ion impacts the surface, as compared to atomic ion bombardment, a significant amount of main-chain scission occurs, even in cross-linking polymers. This is evident, simply by the enhanced sputtering rates of PS observed with C_{60}^+ in comparison with Ga^+ ($\sim$200 times faster for C_{60}).[101] However, in contrast to degrading polymers, Type I polymers leave behind an overwhelming amount of cross-linking damage in the crater bottom. This highly cross-linked layer prevents the detection of any recognizable molecular species. We have shown that when employing increased temperatures, one can promote degradation in polymers. Similar local effects are likely to occur when the energy density is high, such as is the case with cluster ion beams, which cause a local thermal spike. However, in the subsurface layers, where there is a lower energy density (the regions surrounding the crater impact, for example), cross-linking mechanisms will be more prominent, particularly in Type I polymers. The energy density in these surrounding regions will be elevated enough to increase the rate of cross-linking damage but will not be high enough to cause depolymermization and/or enhanced sputtering effect that is inherent in the thermal spike region of the crater. This cross-linked layer will obviously be thicker in samples analyzed at more normal beam angles, where the energy is directed toward the bulk of the material.

The effect of beam angle on depth profiling in the Irganox multilayer reference material was also studied recently[11,85] and is described in more detail in the following section (Fig. 5.35). Recall that the FWHM of each of the peaks in the Irganox depth profile under different sputter conditions is plotted in Figure 5.24, where the grazing incidence angle data is plotted in comparison to other conditions. Note that the data acquired at a more grazing incidence angle of 76° ($\square$) are actually among the lowest values obtained in the study, only slightly higher than the values obtained at low temperatures. It is certainly a considerable improvement over the data taken at 45° $\bigcirc$), in which a dramatic decay of the depth resolution is observed with depth.

5.6.6 Effect of Sample Rotation

It is already well known that reducing the sputter-induced topography results in significant improvements for inorganic depth profiling.[102] The same applies for organic depth profiling.[11,17,85,103,104] When revisiting Figure 5.24b, for example, the FWHM obtained when employing sample rotation is represented ($\triangle$). When comparing this to the values obtained without sample rotation ($\bigcirc$), one can clearly see the improvement in the depth resolution, particularly at greater erosion depths, where the sample topography really starts accumulating under normal sputtering conditions.

Although there is a considerable improvement observed over normal conditions, the values for depth resolution are still not quite as low as what is obtained when employing cryogenic temperatures or at grazing beam incidence angles. This

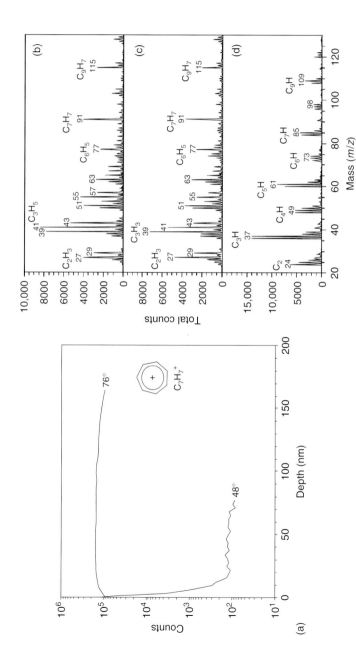

Figure 5.34 Depth profiling in a PS thin film using 20 keV C_{60}^+ primary ions for sputtering and analysis. (a) Signal intensity characteristic of PS ($C_7H_7^+$, m/z 91) monitored as a function of increasing sputtered depth at two different beam angles with respect to the surface normal, and (b, c, d) positive ion mass spectra acquired before (b) and after (c, d) sputtering with C_{60}^+ to a dose of 2×10^{14} ions/cm^2. The incidence angle of the primary ion beam was $76°$ in (c), and $48°$ in (d). Reproduced from Iida et al.,[24] with permission from Wiley.

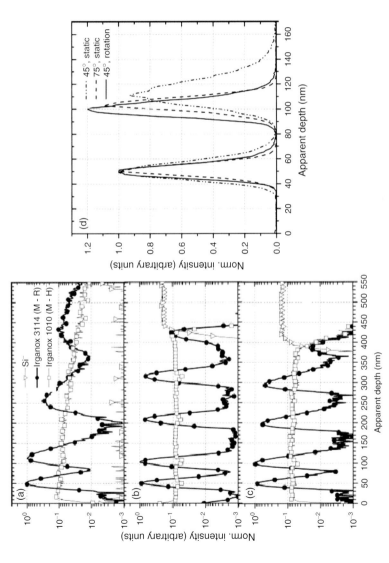

Figure 5.35 Depth profiles obtained in dual-beam mode under different conditions. In all cases, the sample was sputtered with 20 keV C_{60}^{+}, while the analysis was performed using a 25 keV Bi_{3}^{+} ion beam. (a) 45° incident angle, no sample rotation; (b) 75° incident angle, no sample rotation; (c) 45° incident angle, with sample rotation at 14 Hz; and (d) overlay of three profiles acquired from the first two layers (plotted on a linear scale). All profiles shown are depth calibrated and scaled to the first layer. Reproduced from Rading et al.,[104] with permission from Wiley.

result can be misleading, however, as depth resolutions are only one measure of success in organic depth profiling.

Figure 5.35 shows the effect of sample rotation on depth profiling in Irganox delta layers and compares this with the results obtained at glancing beam incidence, and, of course, to that obtained under normal beam conditions.[iv] Figure 5.35a shows the depth profile obtained using 20 keV C_{60}^+ at $45°$ incidence and no rotation. Note that the results are displayed on a log scale so that the differences in depth resolution are readily observed. Furthermore, note that the vertical scaling of the graphs shows a grid line every 50 nm and is therefore representative of the spacing of the Irganox 3114 layers in this example (Fig. 5.24a), which is 50 nm for the first two layers and 100 nm for the last two layers. The sputter rates used to calibrate the depth scale were based solely on the sputter rate of the first layer. Therefore, it should be easy to measure the linearity of the sputter rates, as the peaks will be located right over these grid lines when the sputter rates are constant. As can be seen, under normal conditions (Fig. 5.35), this is certainly not the case, and the sputter rate declines with depth. Furthermore, the depth profile experiences severe degradation in depth resolution with increasing sputtered depth into the sample.

Figure 5.35b shows the depth profile acquired at a more glancing incidence angle ($75°$ with respect to the surface normal). A significant improvement in the profile is observed, with particular reference to the depth resolution. Furthermore, the sputtering rates are relatively constant in comparison to the $45°$ profile. However, note that there is still some nonlinearity of the sputtering rate as the peaks start drifting to the right of the grid lines, thus indicating that the sputter yields are slightly decreasing with dose.

In Figure 5.35c, the depth profile is acquired using an incidence angle of $45°$, but with 14 Hz sample rotation. As can be seen, there is a significant improvement in the FWHM for each layer with respect to that observed with no sample rotation (also at $45°$ incidence) in Figure 5.35a. The FWHM is not as low as was observed using the glancing incidence angle in Figure 5.35b. However, note that the sputter rates are indeed linear throughout all layers when employing sample rotation, unlike what was observed at glancing incidence, indicating that there is decreased accumulation of beam damage with more uniform sputtering.

Figure 5.35d shows an overlay of the first two layers plotted on a linear scale, where the differences between the profiles are readily observed. From these results, it can be summarized that employing sample rotation improves the sputter properties, yielding significant improvements in depth resolutions and sputter yield linearity relative to normal sputtering conditions. On the other hand, the depth resolution is much better when employing grazing incidence angles for depth profiling as opposed to sample rotation. Despite this improvement in depth resolution, however, there is increased damage accumulation when employing grazing incidence angles relative to what is observed when employing sample rotation, causing nonlinear sputtering. It is likely that glancing angles therefore will approach the catastrophic failure region of the profile at a faster rate than samples analyzed at $45°$ incidence with sample rotation, although no evidence has been published yet to substantiate this hypothesis.

The effects of beam energy and sample rotation employed simultaneously were also examined by Rading et al.,[104] where it was found that samples analyzed at lower beam energies with sample rotation provide further improvements in the depth resolution as compared to higher beam energies with sample rotation.

It is expected that the reason for this improvement during sample rotation is similar to that observed for inorganic materials. When employing sample rotation, there will be an increase in the uniformity of the sputtering process that will end in a reduction in the sputter-induced topography formation. It is expected that uniform sputtering of organics will prevent the formation of highly cross-linked "nucleation" sites, where the damage becomes accelerated.

5.6.7 Ion Source Selection

The first thing to consider during any organic depth profiling experiment is which ion source should be used for sputtering and/or analysis of a particular material. Cluster SIMS has been rapidly expanding over the past decade to include several new sources including a high energy C_{60}^+ source, and the more recent development of GCIBs for SIMS applications. This latter example has been taking the organic depth profiling community by storm with its apparent ability to sputter through any material, independent of its chemistry. Although this source appears to be the most promising source to date for molecular depth profiling in organic materials, it is also very likely that over time, limitations will be presented. Here, we review and discuss the benefits and limitations of cluster sources commonly used for depth profiling applications. The discussion is limited to the most commonly employed cluster sources used for depth profiling applications, including C_{60}^+, SF_5^+, $Ar_{n>500}^+$, Bi_3^+, Au_3^+, and large water clusters (electrospray ionization sources). Low energy Cs^+ ions, although not a cluster source, are included in this discussion as they are also employed for depth profiling in cross-linking polymers.

5.6.7.1 SF_5^+ and Other Small Cluster Ions

SF_5^+ is one of the earlier cluster ion source designs and was among the first on the scene for organic depth profiling. However, there are many limitations to utilizing this source for organic depth profiling applications and it is not nearly as versatile as C_{60}^+ and/or GCIB have proven to be. One of the major shortcomings of SF_5^+ and other smaller cluster ion beams is that they tend to accumulate damage much more rapidly than other cluster ion beam sources.[11,12,16,18,85] The reasoning behind this is most likely attributed to its small size. First, there is higher impact energy per constituent atom in the cluster when the clusters are smaller. Secondly, and more importantly, smaller cluster ions tend to have a different sputtering mechanism entirely, exhibiting characteristics of both the collision cascade regime described for atomic ion bombardment and the thermal spike regime described for larger clusters. These factors result in a combination of decreased sputtering yield and increased rates of cross-linking and damage accumulation in organic materials.

The gel point is reached much more rapidly when employing smaller clusters, such as SF_5^+, and therefore they have only a limited utility for in-depth analysis of organic samples. This is readily observed by comparing the C_{60}^+ depth profile of the Irganox standard shown in Figure 5.24b to the SF_5^+ depth profile shown in

Figure 5.36, in which only the first layer was observed before failure occurred. It should also be noted that in the case of SF_5^+, only lower mass secondary ions were observed (e.g., CN^- and CNO^- signals), and the larger molecular species monitored with C_{60} were not observed during the depth profiling experiment. Moreover, the results shown in Figure 5.36 were only obtained when the sample was maintained at -100 °C during the experiment. These increased damage accumulation affects with SF_5^+ as compared to C_{60}^+ have also been observed in PMMA samples[12,16] and PLA-based drug delivery systems.[18]

A particularly surprising result from a recent interlaboratory study appears to show that atomic ion bombardment is much more effective in resolving these layers than SF_5^+.[85] When employing Ar or Cs, for example, all four layers are resolved, whereas, in Figure 5.36, there is only a very broad feature that is indicative of the second layer. However, this is an effect of sputter rate nonlinearities. The initial sputtering yields for SF_5^+ in irganox are orders of magnitude higher than for atomic ion beams. Atomic ion sources, however, have more linear sputtering yields with dose, as the gel point is reached almost immediately and therefore remains low throughout the entire profile. On the contrary, the sputtering yields for SF_5^+ start out high and then decay rapidly after the gel point is reached, to a value comparable to that of conventional atomic sources.

5.6.7.2 C_{60}^{n+} and Similar Carbon Cluster Sources

C_{60}^+ is a much more versatile source than SF_5^+ and with its commercial introduction in the early part of the century[105], it rapidly expanded the field of organic depth profiling, expanding the utility of cluster sources for other surface analytical equipment, such as XPS.[106] Furthermore, it was apparent that these larger cluster sources were much more useful for sputter depth profiling than smaller cluster

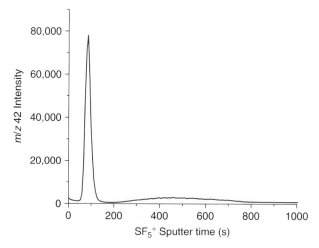

Figure 5.36   Depth profile attempt of Irganox delta layer standard, acquired using 8 keV SF_5^+ for sputtering, combined with 25 keV Bi_3^+ for analysis. The sample temperature was maintained at -100 °C during the experiment. Signal characteristic of Irganox 3114 (m/z 42, CNO^+) is shown as a function of SF_5^+ sputtering time.

sources, yielding significant enhancements in sputter yields and a definite improvement in the quality of the resulting depth profiles; even allowing for 3D analysis in more complex biological systems.[5,85,18,107−116] The vast improvement in the performance of C_{60} over smaller clusters is clearly demonstrated by comparison of Figure 5.24 and Figure 5.36 (depth profiling in Irganox standard using $C_{60}{}^+$ and $SF_5{}^+$, respectively). A further summary of the sputter parameters as reported in the literature (i.e., depth resolutions, range of sputter yield volumes, ability to etch cross-linking polymers, and spatial resolutions) for different ion sources is shown in Table 5.1, where it can be seen that $C_{60}{}^+$ is a vast improvement over $SF_5{}^+$. However, it should also be noted, that GCIB exhibits much improved conditions over C_{60} for molecular depth profiling.

Although $C_{60}{}^+$ and similar carbon-based cluster sources, such as coronene, perform reasonably well for organic depth profiling in many samples, there are still several limitations inherent in the source. First, the depth profile capability in certain materials, such as cross-linking polymers, remains a difficulty and requires specialized conditions for optimum results (e.g., grazing incidence angles, chamber backfilling, etc.). Second, even in materials that are amenable to being depth profiled with $C_{60}{}^+$, the signal is eventually lost in a similar manner as was described earlier (Fig. 5.21). Third, carbon deposition remains a problem for any sputter depth profile experiment performed at lower energies (<10 keV), where the best depth profiles are likely to be obtained.[46,117] This deposition is particularly prominent in Si, typically used as a substrate for many organic depth profiling experiments where strong Si–C bonds are formed.[118−121]

5.6.7.3 The Gas Cluster Ion Beam (GCIB)

So far, the most promising results for organic depth profiling have been obtained through use of GCIB.[19−21,23,122] GCIBs consist of hundreds to thousands of atoms or molecules of gaseous materials. The individual gas atoms are first condensed into neutral clusters, generated through cooling in a supersonic expansion (see Chapter 3). These neutral clusters of varying size are subsequently ionized by electron bombardment and then accelerated.[123−126] The chemistry of these beams is variable and

TABLE 5.1 Summary of Sputter Parameters Using Various Ion Beam Sources

Cluster Source	Maximum Erosion Depth in PMMA at 25 °C	Cross-Linking Polymers?	Spatial Resolution, μm	Depth Resolution in Irganox Standard under Optimized Conditions, nm	Sputter Yield Volumes in Irganox, nm^3/ion
$SF_5{}^+$	~100 nm (5 keV)	No	3–5	11 (8 keV, −100 °C)	12 (8 keV, −100 °C)
$C60_n{}^+$	1.5–2 μm (40 keV)	Limited	3–5	~9 (10 keV, −80 °C)	75–300
$Ar_{(n>500)}{}^+$	>10 μm	Yes	>30	5 (2.5 keV, 14 Hx rotation)	75–300
Low energy Cs^+	Not reported	Yes	3–5	Not reported	0.0195–0.146

they can be comprised of thousands of atoms of Ar, O_2, CO_2, or SF_6 gases, to name a few.[124,127,128] The GCIB source was originally designed for processing of material surfaces,[124,129] but has recently been developed for SIMS applications,[126,130−134] where it has shown particular promise for the analysis of organic materials owing to extremely low damage sputtering processes.[19,21,23,93−97,122,132] Some particularly nice examples of what this source can do are demonstrated in Figure 5.37−5.39.

Figure 5.37 shows SIMS depth profiles acquired using Ar_{700}^{+} operated at 5.5 keV from thin PMMA (Fig. 5.37a) and PS (Fig. 5.37b) films.[19] While the results in Figure 5.37a are expected, owing to the degrading nature of PMMA (Section 5.5), the results in Figure 5.37b are extraordinary, and represent the first successful attempt to depth profile PS, a cross-linking polymer, with a cluster source. No variations in experimental conditions were required (no sample rotation, no glancing incidence angles, or low temperatures) and similar results have been obtained for other cross-linking polymers, such as PC and polyimide.[19,68] An even more exciting result is that there is no apparent depth limitation observed for PMMA or PS. As the literature stands at present, films as thick as 10 μm have been depth profiled without signal loss for both PMMA and PS.[135]

Direct comparisons of the GCIB source with C_{60} can be observed in Table 5.1, where it is clearly seen that the best depth resolutions, maximum erosion depths, and sputter rate linearities have indeed been obtained when employing the GCIB source. Further enhancements in sputter properties are observed when employing sample rotation.[135]

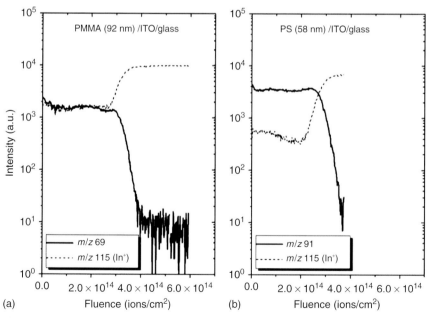

Figure 5.37 Variation of the characteristic secondary ion intensities for (a) PMMA and (b) PS films with incident fluence. Both etching and analysis were carried out using 5.5 keV Ar_{700}^{+} cluster ions. Reproduced from Ninomiya et al.,[19] with permission from Wiley.

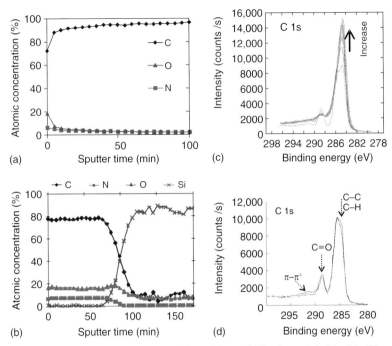

Figure 5.38 XPS depth profiles of C1s, O1s, and N1s for polyimide thin films (100 nm) using (a) 10 keV C_{60}^+ and (b) 10 keV Ar_{2500}^+. (c, d) Corresponding C1s high resolution spectrum acquired as a function of sputtering time using (c) 10 keV C_{60}^+ at varying sputter times; arrow indicates direction of increasing C_{60}^+ bombardment time; and (d) 5 keV Ar_{2500}^+ pristine surface (blue curve) versus after 10 min of sputtering with Ar_{2500}^+ (red curve). Reproduced from Miyayama et al.,[136] with permission from the American Vacuum Society.

Another example of successful depth profiling, where C_{60}^+ has previously failed is shown in Figure 5.38, which depicts XPS depth profiling in a polyimide sample using C_{60}^+ (Fig. 5.38a and b) versus Ar_{2500}^+ (Fig. 5.38c and d). The depth profile for the C_{60}^+ is shown in Figure 5.38a.[136] As can be seen, the C concentration increases while the N and O concentrations decrease as a function of increasing primary ion sputtering time, consistent with structural damage and/or C-deposition. Furthermore, the Si substrate is not reached during the time period of the experiment when employing C_{60}^+ as the sputter source. The corresponding high resolution XPS spectra are consistent with extensive beam-induced damage to the imide components. The C1s spectra as a function of increasing sputter time is displayed in Figure 5.38b. The carbonyl functionality diminishes with sputtering, with a corresponding increase in the adventitious carbon component, consistent with beam-induced chemical damage and/or carbon deposition. Similar chemical damage effects are evident in the N1s spectra (not shown here), indicating structural damage to the imide bond. This confirms that the damage occurring during the sputtering process is not purely a result of carbon deposition.

Figure 5.38c and d shows the corresponding depth profile and high resolution C1s spectra, respectively, for Ar_{2500}^+ analysis. As can be seen, when the GCIB is used, there is no detectable damage to the structure after sputtering, allowing for low damage depth profile analysis as compared to that obtained with C_{60}^+. The depth profile shows constant C, O, and N signals with increasing sputter time, and the entire film is eroded through, as indicated by the rise in Si signal with commensurate decay in the C, O, and N signals. Most impressive is the absolute lack of change in any of the high resolution spectra after sputtering with Ar_{2500}^+ (Fig. 5.38d), indicating there is no chemical damage occurring during the sputtering of the polyimide.

One of the most exciting applications of this source is its ability to depth profile organic electronics components—obtaining profiles that are as yet unattainable using the more conventional cluster ion beam sources. This is a particularly important area of research, as currently there are very few mechanisms for analyzing the molecular compositions at the interfaces in these types of devices. Figure 5.39a shows an example of depth profiling in organic light emitting diodes (OLEDs) using GCIB. The structures for the OLED materials depth profiled, tris(8-hydroxyquinolinato)aluminum (Alq$_3$) and N,N'-di-[(1-naphthyl)-N,N'-diphenyl]-1,1'-biphenyl)-4,4'-diamine (NPD), are given above the figure. While these materials have been particularly successful using the GCIB, they have been less than successful for molecular depth profiling with C_{60}^+ under normal conditions.[44] Figure 5.39b shows the sputtered depth as a function of dose for Alq3 and NPD with Ar_{700}^+ ions. The data for C_{60}^+ acquired from an earlier publication are also shown.[44] Clearly, the sputter yields are linear for the gas cluster ion beam source, while for the C_{60}^+, this is not the case.

A recent study reports on the performance of the GCIB for analysis of the Irganox delta layer[20] and even more recent investigations of this sample under optimized conditions[135] show that GCIB outperforms C_{60}^+ for molecular depth profiling in terms of linearity of sputtering yields and depth resolutions (Table 5.1). This improvement is observed even without any of the experimental modifications discussed above, where depth resolutions of approximately 7 nm have been reported.[135] With sample rotation and low beam energies (2.5 keV), values as low as 5 nm have been reported for depth resolutions in the first delta layer in the Irganox multilayer system, better than any previously reported value for depth resolution in organic depth profiling.[135]

Similar to organic depth profiling with other cluster beams, GCIB sources show considerable improvements when employing sample rotation and lower beam energies. In terms of beam angle, however, it has been shown that increasing the beam angle toward more glancing incidence angles actually results in increased fragmentation in the mass spectrum, contrary to what is observed with C_{60}.[137] This was attributed to more normal bombardment processes promoting the emission of intact ions toward the vacuum by limiting the collisions among molecules in the film. However, similar to what has been observed with C_{60}^+ depth profiling, glancing incidence angles did yield more constant secondary ion signals with increasing dose, results consistent with decreased damage remaining in the crater bottom. The effect of incidence angle on sputtering of inorganic materials has also been studied,

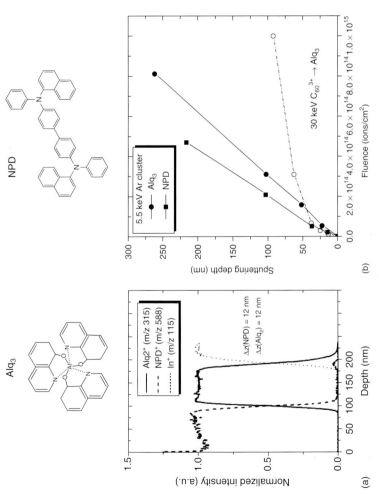

Figure 5.39 (a) Depth profiles of Alq_2^+, NPD^+, and In^+ secondary ions from a glass/ITO/Alq$_3$ (100 nm)/NPD/Alq$_3$ (100 nm)/NPS (100 nm) sample using Ar_{700}^{+125}. The depth resolution was estimated to be between 10 and 20 nm. (b) Fluence-dependent sputtering depth of Alq$_3$ (solid circles) and NPD (solid squares) for 5.5 keV Ar cluster ions (open circles)[21] and 30 keV C_{60}^+ in Alq_3^{+46}. Reproduced from Ninomiya et al.,[21,122] with permission from Wiley.

where it has been shown that glancing angles actually yielded increased topography formation as compared to normal incidence.[124] It is obvious that more research is required for a more thorough understanding of this process.

One of the most interesting phenomenon with gas cluster ion beams is the ability to extend the sputtering out beyond the so-called "sputter threshold," which has been defined as ~10 eV/nucleon.[20,95,134,137,138] This ability is not achievable using C_{60}^+ beams and has been attributed to different mechanistic aspects of sputtering that are observed when employing massive gas clusters as compared to C_{60} (see Chapter 2).[139] Beyond this value, the sputtering yield of material rapidly decays. However, the fragmentation is also significantly suppressed, allowing for extremely low damage sputtering. Enabling low energy per atom sputtering also allows for more controlled sputtering of species in the range of known bond dissociation energies.[95,140] Maintaining the kinetic energy per atom below the bond dissociation energies observed in organic materials (e.g., C–C, C–N, and C–O), which are typically ~3–5 eV/atom reduces fragmentation significantly.[140]

An example of the effect of energy per atom on the fragmentation and sputtering yield of molecular samples is illustrated in Figure 5.40. Figure 5.40a shows the intensity of fragment ions (■, □) and Na-cationized molecules (●, ○) of a small polypeptide consisting of six aspartic acid monomers $(Asp)_6$ plotted as a function of increasing energy per nucleon (increasing cluster size). Two different accelerating energies are plotted for comparison. As can be seen in both cases, below

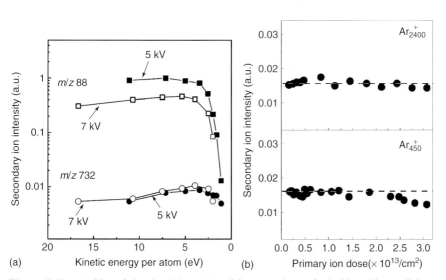

Figure 5.40   (a) Plot of the signal intensity of fragment ions of m/z 88 and intact [M + Na]$^+$ ions at m/z 732 in secondary ion mass spectra of small peptide chains, consisting of six monomer units of aspartic acid $(Asp)_6$ plotted as a function of kinetic energy per atom, at acceleration voltages of 5 and 7.5 kV; (b) plot of the signal intensity of m/z 732 as a function of primary ion dose for two different cluster sizes, at a constant acceleration voltage (10 kV). Reproduced from Mochiji et al.,[95] with permission from Wiley.

5 keV/atom, the sputter properties start to change rapidly, causing a significant decay in both fragment ion and molecular ion intensities. However, the fragment ion signal decreases at a much faster rate, indicating that decreased fragmentation occurs at these subthreshold levels. Therefore, it is ideal to perform analysis at these subthreshold levels, and indeed the most ideal results have been obtained at energies below 5 keV/atom for both depth profiling and analysis.[95,134,135] These results are surprisingly consistent with theoretical data based on molecular dynamics simulations (see Chapter 2).[138,139]

Figure 5.40b shows the corresponding intensity as a function of dose for two different Ar gas cluster sizes, Ar_{450}^+ and Ar_{2400}^+, both operated at 10 kV accelerating voltages. As can be seen in the case of Ar_{450}^+, there is evidence of damage accumulation observed at higher doses, whereby the intensity starts decreasing steadily. This confirms that for molecular depth profiling, larger gas cluster sizes are required for low damage sputtering processes.

The topography formation with different cluster sizes in PMMA has also recently been investigated, where it was shown that sputtering with larger cluster ions resulted in decreased topography formation.[93] The results from this study are shown in Figure 5.41, which plots the surface roughness as measured by AFM, as a function of dose for different gas cluster sizes ranging from 1000 to 16,000 Ar atoms. One can clearly see that the roughness increases with depth, and that the roughening occurs at a much faster rate when employing smaller clusters. However, one needs to balance the effect of low damage with the reduction in sputtering rates observed at lower energies.

It appears, overall, that larger clusters perform better for low damage sputtering applications. This is likely a result of the decreased energy per atom inherent in larger clusters. In order to directly ascertain the effect of cluster size (and not energy per atom) on the depth profiling capabilities of the gas cluster source, one needs to sputter at a constant energy per atom with different cluster sizes. In other words, the acceleration energy should be modified accordingly such that the energy per atom is the same for each cluster. Although no experimental data has

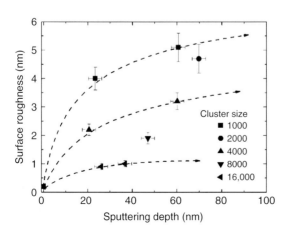

Figure 5.41 Surface roughness monitored as a function of sputtering depth into PMMA using 20 keV Ar gas cluster irradiation of varying sizes. Reproduced from Ichiki et al.,[93] with permission from Wiley.

been obtained to the knowledge of the authors, molecular dynamics simulations have been performed where the energy per atom was kept constant at 5 eV/atom, and the size of the cluster was varied. The calculations predict that the sputtering yield will increase with increasing cluster size, and that there is an almost linear dependence of the sputtering yield on the radius of the cluster.[139] Hence, it is likely that the increased sputtering yields observed during sputtering with smaller cluster sizes under identical accelerating voltages is purely a result of the changes in the energy per constituent atom in the cluster.

Another reason for the inherent success of GCIB for organic depth profiling, particularly in cross-linking materials, is that GCIB is able to efficiently remove highly damaged graphitic layers. A recent experiment studied the ability to remove ion beam damage layers formed with an Ar ion beam, using GCIB (Fig. 5.42).[68] In this study, the authors predamaged a PI surface with an Ar^+ ion beam operated at varying beam energies up to a dose of $\sim 1 \times 10^{16}$ ions/cm^2, thus creating a thick graphitic-like overlayer. In the XPS data shown in Figure 5.42, this creates an increase in C intensity with commensurate decays in N and O signals. Above this initial dose, a GCIB was used to characterize the sample. This damage was subsequently removed, and the chemical structure recovered with GCIB. This ability to remove predamaged regions in a cross-linking polymer has many implications toward the mechanism of depth profiling with GCIB, where damage that would usually result in catastrophic failure when employing conventional sources, is removed when employing GCIB. This means the gel point is not significant in these systems and may no longer be a limitation.

Overall, the GCIB is currently the most promising for use as a cluster source for sputter depth profiling, but is rather limited for imaging applications and high mass resolution analysis because of its extremely large spot size (Table 5.1). Therefore, until better designs are created, one should employ a dual-beam approach for depth profiling when high spatial resolution is required. Another possibility for high spatial resolution imaging with GCIB is employing a magnetic sector–based instrument in the microscope mode.

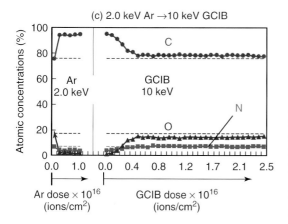

Figure 5.42   About 2 keV Ar^+ ion-induced damage and recovery with 10 keV GCIB, as measured using X-ray photoelectron spectroscopy (XPS). Recreated from Miyayama et al.,[68] with permission from Wiley.

GCIB in reference to SIMS applications is only in its infancy, and much more will likely come from this source as it evolves. Thus far, most of the work published about GCIB-SIMS has been on the analysis of organic samples. Although this is certainly an area where GCIB will likely make its greatest impact, there is still potential for analysis of inorganic materials, as it was originally developed for smoothing and chemical analysis of these surfaces.[124] For example, GCIB has been utilized to dope semiconductors with boron, by bombarding the surface with Ar gas clusters containing a known concentration of boron atoms.[124,129] Moreover, other gases have been employed for sputtering in these systems, where it was found that utilization of reactive gases such as O_2, SF_6, or CO_2 will significantly enhance sputtering rates and change topography formation processes in inorganic samples. Figure 5.43 plots the sputtering rates of W, Au, Si, and SiC under Ar atomic ion bombardment, Ar gas cluster ion bombardment, and SF_6 gas cluster bombardment. In all cases, SF_6 gas clusters yield a dramatic increase in the sputtering rate, relative to the Ar gas clusters.

There is a considerable amount of research to be done in order to discern the effect of different cluster compositions on depth profiling in organic samples. It is very likely that similar effects will be observed for organic materials, and as this is a rapidly expanding field, there are likely to be several publications on this topic, even before this book goes to print.

5.6.7.4 Low Energy Reactive Ion Beams

One method recently developed for improved depth profiling is the use of low energy Cs^+ primary ions for sputter depth profiling. Low energy Cs^+ is particularly useful for characterization of Type I polymers, or cross-linking polymers, which, until recently, have been more difficult to depth profile when using conventional cluster sources such as C_{60}.[141−145] The success of this ion beam for these cases has been attributed to the application of extremely low beam energies (<500 eV) in conjunction with the reactive nature of the Cs, which interacts with free radicals in the polymer and inhibits cross-linking.[144]

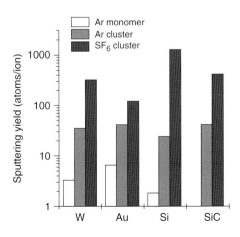

Figure 5.43 Reactive sputtering yield with SF_6 cluster ions as compared to Ar clusters for various materials at a 20 keV acceleration energy. Reproduced from Yamada et al.,[124] with permission from Elsevier.

Figure 5.44 shows a comparison of depth profiling in a PC sample using C_{60}^+ primary ions as compared to low energy Cs^+ ions, both bombarding the surface at $45°$ with respect to the surface normal.[145] When employing C_{60}^+ as a sputter source under normal conditions[§] (Fig. 5.44a), there is a gradual decay in signal with increasing depth into the film, consistent with increased cross-linking and sputter rate nonlinearities. However, when employing the low energy Cs^+ source for the same film (Fig. 5.44b), the molecular signal is retained throughout the depth profile, with a strong Si signal at the interface, unlike what was observed with C_{60}^+.

Although this source does show promise for depth profiling in certain cross-linking materials, it is not as useful for characterization of degrading polymers or organic samples, which perform much better under cluster bombardment. The differences are readily observed when directly comparing the depth profiles of the Irganox layers performed by the round-robin study.[85] First, the sputtering yields are significantly reduced when employing low energy Cs^+ beams (Table 5.1). Furthermore, the shapes of the peaks were not Gaussian in nature and exhibited significant structure in its decay; perhaps consistent with ion beam mixing. Finally, it is important to note that the profile of PC shown for C_{60}^+ in Figure 5.44a will improve greatly when employing glancing incidence angles or sample rotation. Therefore, it is not likely that low energy Cs^+ will be required for these types of analysis. Nonetheless, low energy Cs could be a viable option for those who do not have access to a cluster source or require a greater versatility with regard to inorganic depth profiling.

5.6.7.5 Electrospray Droplet Impact (EDI) Source for SIMS

More recently, an electrospray droplet impact (EDI) source has been developed for SIMS (and XPS), whereby charged droplets of 1 M acetic acid are electrosprayed directly into the main chamber vacuum, forming massive water clusters of the approximate structure, $[(H_2O)_{90,000} + 100H]^{100+}$.[146−148] The difference between the desorption electrospray ionization (DESI) process introduced in Chapter 4 and EDI-SIMS is that the former uses electrospray droplets to dissolve/ionize the samples under atmospheric pressure, while the latter uses the electrospray charged droplets accelerated in a vacuum and measures the secondary ions produced by a high momentum collision between the projectile and the solid sample, similar to what is done during conventional cluster ion beam sputtering.[149]

The EDI source has been shown to be particularly useful in obtaining low damage mass spectra similar to that observed with MALDI without the matrix requirement (see Chapter 4).[149] These large water droplets can also serve as a low damage sputter source for XPS and/or SIMS. EDI has proven to be capable of low damage etching in polymers such as poly(vinyl chloride) (PVC)[150], PS[151−153], PET[154], and PMMA[155], whereby the XPS and SIMS spectra were found to be independent of the irradiation time when employing EDI as a sputter source.[150,152,153,155]

An example of low damage etching as studied by XPS for PET is given in Figure 5.45, which shows the high resolution C1s spectra acquired from a PET

[§]Normal conditions are defined as $45°$ incidence, no sample rotation, and at room temperature.

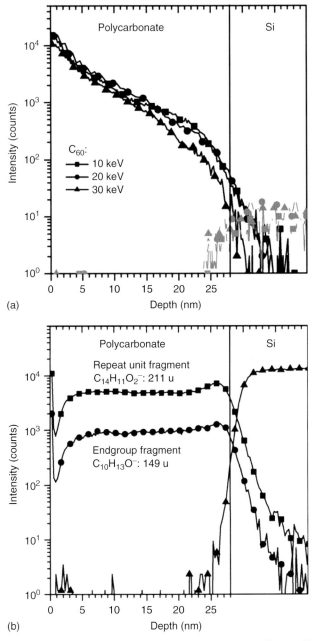

Figure 5.44 Depth profiling in a thin polycarbonate film on Si: (a) acquired using a C_{60} sputter source. The signal shown is representative of the repeat unit fragment characteristic of PC at m/z 211 ($C_{14}H_{11}O_2{}^-$) and (b) acquired using a low energy Cs^+ sputter source operated at 200 eV. Both experiments employ a 25 keV $Bi_3{}^+$ analysis source. Reproduced from Cramer et al.,[145] with permission from Elsevier.

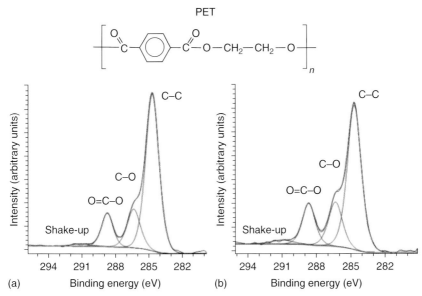

Figure 5.45 C1s chemical-state analysis of PET before (a), and after (b) 60 min etching by EDI. The depth after etching is approximately 12 nm. The structure of PET is indicated above the figure. Reproduced from Sakai et al.,[154] with permission from Wiley.

sample (100 nm) before (Fig. 5.45a) and after (Fig. 5.45b) etching with an EDI source for 60 min. The features of the C1s spectra before etching are consistent with the structure of PET, showing binding energies indicative of the presence of C–O bonds, O–C=O bonds, C–C–C bonds. Furthermore, a shake-up peak is observed, indicating the presence of aromatic rings. After etching, there is virtually no change in the XPS spectra when employing EDI.

EDI has also proven useful for low damage molecular depth profiling of cross-linking polymers with both XPS and SIMS. Figure 5.46 shows some examples of this where the authors employed XPS to monitor the atomic concentrations and fine structure of the C1s spectra acquired from a PS film as a function of increasing EDI etching time. In Figure 5.46a, depth profiling of a very thin layer of PS (8 nm) on Si is demonstrated. Similar depth profiles have also been demonstrated from slightly thicker PS films (~50 nm), employing both XPS and SIMS to acquire the profiles.[151,152]

Similarly, in Figure 5.46b, the ratio of the C1s intensity to the shake-up peak in the PS C1s spectra is monitored as a function of 1 keV Ar^+ sputtering depth and 10 keV EDI etching depth. The shake-up peak is indicative of the resonance in the benzene ring structure and therefore its presence after sputtering indicates that this structure is preserved and no cross-linking has occurred. As can be seen with the Ar^+ source, this ratio increases dramatically, showing that the resonance is being destroyed. However, when employing the EDI source, there is no change with depth, at least for the first 25 nm for which this experiment was performed.

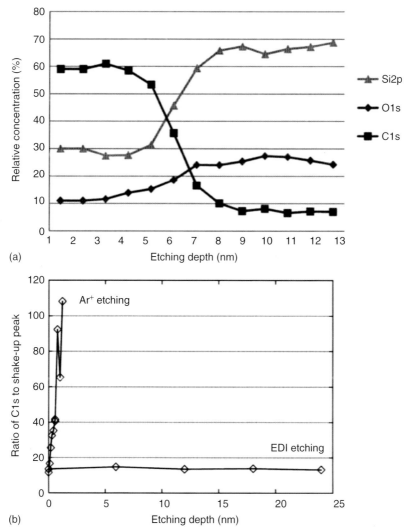

Figure 5.46 EDI-XPS depth profiling in PS. (a) XPS depth profile of a thin PS film (8 nm) and (b) intensity ratio of C1s peak to shake-up peak as a function of etching depth in PS bulk film (50 μm) with 1 keV Ar⁺ and 10 keV EDI. Reproduced from Sakai et al.,[153] with permission from the American Vacuum Society.

Although this source shows promise for low damage molecular depth profiling in organic materials, it is also obvious that this source needs some work if it is to be utilized for analysis of organic samples in lieu of GCIB. From Figure 5.45, we know that it takes 1 h to etch through only 12 nm of PET when using the EDI source, and it took 3 h to acquire the depth profile in Figure 5.46a. This is simply too long in comparison with most cluster ion sources, which can achieve this in a matter of seconds.

5.6.7.6 Liquid Metal Ion Gun Clusters (Bi_3^+ and Au_3^+)

Liquid metal ion gun (LMIG) clusters, such as Bi_n^+ and Au_n are proving to be superior candidates for molecular imaging applications, combining the submicrometer spatial resolution capabilities of the LMIG source with the enhanced molecular signals typical of cluster ion beams.

However, while these sources show great promise for molecular imaging applications, they are severely limited in their ability to be utilized as a sputter source for organic depth profiling, where they typically result in extensive damage to the sample relative to what is typically observed for other cluster sources, including SF_3^+, C_{60}^+, and $Ar_{n>500}^+$.

For example, the damage characteristics of PET under Au_3^+ and C_{60}^+ ion bombardment were recently studied in detail by Conlan et al.[156] In this study, the authors plotted the relative intensity of selected PET secondary ions as a function of increasing primary ion fluence. It was found that damage accumulated under Au_3^+ bombardment, but not under C_{60}^+ bombardment, which maintained relatively constant signals with increasing fluence. Similar results were found in other molecular samples, such as trehalose.[157]

Even when employing LMIG sources for analysis in dual-beam depth profiling, damage will eventually accumulate.[158−160] Therefore, one needs to use care, when employing dual-beam depth profiling approaches, to minimize the dose of the analysis beam. Figure 5.47 depicts depth profiling with 10 keV C_{60}^+, using different analysis beam conditions. The profiles are shown as a function of parameters R, and R', which can be defined as the fraction of the total dose that is attributed to the analysis beam (R) and a normalized version of this variable (R'), which accounts for sputter yield variations. These parameters are defined in Equation 5.25 and Equation 5.26 below.[159,160].

$$R = \frac{\varphi_{Bi}}{\varphi_{Bi} + \varphi_{C_{60}}} \tag{5.25}$$

$$R' = \frac{Y_{Bi}\varphi_{Bi}}{Y_{Bi}\varphi_{Bi} + Y_{C_{60}}\varphi_{C_{60}}} \tag{5.26}$$

where φ represents the per-cycle ion dose of the respective primary ion beam, for Bi_n^+ and C_{60}^{n+} primary ions as indicated by the subscripts, denoted φ_{Bi} and $\varphi_{C_{60}}$, respectively, and Y_{Bi} and $Y_{C_{60}}$ represent the corresponding sputtering yield of that material.

In Figure 5.47a, where Bi_1^+ ions are employed for analysis, one can see that the initial signal intensity drop observed increases with increasing R, indicating that there is an increased amount of damage occurring with increased dose of the analysis beam. When Bi_3^+ primary ions are used, as in Figure 5.47b, this initial signal intensity drop is not as prevalent. However, the signal starts to decay rapidly

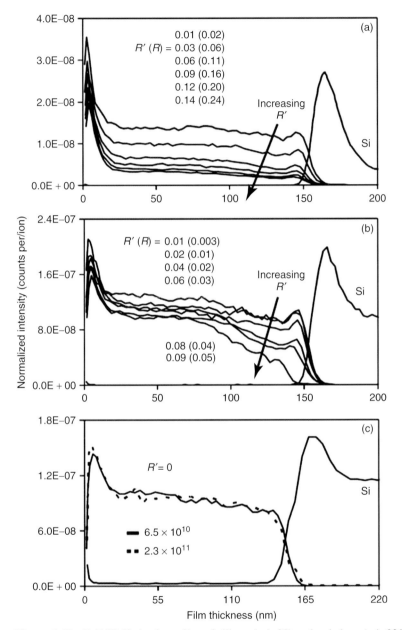

Figure 5.47 ToF-SIMS depth profiles of silicon (m/z 28) and trehalose (m/z 325, [M − OH]$^+$) acquired using different analysis beam conditions. (a) Bi_1^+ ions, (b) Bi_3^+ ions, and (c) C_{60}^+ ions with two different analysis doses as indicated. The sputter source was 10 keV C_{60}^+. In all cases, the profiles are plotted as a function of increasing R and R' values, which are defined in the text. Note that the silicon intensities have been scaled appropriately to fit the figure, and the R values are shown in parenthesis. Reproduced from Muramoto et al.,[160] with permission from Wiley.

under higher primary ion dose conditions, closer to the trehalose/Si interface. This premature decay in signal is likely resultant from an increased cross-link density when using Bi_3^+ as compared to Bi_1^+ ions. When employing C_{60}^+ for both sputtering and analysis (Fig. 5.47c), the depth profiles show the same shape under all R values investigated. One can conclude from this data that the choice of primary ion for analysis does indeed play a very important role in organic depth profiling, and that while LMIG cluster sources are ideal for analysis of molecular samples because of the high spatial resolution capability, there is still an extensive amount of beam damage associated with the LMIG source, whereby the type of damage observed varies according to the cluster size.

For optimum depth profiling, it is recommended that the R-value be maintained below 0.03, meaning that only 3% of the total dose should be attributed to the Bi ion beam.[158−160] Under these conditions, the characteristics of the depth profiles can be superior to single-beam depth profiling with C_{60}^+. When higher analysis doses are required, however, and spatial resolution is not needed, employing a single-beam approach whereby C_{60}^+ is used for both sputtering and analysis may be the best option.

5.6.8 C_{60}^+/Ar^+ Co-sputtering

Recent evidence suggests that co-sputtering with C_{60}^+ and other ion beams, such as Ar^+ can prevent the accumulation of damage associated with carbon deposition,[161] where it has been shown that C_{60}^+/Ar^+ co-sputtering can be used to extend the depth profile range and maintain a more constant sputter rate in polymer depth profiles when employing C_{60}^+ for sputter depth profiling. The theory is that utilizing a low energy atomic beam, such as Ar^+, in conjunction with the C_{60}^+ prevents the deposition of carbon, causing a significant improvement in the quality of the depth profile.

This approach has been very successful as can be seen from the example depicted in Figure 5.48, which shows representative XPS depth profiles acquired from an OLED device. The structures or the organic components comprising the device are indicated below the figure. The depth profile was acquired using a 10 keV C_{60}^+ primary ion source, operated at 10 nA continuous current, alternated with a 200 eV Ar^+ source, operated at higher currents (300 nA) in order to prevent carbon deposition.

As can be seen, the structures of the compounds are very similar in nature to what was profiled in Figure 5.39 with GCIB. These compounds are not capable of being etched using C_{60}^+ alone because of their tendency to cross-link. As can be seen in the figure, however, the in-depth structure of the device is readily observed when using Ar^+/C_{60}^+ co-sputtering. Note that the solid lines in the figure represent depth profiles acquired using Ar^+/C_{60}^+ co-sputtering, while the dashed lines represent depth profiles acquired using 10 keV C_{60}^+ alone. As can be seen, C_{60}^+ depth profile with no co-sputtering results in a catastrophic failure and no 2D structure is observed at all. This result supports the theory that C-deposition plays a very important role in the failure mechanisms of C_{60}^+ during ion bombardment of organics, although the high current, low energy Ar^+ beam may also

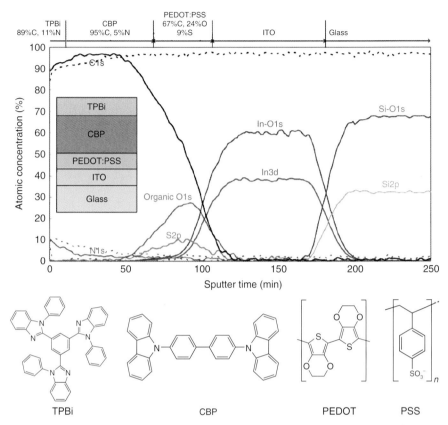

Figure 5.48 XPS depth profiles of organic LED device using 10 kV, 10 nA C_{60}^+ mixed with a 0.2 kV, 300 nA Ar^+ beam (solid lines) and 10 kV, 10 nA C_{60}^+ only (dashed lines.) The structure of the device is indicated above the figure and is illustrated in the inset. The chemical structures of each of the organic components is also indicated below the figure. The device was prepared by spin coating a layer of poly(ethylenedioxythiophene)–poly (styrene sulfonic acid) (PEDOT-PSS) onto 125 nm thick indium tin oxide (ITO) glass. This layer serves as the hole-transporting layer. A light-emissive layer, comprising of 4,4′ -bis(caraxol-9-yl)biphenyl (CBP), doped with Ir-containing dyes was spin coated on the PEDOT:PSS layer. Finally, an electron-transporting layer consisting of 2,2′,2″-(1,3,5-benzinetriyl)-tris(1-phenyl-1-H-benzimidazole) (TPBi) was spin coated on top of the CBP layer. Reproduced from Yu et al.,[161] with permission from ACS publications.

be more effective at removing cross-linking damage as well. Recent investigations by Garrison et al.[162] have suggested that the reason for the improvements observed with co-sputtering is a topography reduction, where it was shown that the Ar ion beam removes some of the ion beam–induced topography created by the C_{60}^+ ion beam. The mechanisms are not well understood at this point and further studies are required.

5.6.9 Chamber Backfilling with a Free Radical Inhibitor Gas

It is well known that some polymers which normally cross-link under irradiation can be degraded if irradiated in the presence of oxygen because the free radicals, which would otherwise lead to cross-linking, react with O_2 to form peroxide structures that eventually decompose and cause oxidative degradation of the main chain. Furthermore, the presence of O_2 serves as a free radical inhibitor that can potentially prevent cross-linking from occurring.

Although SIMS depth profiling in the presence of O_2 has been attempted in some polymers, such as PS, it has been not yet been shown to be effective at the low concentrations required to maintain high vacuum conditions.[163] Meanwhile, NO gas, another free radical inhibitor, has been shown to be very effective for depth profiling in polymers, allowing for depth profiling in PS and several other Type I cross-linking polymers with C_{60} ion beams.[163] The reason NO is successful, where O_2 is not, may be because NO is a more powerful free radical inhibitor, owing to its having a free electron in its native state. Clearly, there is a need for more research in this area, so that we can better understand the mechanisms involved. Although this work has been presented at conferences,[163] there are no currently available publications on this.

5.6.10 Other Considerations for Organic Depth Profiling Experiments

If possible, the electron flood gun fluence should always be limited during the depth profile experiment, as the electron beam will damage organic and polymeric materials under higher fluences. Gilmore et al.[164] defined a useful electron fluence limit of 6×10^{18} electrons/m^2 for organic and polymer samples. This may or may not apply for depth profiling applications, in which the effect of electron beam damage relative to ion beam damage; a study similar to that described in Figure 5.47 but with electron dose instead of primary ion dose, for example, has not yet been investigated. In addition, when employing cluster ion beams, oftentimes charge neutralization is not required.

At this point in time, there is very little evidence that the charge state of the primary ion has any significant effect on the damage processes in polymers. For example, Szymczak and Wittmaack[165] acquired positive and negative TOF mass spectra of Si sputter cleaned with SF_6^0, SF_5^-, SF_5^+, and Xe^+. Charge was found to play little, if any, role in the ion yields. More recent investigations use C_{60}^+, C_{60}^{2+}, and C_{60}^{3+} interchangeably for depth profiling applications in order to double or triple the acceleration energy of the beam.[39,25,44,85]

5.6.11   Molecular Depth Profiling: Novel Approaches and Methods

In recent years, several new techniques for surface mass spectrometry at atmospheric pressures have been developed, including atmospheric pressure matrix-assisted laser desorption ionization (AP-MALDI),[166] DESI and related methods, and plasma desorption ionization (PDI) methods. While DESI shows great promise for on-site, real-time mass spectral data acquisitions (see Chapter 4), there are very few examples of molecular depth profiling using DESI. One example was recently published by Green et al.,[167] which demonstrated that DESI was able to depth profile through a Rhodamine B film under optimized conditions.

PDI techniques, such as direct analysis in real time (DART), atmospheric pressure glow discharge ionization (APGDI), and low temperature plasma ionization (LTP) are also promising ambient pressure surface mass spectrometry techniques for characterization of polymeric materials. While methods employing plasmas do not extend to the analysis of larger biomolecules,[168] they show particular promise as a fragment ion depth profiling tool. Depth profiling in various polymers was recently demonstrated by Tuccito et al.,[169] for example, where the authors utilized pulsed radiofrequency glow discharge (RFGD) plasma time-of-flight mass spectrometry (ToF-MS) for depth profiling in thin films (300–400 nm) of PMMA, PS, PET, and PAMS. The results are promising, particularly for Type I polymers that tend to cross-link when irradiated (Fig. 5.49). Care must be taken of course to find the appropriate plasma conditions for low damage etching (i.e., plasma power and pressures).

5.7   CONCLUSIONS

Over the past twenty years, SIMS has come a long way toward realizing its goals for molecular depth profiling; starting early on with the in-depth characterization of thin amino acid films, moving toward more complex polymer coatings, and finally developing SIMS for 3D imaging in increasingly complex biological systems. Furthermore, quantitative depth profiling in molecular systems has been demonstrated.

We have also obtained a better understanding of the physical and chemical sputtering processes that occur during cluster ion bombardment of molecular samples, allowing one to avoid, or, at the very least, minimize the effects of beam-induced damage accumulation by employing optimized parameters. These parameters include sample temperature, ion beam angle, ion beam energy, ion beam size, ion beam composition, and variation of the ambient environment during the depth profile process. As a result, cluster SIMS has been proven to be useful for characterization of real-world samples, including pharmaceuticals, organic electronics, coatings and films, and complex cellular structures.

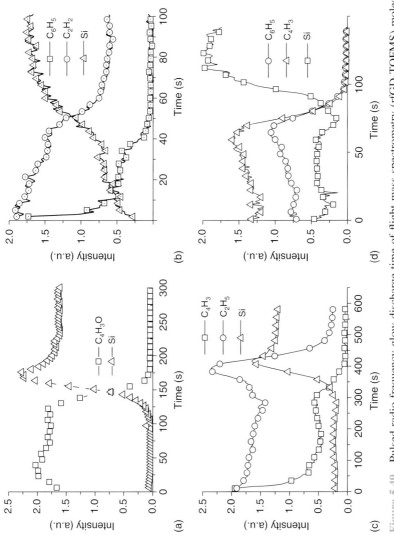

Figure 5.49 Pulsed radio frequency glow discharge time-of-flight mass spectrometry (rfGD-TOFMS) molecular depth profiles from thin films (300–400 nm) of (a) PMMA, (b) PS, (c) PET, and (d) PAMS. Reproduced from Tuccitto et al.,[169] reproduced with permission from Wiley.

Although we have come a long way, it is expected that further growth will occur over the next few decades, as new sources are developed and optimized, and additional parameters are defined. The newly developed GCIB source for SIMS, for example, is only in its infancy and therefore optimum parameters are only now being explored. In addition, there are several additional methods, including FIB-SIMS, PDI, DESI, and EDI-SIMS, which may also prove useful for molecular depth profiling applications.

This chapter has given an overview of the molecular depth profiling basics. We now move on to discuss the more advanced aspects of depth profiling in the following chapters. Much of the ensuing discussions, particularly in Chapters 6 and 8, will be equally important for molecular depth profiling applications.

REFERENCES

1. Gillen, G.; Roberson, S. *Rapid Commun. Mass Spectrom.* **1998**, 12, 1303–1312.
2. Gillen, G.; Simons, D. S.; Williams, P. *Anal. Chem.* **1990**, 62 (19), 2122–2130.
3. Wucher, A. *Surf. Interface Anal.* **2008**, 40 (12), 1545–1551.
4. Cheng, J.; Wucher, A.; Winograd, N. *J. Phys. Chem. B.* **2006**, 110 (16), 8329–8336.
5. Mahoney, C. M. *Mass Spectrom. Rev.* **2010**, 29 (2), 247–293.
6. Mahoney, C. M. *Surf. Interface Anal.* **2010**, 42, 1393–1401.
7. Cornett, D. S.; Lee, T. D.; Mahoney, J. F. *Rapid Commun. Mass Spectrom.* **1994**, 8 (12), 996–1000.
8. Brox, O.; Hellweg, S.; Benninghoven, A. Dynamic SIMS of Polymers? In *Proceedings of 12th International Conference on Secondary Ion Mass Spectrometry; Brussels, Belgium,* 2000; pp 777–780.
9. Mahoney, C. M.; Roberson, S.; Gillen, G. *Appl. Surf. Sci.* **2004**, 231–2, 174–178.
10. Mahoney, C. M.; Roberson, S. V.; Gillen, G. *Anal. Chem.* **2004**, 76 (11), 3199–3207.
11. Shard, A. G.; Goster, I. S.; Gilmore, I. S.; Lee, L. S.; Ray, S.; Yang, L. *Surf. Interface Anal.* **2011**, 43 (1–2), 510–513.
12. Mahoney, C. M.; Fahey, A. J.; Gillen, G.; Xu, C.; Batteas, J. D. *Appl. Surf. Sci.* **2006**, 252 (19), 6502–6505.
13. Mahoney, C. M.; Patwardhan, D. V.; McDermott, M. K. *Appl. Surf. Sci.* **2006**, 252 (19), 6554–6557.
14. Mahoney, C. M.; Fahey, A. J.; Gillen, G. *Anal. Chem.* **2007**, 79 (3), 828–836.
15. Mahoney, C. M.; Fahey, A. J.; Gillen, G.; Xu, C.; Batteas, J. D. *Anal. Chem.* **2007**, 79 (3), 837–845.
16. Moellers, R.; Tuccitto, N.; Torrisi, V.; Niehuis, E.; Licciardello, A. *Appl. Surf. Sci.* **2006**, 252 (19), 6509–6512.
17. Sjovall, P.; Rading, D.; Ray, S.; Yang, L.; Shard, A. G. *J. Phys. Chem. B.* **2010**, 114 (2), 769–774.
18. Fisher, G. L.; Belu, A. M.; Mahoney, C. M.; Wormuth, K.; Sanada, N. *Anal. Chem.* **2009**, 81 (24), 9930–9940.
19. Ninomiya, S.; Ichiki, K.; Yamada, H.; Nakata, Y.; Seki, T.; Aoki, T.; Matsuo, J. *Rapid Commun. Mass Spectrom.* **2009**, 23 (11), 1601.
20. Lee, J. L. S.; Ninomiya, S.; Matsuo, J.; Gilmore, I. S.; Seah, M. P.; Shard, A. G. *Anal. Chem.* **2009**, 82 (1), 98–105.
21. Ninomiya, S.; Ichiki, K.; Yamada, H.; Nakata, Y.; Seki, T.; Aoki, T.; Matsuo, J. *Rapid Commun. Mass Spectrom.* **2009**, 23 (20), 3264.
22. Kozole, J.; Wucher, A.; Winograd, N. *Anal. Chem.* **2008**, 80 (14), 5293–5301.
23. Ninomiya, S.; Ichiki, K.; Yamada, H.; Nakata, Y.; Seki, T.; Aoki, T.; Matsuo, J. *Surf. Interface Anal.* **2011**, 43 (1–2), 221–224.

24. Iida, S.; Miyayama, T.; Sanada, N.; Suzuki, M.; Fisher, G. L.; Bryan, S. R. *Surf. Interface Anal.* **2011**, 43 (1–2), 214–216.

25. Fisher, G. L.; Dickinson, M.; Bryan, S. R.; Moulder, J. *Appl. Surf. Sci.* **2008**, 255 (4), 819–823.

26. Mahoney, C. M.; Kushmerick, J. G.; Steffens, K. L. *J. Phys. Chem. C.* **2010**, 114 (34), 14510–14519.

27. Shard, A. G.; Green, F. M.; Gilmore, I. S. *Appl. Surf. Sci.* **2008**, 255 (4), 962–965.

28. Shard, A. G.; Green, F. M.; Brewer, P. J.; Seah, M. P.; Gilmore, I. S. *J. Phys. Chem. B.* **2008**, 112 (9), 2596–2605.

29. Belu, A.; Mahoney, C.; Wormuth, K. *J. Contr. Release* **2008**, 126 (2), 111–121.

30. Mahoney, C. M.; Fahey, A. J.; Belu, A. M. *Anal. Chem.* **2008**, 80 (3), 624–632.

31. Mahoney, C. M.; Fahey, A. J.; Belu, A. M. *J. Surf. Anal.* **2011**, 17 (3), 299–304.

32. Mahoney, C. M.; Yu, J.; Gardella, J. A., Jr. *Anal. Chem.* **2005**, 77 (11), 3570–3578.

33. Zheng, L.; Wucher, A.; Winograd, N. *Anal. Chem.* **2008**, 80 (19), 7363–7371.

34. Zheng, L.; Wucher, A.; Winograd, N. *J. Am. Soc. Mass Spectrom.* **2008**, 19 (1), 96–102.

35. Py, M.; Barnes, J. P.; Charbonneau, M.; Tiron, R.; Buckley, J. *Surf. Interface Anal.* **2011**, 43 (1–2), 179–182.

36. Mahoney, C. M.; Yu, J. X.; Fahey, A.; Gardella, J. A. *Appl. Surf. Sci.* **2006**, 252 (19), 6609–6614.

37. Wucher, A.; Cheng, J.; Winograd, N. *Anal. Chem.* **2007**, 79 (15), 5529–5539.

38. Jones, E. A.; Lockyer, N. P.; Vickerman, J. C. *Int. J. Mass Spectrom.* **2007**, 260 (2–3), 146–157.

39. Wucher, A.; Cheng, J.; Winograd, N. *J. Phys. Chem. C.* **2008**, 112 (42), 16550–16555.

40. Conlan, X. A.; Lockyer, N. P.; Vickerman, J. C. *Rapid Commun. Mass Spectrom.* **2006**, 20 (8), 1327–1334.

41. Winograd, N.; Postawa, Z.; Cheng, J.; Szakal, C.; Kozole, J.; Garrison, B. J. *Appl. Surf. Sci.* **2006**, 252 (19), 6836–6843.

42. Ho, P. S.; Lewis, J. E.; Howard, J. K. *J. Vac. Sci. Technol.* **1977**, 14 (1), 322–325.

43. Ho, P. S.; Lewis, J. E.; Chu, W. K. *Surf. Sci.* **1979**, 85 (1), 19–28.

44. Shard, A. G.; Brewer, P. J.; Green, F. M.; Gilmore, I. S. *Surf. Interface Anal.* **2007**, 39 (4), 294–298.

45. Benninghoven, A.; Rudenauer, F. G.; Werner, H. W. Secondary Ion Mass Spectrometry: Basic Concepts, Instrumental Aspects and Trends; Wiley: New York; **1987**.

46. Gillen, G.; Batteas, J.; Michaels, C. A.; Chi, P.; Small, J.; Windsor, E.; Fahey, A.; Verkouteren, J.; Kim, K. J. *Appl. Surf. Sci.* **2006**, 252 (19), 6521–6525.

47. Wucher, A. *Appl. Surf. Sci.* **2006**, 252 (19), 6482–6489.

48. Seah, M. P. *Surf. Interface Anal.* **2007**, 39 (7), 634–643.

49. Seah, M. P.; Gilmore, I. S. *Surf. Interface Anal.* **2011**, 43, 228–235.

50. Willingham, D.; Brenes, D. A.; Wucher, A.; Winograd, N. *J. Phys. Chem. C.* **2010**, 114 (12), 5391–5399.

51. Rabbani, S.; Barber, A. M.; Fletcher, J. S.; Lockyer, N. P.; Vickerman, J. C. *Anal. Chem.* **2011**, 83 (10), 3793–3800.

52. Ivanov, V. S. Radiation Chemistry of Polymers; VSP, Netherlands; **1992**.

53. Chapiro, A. *Radiat. Res. Suppl.* **1964**, 4, 179–191.

54. Licciardello, A.; Puglisi, O.; Calcagno, L.; Foti, G. *Nucl. Instrum. Meth. Phys. Res. B.* **1988**, 32 (1–4), 131–135.

55. Calcagno, L. *Nucl. Instrum. Meth. Phys. Res. B.* **1995**, 105 (1–4), 63–70.

56. Licciardello, A.; Fragal, M. E.; Foti, G.; Compagnini, G.; Puglisi, O. *Nucl. Instrum. Meth. Phys. Res. B.* **1996**, 116 (1–4), 168–172.

57. Koval, Y. *J. Vac. Sci. Technol. B.* **2004**, 22, 843.

58. Puglisi, O.; Licciardello, A.; Calcagno, L.; Foti, G. *Nucl. Instrum. Meth. Phys. Res. B.* **1987**, 19, 865–871.

59. Chilkoti, A.; Lopez, G. P.; Ratner, B. D.; Hearn, M. J.; Briggs, D. *Macromolecules* **1993**, 26 (18), 4825–4832.

60. Wagner, M. S.; Lenghaus, K.; Gillen, G.; Tarlov, M. J. *Appl. Surf. Sci.* **2006**, 253 (5), 2603–2610.

61. Calcagno, L.; Foti, G.; Licciardello, A.; Puglisi, O. *Appl. Phys. Lett.* **1987**, 51 (12), 907–909.

62. Licciardello, A.; Puglisi, O. *Nucl. Instrum. Meth. Phys. Res. B.* **1994**, 91 (1–4), 436–441.
63. Puglisi, O.; Licciardello, A.; Calcagno, L.; Foti, G. *J. Mat. Res.* **1988**, 3 (6), 1247–1252.
64. Calcagno, L.; Foti, G. *Nucl. Instrum. Meth. Phys. Res. B.* **1991**, 59, 1153–1158.
65. Licciardello, A.; Wenclawiak, B.; Boes, C.; Benninghoven, A. *Surf. Interface Anal.* **1994**, 22 (1–12), 528–531.
66. Puglisi, O.; Marletta, G.; Pignataro, S.; Foti, G.; Trovato, A.; Rimini, E. *Chem. Phys. Lett.* **1981**, 78 (2), 207–211.
67. Fuoco, E. R.; Gillen, G.; Wijesundara, M. B. J.; Wallace, W. E.; Hanley, L. *J. Phys. Chem. B.* **2001**, 105 (18), 3950–3956.
68. Miyayama, T.; Sanada, N.; Bryan, S. R.; Hammond, J. S.; Suzuki, M. *Surf. Interface Anal.* **2010**, 42 (9), 1453–1457.
69. Lazareva, I.; Koval, Y.; Alam, M.; Stromsdorfer, S.; Muller, P. *Appl. Phys. Lett.* **2009**, 90 (26), 262108.
70. Davenas, J.; Boiteux, G.; Xu, X. L.; Adem, E. *Nucl. Instrum. Meth. Phys. Res. B.* **1988**, 32 (1–4), 136–141.
71. Venkatesan, T.; Forrest, S. R.; Kaplan, M. L.; Murray, C. A.; Schmidt, P. H.; Wilkens, B. J. *J. Appl. Phys.* **2009**, 54 (6), 3150–3153.
72. Forrest, S. R.; Kaplan, M. L.; Schmidt, P. H.; Venkatesan, T.; Lovinger, A. J. *Appl. Phys. Lett.* **1982**, 41 (8), 708–710.
73. Venkatesan, T. *Nucl. Instrum. Meth. Phys. Res. B.* **1985**, 7, 461–467.
74. Davenas, J.; Thevenard, P.; Boiteux, G.; Fallavier, M.; Lu, X. L. *Nucl. Instrum. Meth. Phys. Res. B.* **1990**, 46 (1–4), 317 323.
75. Netcheva, S.; Bertrand, P. *Nucl. Instrum. Meth. Phys. Res. B.* **1999**, 151 (1–4), 129–134.
76. Marletta, G. *Nucl. Instrum. Meth. Phys. Res. B.* **1990**, 46 (1–4), 295–305.
77. Marletta, G.; Licciardello Lucia, A. *Nucl. Instrum. Meth. Phys. Res. B.* **1989**, 37, 712–715.
78. Marletta, G.; Pignataro, S.; Oliveri, C. *Nucl. Instrum. Meth. Phys. Res. B.* **1989**, 39 (1–4), 792–795.
79. Puglisi, O.; Licciardello, A.; Pignataro, S.; Calcagno, L.; Foti, G. *Radiat. Eff. Defects Solids* **1986**, 98 (1), 161–170.
80. Dowsett, M. G.; Rowlands, G.; Allen, P. N.; Barlow, R. D. *Surf. Interface Anal.* **1994**, 21 (5), 310–315.
81. Lu, C.; Wucher, A.; Winograd, N. *Anal. Chem.* **2010**, 83 (1), 351–358.
82. Zheng, L.; Wucher, A.; Winograd, N. *Appl. Surf. Sci.* **2008**, 255 (4), 816–818.
83. Piwowar, A.; Fletcher, J.; Lockyer, N.; Vickerman, J. *Surf. Interface Anal.* **2011**, 43 (1–2), 207–210.
84. Piwowar, A. M.; Fletcher, J. S.; Kordys, J.; Lockyer, N. P.; Winograd, N.; Vickerman, J. C. *Anal. Chem.* **2010**, 82 (19), 8291–8299.
85. Shard, A. G.; Ray, S.; Seah, M. P.; Yang, L. *Surf. Interface Anal.* **2011**, 43 (9), 1240–1250.
86. Fragal, M.; Compagnini, G.; Torrisi, L.; Puglisi, O. *Nucl. Instrum. Meth. Phys. Res. B.* **1998**, 141 (1–4), 169–173.
87. Wagner, M. S. *Anal. Chem.* **2004**, 76 (5), 1264–1272.
88. Papaleo, R. M.; de Oliveira, L. D.; Farenzena, L. S.; Livi, R. P. *Nucl. Instrum. Meth. Phys. Res. B.* **2001**, 185 (1–4), 55–60.
89. Papaleo, R. M.; Hasenkamp, W.; Barbosa, L. G.; Leal, R. *Nucl. Instrum. Meth. Phys. Res. B.* **2006**, 242 (1–2), 190–193.
90. Fletcher, J. S.; Conlan, X. A.; Lockyer, N. P.; Vickerman, J. C. *Appl. Surf. Sci.* **2006**, 252 (19), 6513–6516.
91. Wagner, M. S.; Gillen, G. *Appl. Surf. Sci.* **2004**, 231–2, 169–173.
92. Bolotin, I. L.; Tetzler, S. H.; Hanley, L. *Appl. Surf. Sci.* **2006**, 252 (19), 6533–6536.
93. Ichiki, K.; Ninomiya, S.; Nakata, Y.; Yamada, H.; Seki, T.; Aoki, T.; Matsuo, J. *Surf. Interface Anal.* **2010**, 43 (1–2), 120–122.
94. Ichiki, K.; Ninomiya, S.; Nakata, Y.; Honda, Y.; Seki, T.; Aoki, T.; Matsuo, J. *Appl. Surf. Sci.* **2008**, 255 (4), 1148–1150.
95. Mochiji, K.; Hashinokuchi, M.; Moritani, K.; Toyoda, N. *Rapid Commun. Mass Spectrom.* **2009**, 23 (5), 648–652.

96. Ninomiya, S.; Nakata, Y.; Honda, Y.; Ichiki, K.; Seki, T.; Aoki, T.; Matsuo, J. *Appl. Surf. Sci.* **2008**, 255 (4), 1588–1590.

97. Ninomiya, S.; Nakata, Y.; Ichiki, K.; Seki, T.; Aoki, T.; Matsuo, J. *Nucl. Instrum. Meth. Phys. Res. B.* **2007**, 256 (1), 493–496.

98. Feld, H.; Leute, A.; Rading, D.; Benninghoven, A.; Chiarelli, M. P.; Hercules, D. M. *Anal. Chem.* **1993**, 65 (15), 1947–1953.

99. Miyayama, T.; Sanada, N.; Iida, S.; Hammond, J. S.; Suzuki, M. *Appl. Surf. Sci.* **2008**, 255 (4), 951–953.

100. Ryan, K. E.; Smiley, E. J.; Winograd, N.; Garrison, B. J. *Appl. Surf. Sci.* **2008**, 255 (4), 844–846.

101. Nieuwjaer, N.; Poleunis, C.; Delcorte, A.; Bertrand, P. *Surf. Interface Anal.* **2009**, 41 (1), 6–10.

102. Zalar, A. *Thin Solid Films* **1985**, 124 (3–4), 223–230.

103. Garrison, B. J.; Postawa, Z. *Chem. Phys. Lett.* **2011**, 506, 129–134.

104. Rading, D.; Moellers, R.; Kollmer, F.; Paul, W.; Niehuis, E. *Surf. Interface Anal.* **2011**, 43 (1–2), 198–200.

105. Weibel, D.; Wong, S.; Lockyer, N.; Blenkinsopp, P.; Hill, R.; Vickerman, J. C. *Anal. Chem.* **2003**, 75 (7), 1754–1764.

106. Nobuta, T.; Ogawa, T. *J. Mater. Sci.* **2009**, 44 (7), 1800–1812.

107. Fletcher, J. S.; Vickerman, J. C. *Anal. Bioanal. Chem.* **2010**, 396 (1), 85–104.

108. Fletcher, J. S. *Analyst* **2009**, 134 (11), 2204–2215.

109. McDonnell, L. A.; Heeren, R. M. A. *Mass Spectrom. Rev.* **2007**, 26 (4), 606–643.

110. Solon, E. G.; Schweitzer, A.; Stoeckli, M.; Prideaux, B. *AAPS J.* **2010**, 12 (1), 11–26.

111. Breitenstein, D.; Rommel, C. E.; Mollers, R.; Wegener, J.; Hagenhoff, B. *Angew. Chem. Int. Ed.* **2007**, 46 (28), 5332–5335.

112. Jones, E. A.; Lockyer, N. P.; Vickerman, J. C. *Anal. Chem.* **2008**, 80 (6), 2125–2132.

113. Yamada, H.; Ichiki, K.; Nakata, Y.; Ninomiya, S.; Seki, T.; Aoki, T.; Matsuo, J. *Nucl. Instrum. Meth. Phys. Res. B.* **2010**, 268 (11–12), 1736–1740.

114. Fletcher, J. S.; Rabbani, S.; Henderson, A.; Blenkinsopp, P.; Thompson, S. P.; Lockyer, N. P.; Vickerman, J. C. *Anal. Chem.* **2008**, 80 (23), 9058–9064.

115. Fletcher, J. S.; Lockyer, N. P.; Vaidyanathan, S.; Vickerman, J. C. *Anal. Chem.* **2007**, 79 (6), 2199–2206.

116. Fletcher, J. S.; Lockyer, N. P.; Vickerman, J. C. *Mass Spectrom. Rev.* **2011**, 30 (1), 142–174.

117. Lee, J. L. S.; Seah, M. P.; Gilmore, I. S. *Appl. Surf. Sci.* **2008**, 255 (4), 934–937.

118. Krantzman, K. D.; Kingsbury, D. B.; Garrison, B. J. *Appl. Surf. Sci.* **2006**, 252 (19), 6463–6465.

119. Krantzman, K. D.; Kingsbury, D. B.; Garrison, B. J. *Nucl. Instrum. Meth. Phys. Res. B.* **2007**, 255 (1), 238–241.

120. Krantzman, K. D.; Garrison, B. J. *Surf. Interface Anal.* **2011**, 43, 123–125.

121. Krantzman, K. D.; Garrison, B. J. *Nucl. Instrum. Meth. Phys. Res. B.* **2009**, 267 (4), 652–655.

122. Ninomiya, S.; Ichiki, K.; Yamada, H.; Nakata, Y.; Seki, T.; Aoki, T.; Matsuo, J. *Surf. Interface Anal.* **2011**, 43 (1–2), 95–98.

123. Yamada, I.; Takaoka, G. H. *Jpn. J. App. Phys.* **1993**, 32, 2121–2141.

124. Yamada, I.; Matsuo, J.; Toyoda, N.; Kirkpatrick, A. *Mater. Sci. Eng. R Rep.* **2001**, 34 (6), 231–295.

125. Hagena, O. F. *Rev. Sci. Instrum.* **1992**, 63 (4), 2374–2379.

126. Seki, T.; Matsuo, J.; Takaoka, G. H.; Yamada, I. *Nucl. Instrum. Meth. Phys. Res. B.* **2003**, 206, 902–906.

127. Toyoda, N.; Kitani, H.; Matsuo, J.; Yamada, I. *Nucl. Instrum. Meth. Phys. Res. B.* **1997**, 121 (1–4), 484–488.

128. Toyoda, N.; Hagiwara, N.; Matsuo, J.; Yamada, I. *Nucl. Instrum. Meth. Phys. Res. B.* **1999**, 148 (1–4), 639–644.

129. Yamada, I.; Matsuo, J.; Insepov, Z.; Takeuchi, D.; Akizuki, M.; Toyoda, N. *J. Vac. Sci. Technol.* **1996**, 14 (3), 781–785.

130. Matsuo, J.; Okubo, C.; Seki, T.; Aoki, T.; Toyoda, N.; Yamada, I. *Nucl. Instrum. Meth. Phys. Res. B.* **2004**, 219, 463–467.

131. Toyoda, N.; Matsuo, J.; Aoki, T.; Yamada, I.; Fenner, D. B. *Nucl. Instrum. Meth. Phys. Res. B.* **2002**, 190 (1–4), 860–864.

132. Matsuo, J.; Ninomiya, S.; Nakata, Y.; Honda, Y.; Ichiki, K.; Seki, T.; Aoki, T. *Appl. Surf. Sci.* **2008**, 255 (4), 1235–1238.

133. Fenner, D. B.; Shao, Y. *J. Vac. Sci. Technol. A.* **2003**, 21, 47.

134. Moritani, K.; Hashinokuchi, M.; Nakagawa, J.; Kashiwagi, T.; Toyoda, N.; Mochiji, K. *Appl. Surf. Sci.* **2008**, 255 (4), 948–950.

135. Rading, D.; Moellers, R.; Cramer, H.-G.; Niehuis, E. *Surf. Interface Anal.* **2012**, 45 (1), 171–174.

136. Miyayama, T.; Sanada, N.; Suzuki, M.; Hammond, J. S.; Si, S. Q. D.; Takahara, A. *J. Vac. Sci. Technol. A.* **2010**, 28 (2), L1–L4.

137. Oshima, S.; Kashihara, I.; Moritani, K.; Inui, N.; Mochiji, K. *Rapid Commun. Mass Spectrom.* **2011**, 25 (8), 1070–1074.

138. Delcorte, A.; Garrison, B. J.; Hamraoui, K. *Surf. Interface Anal.* **2011**, 43 (1–2), 16–19.

139. Rzeznik, L.; Czerwinski, B.; Garrison, B. J.; Winograd, N.; Postawa, Z. *J. Phys. Chem. C.* **2008**, 112 (2), 521–531.

140. Moritani, K.; Houzumi, S.; Takeshima, K.; Toyoda, N.; Mochiji, K. *J. Phys. Chem. C.* **2008**, 112 (30), 11357–11362.

141. Mine, N.; Douhard, B.; Brison, J.; Houssiau, L. *Rapid Commun. Mass Spectrom.* **2007**, 21 (16), 2680–2684.

142. Houssiau, L.; Douhard, B.; Mine, N. *Appl. Surf. Sci.* **2008**, 255 (4), 970–972.

143. Houssiau, L.; Mine, N. *Surf. Interface Anal.* **2010**, 42 (8), 1402–1408.

144. Houssiau, L.; Mine, N. *Surf. Interface Anal.* **2011**, 43 (1–2), 146–150.

145. Cramer, H. G.; Grehl, T.; Kollmer, F.; Moellers, R.; Niehuis, E.; Rading, D. *Appl. Surf. Sci.* **2008**, 255 (4), 966–969.

146. Hiraoka, K.; Mori, K.; Asakawa, D. *J. Mass Spectrom.* **2006**, 41 (7), 894–902.

147. Hiraoka, K.; Asakawa, D.; Fujimaki, S.; Takamizawa, A.; Mori, K. *Eur. Phys. J.Atom. Mol. Opt. Phys.* **2006**, 38 (1), 225–229.

148. Asakawa, D.; Fujimaki, S.; Hashimoto, Y.; Mori, K.; Hiraoka, K. *Rapid Commun. Mass Spectrom.* **2007**, 21 (10), 1579–1586.

149. Asakawa, D.; Chen, L. C.; Hiraoka, K. *J. Mass Spectrom.* **2009**, 44 (6), 945–951.

150. Hiraoka, K.; Iijima, Y.; Sakai, Y. *Surf. Interface Anal.* **2011**, 43 (1–2), 236–240.

151. Sakai, Y.; Iijima, Y.; Mukou, S.; Hiraoka, K. *Surf. Interface Anal.* **2011**, 43 (1–2), 167–170.

152. Sakai, Y.; Iijima, Y.; Asakawa, D.; Hiraoka, K. *Surf. Interface Anal.* **2010**, 42 (6–7), 658–661.

153. Sakai, Y.; Iijima, Y.; Takaishi, R.; Asakawa, D.; Hiraoka, K. *J. Vac. Sci. Technol. A.* **2009**, 27, 743.

154. Sakai, Y.; Iijima, Y.; Mori, K.; Hiraoka, K. *Surf. Interface Anal.* **2008**, 40 (13), 1716–1718.

155. Hiraoka, K.; Takaishi, R.; Asakawa, D.; Sakai, Y.; Iijima, Y. *J. Vac. Sci. Technol. A.* **2009**, 27, 748.

156. Conlan, X. A.; Gilmore, I. S.; Henderson, A.; Lockyer, N. P.; Vickerman, J. C. *Appl. Surf. Sci.* **2006**, 252 (19), 6562–6565.

157. Cheng, J.; Kozole, J.; Hengstebeck, R.; Winograd, N. *J. Am. Soc. Mass Spectrom.* **2007**, 18 (3), 406–412.

158. Szakal, C.; Hues, S. M.; Bennett, J.; Gillen, G. *J. Phys. Chem. C.* **2009**, 114 (12), 5338–5343.

159. Brison, J.; Muramoto, S.; Castner, D. G. *J. Phys. Chem. C.* **2010**, 114 (12), 5565–5573.

160. Muramoto, S.; Brison, J.; Castner, D. G. *Surf. Interf. Anal.* **2011**, 43 (1–2), 58–61.

161. Yu, B. Y.; Chen, Y. Y.; Wang, W. B.; Hsu, M. F.; Tsai, S. P.; Lin, W. C.; Lin, Y. C.; Jou, J. H.; Chu, C. W.; Shyue, J. J. *Anal. Chem.* **2008**, 80 (9), 3412–3415.

162. Schiffer, Z. J.; Kennedy, P. E.; Postawa, Z.; Garrison, B. J. *J. Phys. Chem. Lett.* **2011**, 2, 2635–2638.

163. Licciardello, A. Depth Profiling in PS Using NO Backfilling; *Seventeenth International Conference on Secondary Ion Mass Spectrometry*; Toronto, Ontario, Canada, **2011**.

164. Gilmore, I. S.; Seah, M. P. *Appl. Surf. Sci.* **2011**, 187, 89–100.

165. Szymczak, W.; Wittmaack, K. *Nucl. Instrum. Meth. Phys. Res. B.* **1994**, 88 (1–2), 149–153.

166. Laiko, V. V.; Baldwin, M. A.; Burlingame, A. L. *Anal. Chem.* **2000**, 72 (4), 652–657.

167. Green, F. M.; Stokes, P.; Hopley, C.; Seah, M. P.; Gilmore, I. S.; O'Connor, G. *Anal. Chem.* **2009**, 81 (6), 2286–2293.

168. Weston, D. J. *Analyst* **2010**, 135 (4), 661–668.

169. Tuccitto, N.; Lobo, L.; Tempez, A.; Delfanti, I.; Chapon, P.; Canulescu, S.; Bordel, N.; Michler, J.; Licciardello, A. *Rapid Commun. Mass Spectrom.* **2009**, 23 (5), 549–556.

170. Mahoney, C. M.; Weidner, S. M. Surface Analysis and Imaging Techniques. In Mass Spectrometry in Polymer Chemistry; Barner-Kowollik, C., Gruendling, T., Falkenhagen, J., Weidner, S., Eds.; Wiley-VCH; **2011**; pp 149–208.

CHAPTER 6

THREE-DIMENSIONAL IMAGING WITH CLUSTER ION BEAMS

Andreas Wucher, Gregory L. Fisher, and Christine M. Mahoney

6.1 INTRODUCTION

One of the most challenging tasks in surface and thin-film technology is the ability to perform three-dimensional chemical characterization of surface and subsurface structures with high spatial and depth resolution. Such information is, for instance, ultimately needed in semiconductor technology, where devices are highly integrated and stacked with vertical dimensions approaching the nanometer scale. While this was, in the past, mainly done with inorganic materials, designs are now shifting to organic devices, where it is important to analyze the *molecular* structure rather than just the elemental composition. Another demand is triggered by the immensely growing field of biological applications, where it becomes important to localize specific molecules within samples such as tissue and single cells.

In principle, 3D chemical characterization requires an imaging analysis technique that allows both lateral (i.e., parallel to the surface) and vertical (i.e., perpendicular to the surface) resolution on the nanometer scale. Lateral resolution can be obtained by microscope or microprobe imaging techniques which may either employ stigmatic imaging of species carrying the spectroscopic information or by raster scanning a finely focused excitation probe in order to obtain chemically specific information. Vertical resolution, on the other hand, requires the applied imaging technique to be surface sensitive, that is, to probe chemical information only from the uppermost layers of an investigated sample. Useful analytical techniques are therefore restricted to methods such as electron spectroscopy or surface mass spectrometry, delivering chemical information with nanometer scale

Cluster Secondary Ion Mass Spectrometry: Principles and Applications, First Edition.
Edited by Christine M. Mahoney.
© 2013 John Wiley & Sons, Inc. Published 2013 by John Wiley & Sons, Inc.

information depth. However, only surface mass spectrometric-based methods have the ability to obtain molecular information without the need of specific tags.

Surface mass spectrometry utilizes a primary excitation beam to energize the surface, thereby generating surface-specific particles that are then analyzed with respect to their mass-to-charge ratio. Possible primary excitation mechanisms include laser desorption (laser ablation mass microanalysis, LAMMA; laser ionization microanalysis, LIMA; laser desorption and ionization, LDI; matrix-assisted laser desorption/ionization, MALDI), ion sputtering (secondary ion mass spectrometry/secondary neutral mass spectrometry, SIMS/SNMS), electron-stimulated desorption (ESD) or bombardment of the surface with energetic macrodroplets (desorption electrospray ionization, DESI). Among these, only laser desorption and ion sputtering allow for high resolution lateral imaging with molecule-specific information. However, while the laser beam focus is restricted by the diffraction limit, limiting the maximum spatial resolution to the micrometer scale, ion beams can, in principle, be focused down to a spot size of several nanometers,[1] allowing for molecular imaging with spatial resolutions in the nanoscale range. Moreover, the information depth of laser desorption (micrometer range) is much larger than that of ion sputtering (~1 nm), leaving SIMS (and SNMS) as the only viable method for high resolution three-dimensional chemical imaging on the nanometer scale.

A prominent strategy for 3D surface and thin-film characterization is to combine laterally resolved surface analysis with layer-by-layer sample erosion, thereby gradually exposing the subsurface material to a method for surface-sensitive chemical analysis. Among the various possible approaches for controlled erosion, sputter depth profiling has evolved as one of the most versatile tools to obtain chemical information as a function of depth below the initial surface. Here, the solid surface is bombarded by an energetic ion beam, thereby removing surface material because of ion sputtering. Combined with imaging surface analysis, this strategy, although destructive, allows one to reconstruct the three-dimensional structure of the investigated sample by means of a tomography approach, that is, stacking successive 2D slices as illustrated in Figure 6.1a.

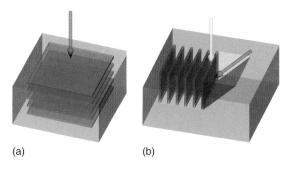

(a)   (b)

Figure 6.1 Schematic view of (a) 3D sputter depth profiling and (b) 3D focused ion beam (FIB) tomography. Red arrows indicate the direction of analysis. The yellow arrow in (b) indicates the FIB milling direction. The blue planes represent the successive image stacks.

In principle, every imaging method can be used to analyze the momentarily exposed surface, as long as it is sufficiently surface sensitive and delivers chemistry-specific information at a meaningful level. Mass spectrometry techniques such as SIMS or SNMS are capable of providing molecule-specific information about the surface chemistry and may therefore directly be used for the analysis of molecular samples, provided the chemical integrity of the surface can be preserved under high fluence ion bombardment as is necessary for sputter erosion. While this was originally deemed impossible when atomic projectile ions were more convention-ally used for sputtering, the advent of cluster ion beams has dramatically changed this picture. As shown in Chapter 5, molecular sputter depth profiling using cluster projectiles such as SF_5^+, C_{60}^+, or Ar_n^+ (with n up to several thousand) is now regarded as routine, with an achievable depth resolution of the order of 10 nm or below.

Although there have been quite a few successful demonstrations of 3D SIMS analysis using this approach, there are many potential artifacts that can arise in the data because of inaccurate assumptions (e.g., nonlinear sputter rates and topographic artifacts). Moreover, a fundamental problem in nanoscale 3D imaging, particularly of molecular systems, is related to detection sensitivity. Typical number densities of such systems are of the order of several molecules per cubic nanometer of sample volume. If one would set out to measure the chemical composition of such a sample with a spatial resolution of $10 \times 10 \times 10$ nm^3, such a "voxel" would only contain a few thousand (matrix) molecules that are available for analysis. If the analyte molecule of interest is not the matrix itself but only present in minor concentration, this number further reduces accordingly. The probability of a sputtered molecule leaving the surface as an intact, unfragmented molecular secondary ion is typically rather small (of the order of $10^{-5} \cdots 10^{-3}$), and even for characteristic fragments this number is not significantly larger. This means that (much) less than one single ion would, on average, be detectable from each voxel in a SIMS experiment. Therefore, one needs to find a useful spatial resolution that can be defined by the condition that at least one ion should be detectable from a voxel. From the above estimates, it is clear that this may require the voxel volume to be increased by orders of magnitude, which can be accomplished by sacrifices either in lateral (parallel to the surface) or in-depth (along the surface normal) resolution.

This chapter outlines the considerations that must be taken into account when obtaining and analyzing 3D molecular imaging data. A series of guidelines for beam conditions during sputter depth profiling are described, as well as important data analysis considerations. It is shown that 3D imaging should always be correlated with sample topography in order to display true 3D representations of the data. Furthermore, correction factors accounting for differential sputtering, initial surface topography, and low data density are introduced. Finally, although the primary focus of this chapter will be on sputter depth profiling with cluster ion beams, other means of obtaining in-depth information from soft samples will also be discussed (e.g., Focused Ion Beam SIMS (FIB-SIMS) and physical cross sectioning methods). Overall it is expected that this chapter will serve as a set of guidelines for 3D organic analysis using SIMS.

6.2   GENERAL STRATEGIES

There are several approaches to obtaining 3D molecular information as a function of depth using SIMS or SNMS, based both on physical cross sectioning and ion bombardment methods. Each of these approaches will now be discussed in turn.

6.2.1   Three-Dimensional Sputter Depth Profiling

The standard strategy of 3D sputter depth profiling involves acquiring a stack of mass spectrometric images that are separated by intermediate sputter erosion cycles. One example of experimental data acquired using this strategy was shown in Chapter 5 (Fig. 5.7). Other examples of data acquired in this manner are shown in Figures 6.2, through 6.6, and Figure 6.8. In fact, most of the data published to date have been obtained in this way. Mass spectral data are usually acquired using time-of-flight (ToF) spectrometry in order to allow parallel detection of all masses and avoid any restriction of the acquired mass range. During the acquisition of an image, the surface is kept in a "static" mode where the applied projectile ion fluence is low enough to prevent damage or erosion of more than a small fraction (typically less than 1%) of a monolayer. As an important feature of this strategy, data acquisition and sputter erosion are decoupled and can, in principle, be performed using different ion beams using, for instance, the high lateral resolution of a liquid metal ion gun (LMIG) beam for image acquisition along with the low damage cross section of a cluster ion beam for sputter erosion. The projectile ion beam used for ToF data acquisition is operated in a pulsed mode and digitally rastered across the desired analysis field-of-view.

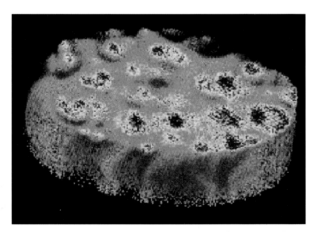

Figure 6.2   Pseudo-3D visualization of SIMS data at m/z 152 acquired on an acetaminophen-doped PLA polymer film using a 250 µm field-of-view. Sputter erosion and data acquisition was performed using a 10 keV SF_5^+ cluster ion beam, and image acquisition was performed in microscope mode. Reproduced from Gillen et al.,[23] with permission from Elsevier.

One hour

One day

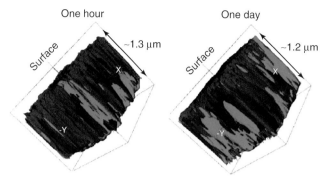

Figure 6.3 Pseudo-3D visualization of ToF-SIMS data at m/z 84 (sirolimus, red), m/z 56 (PLGA, green) and m/z 23 (Na, purple) acquired on a coronary stent coating for two different elution times (as indicated). Sputter erosion was performed using a 20 keV C_{60}^+ cluster ion beam and images were acquired using a 30 keV Au^+ ion beam. Reproduced from Fisher et al.,[24] with permission from ACS Publications.

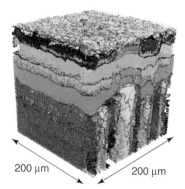

200 μm   200 μm

Figure 6.4 Pseudo-3D visualization of ToF-SIMS data of various molecular ions at m/z 500–850 acquired on an OLED display structure. Sputter erosion was performed using a 5 keV Ar_{1700}^+ cluster ion beam and images were acquired using a 25 keV Bi_3^+ beam. Data courtesy of ION-TOF GmbH, Germany.

During an erosion cycle, the sputtering ion beam is operated in a continuous mode and rastered over the desired erosion area for a preselected amount of time, applying a certain ion fluence per sputtering cycle. It should be noted that this method of depth profiling has a more mature standing for elemental analysis of inorganic samples,[2,3] where issues such as the relation between useful lateral image resolution, detection sensitivity, and statistical signal noise have already been discussed in great detail.[1,4−7] There are also a number of studies where molecules have been localized using this method, for instance, imaging inside single cells via elemental and isotopic tagging of the molecule of interest.[8−14] In a few cases, elemental imaging using cluster ion beams was attempted with the intention of optimizing the depth resolution as compared to atomic projectile ions.[15−23] We will restrict the discussions here to the more recent applications of *molecular* 3D analysis, which has only been made possible by the use of cluster ion beams.

From the body of data that has been published during the past decade, we revert to a few recent examples illustrating the application. One of the more promising applications of the cluster ion beam is for imaging and depth profiling in drug

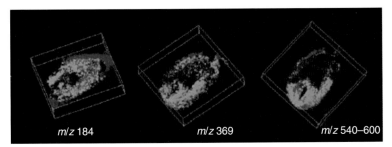

Figure 6.5 Pseudo-3D visualization of ToF-SIMS data at *m/z* 184 (phosphocholine head group), *m/z* 369 (cholesterol), and *m/z* 540–600 (assigned to phosphatidylcholine lipids) acquired on a frog's egg cell (*Xenopus laevis* oocyte). Image acquisition and sputter erosion were performed using a 40 keV C_{60}^{+} cluster ion beam. Reproduced from Fletcher et al.,[70] with permission from Springer.

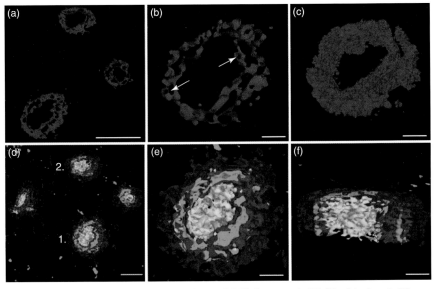

Figure 6.6 Pseudo-3D visualization of ToF-SIMS data at *m/z* 23 (Na, blue), *m/z* 39 (K, green), *m/z* 184 (phosphocholine head group, red), and *m/z* 86 (phosphocholine fragment, yellow) acquired on single thyroid tumor cells. Sputter erosion was performed using a 10 keV C_{60}^{+} cluster ion beam and images were acquired using a 25 keV Bi_{3}^{+} beam. Reproduced from Nygren et al.,[74] with permission from Wiley.

delivery systems. An example of this was given in Chapter 5 for protein delivery systems (Fig. 5.7). Figure 6.2 shows an earlier example with the analysis of a poly(lactic acid) (PLA) film doped with 20% (weight) of a drug (acetaminophen).[23] Sputter erosion and data acquisition were performed using a 10 keV SF_{5}^{+} ion beam and images of the drug-specific molecular ion signal at *m/z* 152 were acquired in microscope mode (see section 6.3.1 for definition of microscope mode). It is clearly evident that the drug distribution is highly inhomogeneous.

Figure 6.3 shows another example of image depth profiling of a drug-eluting coronary stent coating that is composed of 25% (w/w) of a drug (sirolimus) in a poly(lactic-*co*-glycolic acid) (PLGA) matrix.[24] The data were taken in a dual beam configuration using a 30 keV Au^+ ion beam for image acquisition and a 20 keV C_{60}^+ ion beam for sputter erosion. The figure displays a 3D visualization of the resulting image stack showing characteristic molecular fragment ion signals representative of the drug (red) and the PLGA matrix (green) along with that of sodium (purple) after two different elution times. The study shows that the spatial distribution of the drug is related to the elution rate.

With the more recent advent of the gas cluster ion beam source (GCIB), the ability of cluster SIMS for depth profiling in organic electronics has improved dramatically, opening up new doors for characterization of photovoltaics, organic light emitting diodes (OLED), and other organic-based electronics devices. Figure 6.4 depicts an example of 3D analysis in an OLED structure as part of a modern computer display.[25] As is typical for integrated circuits, the sample is highly stacked and structured both in lateral and vertical dimensions. Using a rare gas cluster ion beam (5 keV Ar_{1700}^+) for sputter erosion and a 25 keV Bi_3^+ LMIG ion beam for data acquisition, one can easily identify the different regions using molecule-specific SIMS signals; a feat previously unattainable using any other cluster beam.

One of the most important impacts of the cluster source however, is the ability to obtain 3D molecular information from biological systems on the single cell level. Figure 6.5 shows the first published example of three-dimensional molecular analysis in a biological single cell system.[26] A relatively large (about 1 mm diameter) frog's egg cell (*Xenopus laevis* oocyte) was freeze dried and a 3D depth profile of the entire cell was performed using a 40 keV C_{60}^+ ion beam both for sputter erosion and data acquisition. The important result of this study was that (i) it is possible to erode several micrometers deep into the cell material and (ii) the spatial distribution of molecular ions specific to certain cell components can clearly be discerned.

This pioneering work was later followed by 3D profiles of other biological cells. For example, Figure 6.6 depicts data acquired on freeze-fractured and freeze-dried anaplastic thyroid carcinoma cells.[27] The displayed 3D visualization of the resulting data set clearly shows the structure of a single cell, featuring an enrichment of sodium (blue) from the dry culture medium around the cell periphery, followed by a correlated appearance of potassium (green), and the phosphocholine head group (a characteristic fragment of phospholipids at m/z 184, red), which are taken to be representative for the cell membrane. In the inner region of the cells, an enhanced signal at m/z 86 is found, which is assumed to represent a characteristic fragment of the phosphocholine head group in phospholipids.

In a different instrumental approach, a continuous (DC) ion beam is used for data acquisition and sputter erosion at the same time.[28,29] The beam is rastered across the desired erosion area and mass spectral data are acquired using either a quadrupole mass filter, a pulsed ToF spectrometer, or a combination of both.[28,29] Images are acquired by registering the mass spectrometric signal as a function of the momentary beam position on the sample, which can be varied by either scanning the beam[29] or translating the sample.[28] This way, each measured data point represents an integral over a voxel determined by the lateral beam spot size and the depth eroded during the dwell time of the beam on each pixel. Using the shortest feasible dwell time of about 10 μs along with a cluster ion beam current of 1 nA into a spot size of 1 μm, a typical sputter yield volume of the order of 100 nm^3/impact would correspond to a voxel depth of the order of 3 nm eroded per acquired image. The great advantage of this technique is that practically all sputtered material is used to generate an image, thereby greatly enhancing the detection sensitivity as compared to the conventional alternating cycle method. Furthermore, the analysis times are significantly decreased when employing continuous beams. Again, the lateral voxel (pixel) dimension is restricted by the spot size of the eroding cluster ion beam, which presently limits the achievable lateral resolution to about 1 μm.

A prominent example of a data set acquired using this method is shown in Figure 6.7. The data were taken on formalin-fixed benign hyperplasia prostate cells.[30] The displayed 3D visualization using isosurface rendering along with cross-sectional images clearly shows a localization of the adenine molecular ion signal (*m/z* 136, left panel) representing the DNA in the center of the cells, while the phosphocholine head group fragment signal (*m/z* 184, right panel) representing the lipid cell membrane appears to be localized at the circumference of each cell.

The above examples clearly illustrate that meaningful 3D imaging analysis of complex molecular systems can be performed if cluster ion beams are used for sputter erosion. In order to construct a 3D visualization of the acquired data set, the measured ToF-SIMS images were in all cases simply stacked in the vertical direction with arbitrary equidistant spacing. It should be stressed that such a visualization relies on three implicit assumptions:

1. each image represents a flat slice across the analyzed sample;

2. the original sample surface is flat;

3. there is a linear relation between applied ion fluence and eroded depth, which is the same everywhere in the analyzed volume.

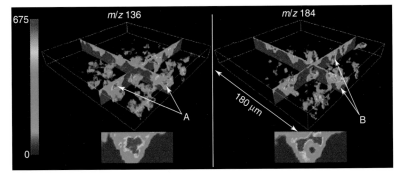

Figure 6.7 Pseudo-3D visualization of ToF-SIMS data at *m/z* 136 (the protonated molecular ion from adenine) and *m/z* 184 (the phosphocholine lipid head group) acquired on benign prostatic hyperplasia cells. Sputter erosion and image acquisition was performed using a 40 keV C$_{60}^+$ cluster ion beam operated in DC mode. Reproduced from Fletcher et al.,[30] with permission from ACS Publications.

In cases where one (or more) of these assumptions is violated, the resulting visualization may greatly deviate from the true three-dimensional sample structure. Examples for such conditions will be shown below. While visualizations of this kind may be extremely valuable to identify the *presence* of different chemical regions in the analyzed sample, they may at the same time be greatly misleading as to the *exact location* of these regions as well as their *spatial dimensions*. In order to illustrate this statement, we refer to an example of a specially prepared test structure consisting of a 300 nm peptide-doped trehalose film deposited on a silicon substrate that was prestructured using a 15 keV Ga$^+$ focused ion beam (FIB) in order to generate significant surface topographical features. This sample was then analyzed by an imaging sputter depth profile using a 20 keV C$_{60}{}^+$ ion beam for erosion and data acquisition. The simple 3D visualization of the resulting data set is depicted in Figure 6.8, where an overlay of ToF-SIMS signals representative for the molecular overlayer (the peptide molecular ion at m/z 425, blue), the FIB projectile material (Ga at m/z 69, red), and the silicon substrate (Si at m/z 28, green) is shown. The image reveals that Ga$^+$ bombardment of the molecular film

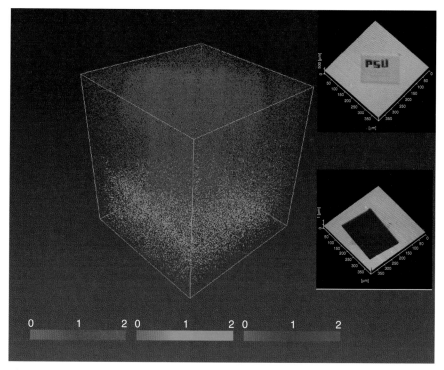

Figure 6.8 Pseudo-3D visualization of ToF-SIMS data at m/z 425 (molecular ion of GGYR peptide, blue), m/z 69 (Ga, red), and m/z 28 (Si, green) acquired on a 300 nm trehalose film deposited on Si and prestructured using a 15 keV Ga$^+$ FIB beam. Image acquisition and sputter erosion were performed using a 20 keV C$_{60}{}^+$ cluster ion beam. Insert: AFM images of the surface taken before and after depth profile analysis. Recreated from Wucher et al.,[58] with permission from ACS publications.

generates a gallium-implanted subsurface region, where the molecular integrity of the sample is practically destroyed. However, the acquired SIMS data do not contain any information regarding the topography of the surface analyzed in each individual ToF-SIMS image. As a consequence, the visualization cannot deliver information about the thickness of the damaged layer, not even of the undamaged parts of the film. In the same way, the data presented in Figure 6.2 and Figure 6.3 reveal an inhomogeneous drug distribution within the analyzed volume; however, the true location of the drug-enriched regions cannot be discerned because the depth information is only qualitative.

Similarly, Figure 6.4 shows that there are different structures stacked in the analyzed volume, but the data deliver no information as to the thickness or depth of the individual films, so it remains unclear which structure is neighbored to which. This information, however, is vital to understanding the electrical functionality of the device. Finally, the data shown in Figure 6.5 reveal the presence of specific molecules such as cholesterol in the interior of the analyzed cells, but it does not provide information on where exactly these molecules are located. For that reason, we will, in the following, call such a simple image stacking procedure a *pseudo-3D* visualization of a measured data set.

In principle, the ion fluence applied between the acquisition of subsequent images must be converted into eroded depth. As will be discussed below, the assumption of a constant erosion rate across the entire sample volume is not always justified, so that this conversion will in general be nonlinear and require a careful depth scale calibration. The data processing from a pseudo-3D visualization as shown in Figures 6.2 through 6.8 to a true three-dimensional representation of the sample chemistry therefore ultimately requires the combination of 3D imaging with topography data taken on the investigated sample before, during, and after the sputter depth profile analysis. Possible methods to obtain such data are stylus profilometry, atomic force microscopy (AFM), or optical interferometry. Conceptually, the best strategy would be to acquire such data *in situ* during the acquisition of the sputter depth profile. While such instrumentation is not currently available, one can at the very least measure the surface topography *ex situ* before and after the depth profile using, for instance, AFM. Examples for such a protocol will be shown in the following section.

6.2.2 Wedge Beveling

Another strategy used to obtain 3D information is by creating a bevel cut across the sample volume of interest and performing an image analysis of the resulting cross section. The bevel cut can be generated either mechanically,[31] via chemical etching,[32] or via ion sputtering using a controlled lateral variation of the applied ion fluence.[33] The latter technique was introduced many years ago as a means of depth scale calibration during inorganic sputter depth profiling.[34]

An example for such an analysis of a molecular film has recently been published by Mao et al.,[35,36] who investigated a system of organic delta layers (Irganox 3114) embedded into an Irganox 1010 matrix. Figure 6.9 shows a ToF-SIMS image

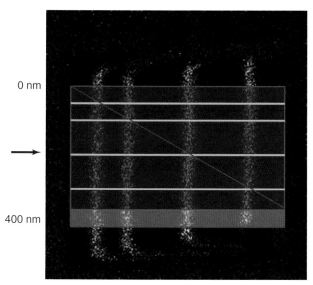

Figure 6.9   ToF-SIMS image of a wedge bevel crater eroded into Irganox 1010 doped with four delta layers of Irganox 3114. The image shows the intensity of the m/z 42 molecular fragment signal representing the delta layers with the arrow indicating the impact direction of the C_{60} ion beam. Insert: schematic side view of wedge (red line) across Irganox 1010 matrix (blue) and Irganox 3114 layers (yellow) down to silicon substrate (green). Recreated from Mao et al.,[36] with permission from ACS Publications.

of a wedge crater eroded into the sample using a 40 keV C_{60}^+ cluster ion beam impinging under $71°$ with respect to the surface normal. As pointed out in the original publication, it is important that the sputter erosion was performed with the sample at cryogenic temperature. The image was acquired using the same C_{60}^+ ion beam, but now impinging under $40°$ with respect to the surface normal. The shape of the eroded crater was measured with AFM and is schematically depicted in the insert as a red line, spanning a range from zero depth (the original surface) on the left hand side to erosion of the entire molecular overlayer down to the silicon substrate on the right hand side. The position of the four delta layers are indicated as yellow lines. It is evident that the displayed SIMS image of the m/z 42 fragment signal of Irganox 3114 exhibits pronounced maxima right where the bevel cut crosses the delta layers, thereby clearly demonstrating that it is possible to localize the embedded Irganox 3114 molecules with a nanometer precision and a depth resolution of about 10 nm.

6.2.3   Physical Cross Sectioning

An alternative approach, one that avoids the issues inherent during sputtering, is to use secondary methods to slice the sample into thin sections, analyze each of the individual slices in succession, and subsequently stack the images for 3D data representation. Such processes have already proven to be successful for analysis

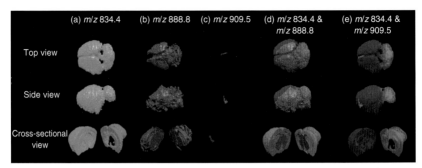

Figure 6.10 Three-dimensional models of the mouse brain by DESI-MS. The signals imaged are characteristic of phosphatidylserines (PS), phosphatidylinositols (PI), and sulfatides (ST) where PS 18:0/22:6 is the major lipid found in the gray matter and ST 24:1 is the major lipid found in the white matter. Top, side, and cross-sectional views are shown for the 3D reconstruction of the distribution of (a) PS 18:0/22:6 in green, (b) ST 24:1 in red, and (c) PI 18:0/22:6 in blue. The same views are shown for the transparent overlaid distributions of the lipids, (d) PS18:0/22:6 and ST24:1 and (e) PS 18:0/22:6 and PI 18:0/22:6. Reproduced from Eberlin et al.,[38] with permission from Wiley.

and 3D imaging with MALDI[37] and DESI.[38] An example is depicted in Figure 6.10 that shows the 3D molecular composition of lipids in a mouse brain, obtained using DESI to characterize a series of 20 μm cryosectioned slices.

Another means of obtaining 3D information without the difficulties of sputtering artifacts is by creating a bevel cut such as was described earlier, except by physically cutting the sample using a method such as ultralow angle microtomy (ULAM). An example is given by Hinder et al.,[39] where the authors used ULAM to cut into a sample of acrylic paint. The slice was created at an angle and subsequently imaged using SIMS in order to obtain in-depth information from the sample.

Figure 6.11 shows the resulting SIMS images of the various components in the paint coating, acquired from a beveled crater bottom. The negative ion images

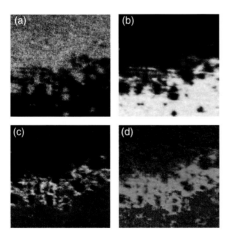

Figure 6.11 Negative ion ToF-SIMS images (500 μm × 500 μm) of a buried polymer/polymer interface exposed by ultralow angle microtomy. (a) *m/z* 66, polyurethane (PU) primer (located 20 μm below surface), (b) *m/z* 19, poly(vinylidene difluoride) (PVDF) coating, (c) *m/z* 85, acrylic copolymers, and (d) overlay of PU (blue), PVDF (red), and acrylics (green). Findings confirm that the acrylic copolymers segregate to the topcoat/primer interface where they enhance the adhesive properties. Recreated from Hinder et al.,[39] with permission from Wiley.

displayed in the figure were acquired using a $Au_3{}^+$ cluster source and show intense molecular and atomic ion images from three different polymeric components in the film: (i) the polyurethane (PU) primer coat located ~20 μm below the surface (m/z 66), (ii) the poly(vinylidene difluoride) (PVDF) topcoat (m/z 19), and (iii) acrylic PMMA/poly(ethyl methacrylate) (PEMA) copolymers that are shown in the figure to segregate the interface region (m/z 85).[39] The data are summarized in an overlay of the three regions, Fig. 6.11(d), created subsequently by the authors of this chapter. This result helped to understand the mechanisms of adhesion in this particular coating, whereby the acrylic copolymer promotes adhesion between the primer and topcoat by segregating the interface. More important, however, is that this method does not have the problems inherent with sputtering and can be useful for analysis of thicker coatings.

6.2.4   FIB-ToF Tomography

A conceptually different approach to obtain 3D chemical information involves a combination of FIB etching and imaging ToF mass spectrometry. A FIB is a highly focused Ga^+ beam that can be used to etch small features into a surface (e.g., lithography), and/or create cross sections and lift-out sections in materials for subsequent analysis with other technologies. For example, FIB is often used in conjunction with secondary electron microscopy (FIB-SEM) to obtain 3D structural information.[40−42] FIB-SEM may also be used in conjunction with energy dispersive X-ray spectroscopy (EDS) capabilities or with wavelength dispersive X-ray spectroscopy (WDS) to obtain elemental information in three dimensions.[41] While the results are promising for 3D analysis, however, there are several limitations to these analytical methods which include low sensitivity, poor spectroscopic resolution, limited chemical information, and no information concerning molecular or structural chemistry—limitations that are readily overcome by employing a SIMS analysis approach.

FIB has been used in conjunction with a quadrupole-based secondary ion mass spectrometer to obtain 3D elemental maps of inorganic samples, in a process referred to as *FIB-SIMS*.[43,44] More recently, an *in situ* FIB option in conjunction with ToF-SIMS imaging mass spectrometry for 3D characterization has been developed—an analytical approach that is called *FIB-ToF*. In this approach, the FIB beam is first used to mill away the sample material around the volume of interest. As indicated in Figure 6.1b, this will lead to "vertical" cutting planes that represent cross sections of the analyzed volume in the direction parallel to the FIB beam. These cross sections can then be imaged using ToF SIMS.

An example of such a strategy is shown in Figures 6.12 and 6.13. A specimen of sintered CuW composite* was used to demonstrate the efficacy of the FIB-ToF approach for 3D characterization wherein successive FIB milling and ToF-SIMS acquisition cycles were employed to image a 10 μm deep volume at a 20 μm × 20 μm analytical field-of-view. Before depth profile analysis, a FIB cross section

---

*Dr. Noel Smith of Oregon Physics and Sherbrooke Metals is acknowledged for providing the CuW composite material.

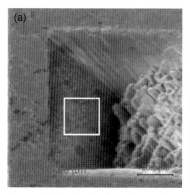 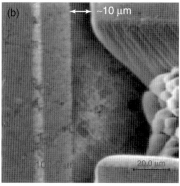

Figure 6.12 Ion-induced secondary electron (SE) images from the CuW composite. (a) An SE image of the initial FIB-milled cross section. The approximate location of the 20 μm × 20 μm analytical field-of-view is indicated. (b) An SE image of the imaged surface following the FIB-ToF tomography. The measured milled depth is approximately 10 μm. Each image is a 100 μm × 100 μm field-of-view (scale bar is 20 μm). The depth of the cross section at the vertical back wall is approximately 50 μm. Data courtesy of Physical Electronics (2010).[45]

was milled into the central region of the sample. The initial cross section for the CuW composite is depicted in Figure 6.12a that shows a secondary electron (SE) image of the initial cross section crater created by the FIB, where the direction of the FIB ion beam cut was 45° with respect to the surface normal, and the LMIG analysis was performed at 90° with respect to the FIB cut, such that the vertical back wall of the FIB crater was in the direct line-of-sight of the LMIG (refer to Fig. 6.1b).

This image reveals the slope of the cross section crater and the slight curtaining[46] as a result of the FIB milling. Additionally, the SE image reveals a large degree of topography at the crater bottom. This topography is the result of differential sputtering and is produced at least in part, when certain components of a composite matrix are sputtered more readily than others; in the present case, the tungsten sputters more readily than does the copper. Since in this design, the analyst is characterizing the vertical back wall of the FIB cross section, the slight topography and/or curtaining does not interfere with analysis as it would during SIMS depth profiling, and a nominally smooth surface is revealed after each slice.

An SE image was also obtained following 20 FIB line cuts and ToF-SIMS acquisition cycles, and is shown in Figure 6.12b where the measured depth of milling in the direction of the line cuts parallel to the surface is 10 μm. The image reveals that after 20 individual line cuts, the back wall of the crater exposed to the LMIG remains sharp, the W grains remain well-defined, and the curtaining has not worsened. It is also observed that the sputtered material from the line cuts has partially deposited onto the crater bottom and side walls.

The 20 individual raw data stream files from the ToF-SIMS analyses of the FIB-sectioned CuW composite were compiled, or concatenated, into a single raw file for the purposes of extracting depth profiles and 3D images, where the resulting

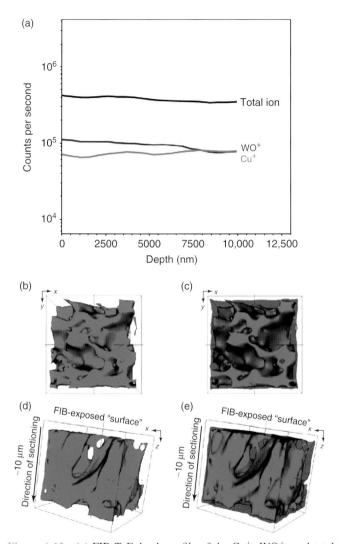

Figure 6.13   (a) FIB-ToF depth profile of the Cu^+, WO^+, and total ion signals that were extracted following concatenation of the 20 individual raw data stream files. The depth scale is based on the total measured depth of the FIB line cuts. For each secondary ion species that is monitored, each data point of the depth profile is a sum of the counts of that species across the entire raster frame, that is 20 μm × 20 μm. (b–e) 3D isosurface models of Cu^+ (green) and WO^+ (blue; reduced opacity) that were generated from the concatenated raw data stream file. (b) Surface view of the Cu^+ isosurface model; (c) surface view of the Cu^+ and WO^+ isosurface overlay; (d) side view of the Cu^+ isosurface model; (e) side view of the Cu^+ and WO^+ isosurface overlay. In each 3D image, the total volume is 20 μm × 20 μm × 10 μm. Data courtesy of Physical Electronics (2010).[45]

depth profiles of Cu^+ and WO^+ are rendered in Figure 6.13a. For comparison, the total ion profile is also shown. What is noteworthy is that other than minor undulation of the Cu^+ and WO^+ signals, the profiles are quite unremarkable. There are no pronounced fluctuations of either Cu^+ and WO^+. The profiles of W^+ and WO_2^+ show the same trend but have reduced signals.

The corresponding three-dimensional distribution of Cu^+ and WO^+ data are provided in Figure 6.13b–e where the Cu^+ (green) and the WO^+ (blue) isosurfaces[47,48] are presented in an overlay. The intensity thresholds of the Cu^+ and WO^+ 3D images were adjusted to produce isosurface models such that nominally pure copper phases and all tungsten oxide grains are visualized. These areas of copper are discernible in Figure 6.13b and 6.13d where, in both images, the cavities occupied by tungsten oxide are evident. However, there are large quantities of copper that exist in mixed phases with tungsten oxide. The corresponding overlay images in Figure 6.13c and 6.13e, wherein the opacity of the WO^+ isosurface has been reduced to expose the internal structure, reveal the 3D distribution of both copper and tungsten oxide. Using the isosurface overlays, the full 3D distribution of each imaged component may be readily observed and fully explored. This level of 3D materials characterization would not be possible via conventional ToF-SIMS depth profiling with a low voltage sputter ion beam because of the artifacts produced as a result of differential sputtering.

Of course, a difficulty of ion beam milling in general is the resulting damage that is imparted to organic molecules. A paramount concern regarding the efficacy of 3D imaging by FIB-ToF tomography is whether the characteristic organic chemistry can be observed, or recovered, following FIB sectioning. Preliminary data indicate that following FIB milling, characteristic and molecular organic signals may be recovered for ToF-SIMS imaging. Before image acquisition, the cross-sectional surface must be subjected to a "polishing" step in order to remove the damaged layer produced by the lateral straggling of the cutting FIB beam. For molecular analysis, a suitable cluster ion beam (C_{60}^+ or Ar_n^+) can be used for that purpose in order to remove the chemical damage produced by the high energy Ga^+ ions. Then, image acquisition can be performed in a static mode. This is demonstrated in Figure 6.14 that presents SE images of a 20 μm PLGA sphere doped with cocaine at a mass fraction of 10%.[†] The sphere was then cross-sectioned by FIB, where the top portion of the sphere was removed (Fig. 6.14a). The FIB milling was performed such that the straggle, or residual damage layer of graphitic carbon produced by FIB milling, is on the order of 10 nm thickness.[49] A 20 keV C_{60}^+ ToF-SIMS depth profile was performed on the cross-sectional surface with the objective of determining whether the characteristic molecular and fragment ion signals of the PLGA matrix and the cocaine dopant could be recovered. The molecular formula of cocaine is $C_{17}H_{21}NO_4$ with a molecular weight of 303.35 g/mol; the molecular ion $[M + H]^+$ is observed at 304.36 m/z. However, the initial ToF-SIMS analysis revealed little but carbon on the as-received surface.

[†]Dr. Christopher Szakal, Matt Staymates, and Keana Scott of NIST are acknowledged for their efforts in producing and FIB sectioning of the polymer sphere specimens.

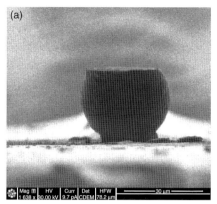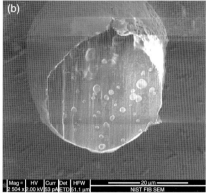

Figure 6.14   Secondary electron (SE) images of a FIB-milled, 20 μm diameter polymer sphere comprised of cocaine (10 wt.%) in PLGA. (a) Side view showing that the surface of the sphere has been removed by FIB milling. The field-of-view is approximately 100 μm. (b) Surface view revealing the morphology of the FIB-milled sphere. The field-of-view is approximately 40 μm. Data courtesy of Physical Electronics (2010).[45]

The depth profile of the FIB-milled sphere is rendered in Figure 6.15a. The depth scale has been approximated based on the sputter rates of bulk and spin cast polymers including PLGA ($\sim$15.1 nm^3/C$_{60}^+$).[47,50] Therefore, the depth scale in the region of the surface that is comprised primarily of FIB-damaged polymer is exaggerated by approximately 1 order of magnitude. Additionally, there is likely some adventitious surface contamination from transportation of the sample at ambient conditions between FIB milling and ToF-SIMS analysis.

The depth profile analysis reveals that after the adventitious surface contamination and the FIB-induced damage layers have been removed via C$_{60}^+$ ion sputtering, the characteristic ion signals of PLGA and the molecular ion signal of cocaine are dramatically improved. At each analytical cycle of the depth profile, the lateral distribution of each chemical constituent is imaged, and these 2D images-at-depth may be extracted from the raw data file at any specified depth interval. The 2D images-at-depth in Figure 6.15b reveal the lateral distribution of cocaine molecules, that is [M + H]$^+$ at m/z 304.36, from the surface of the as-received sample (top image) and at a profiled depth of 1 μm from the surface (bottom image). The cocaine molecular ion image acquired before and after C$_{60}$ sputtering demonstrates that the cocaine signal arises only from the FIB-milled surface of the sphere after the FIB-induced surface damage has been removed. At this time, the source of the heterogeneity in the cocaine signal is unknown, but is presumed to arise via phase segregation after production of the spheres, as opposed to a topography affect.

As explained above, three-dimensional analysis is achieved by subsequently FIB-cutting more and more material away from the edge of the analyzed volume and stacking the resulting cross-sectional images as was indicated earlier in Figure 6.1b. Note that, as vertical cross sections are being imaged, the resulting "depth" resolution in the direction perpendicular to the original sample surface is

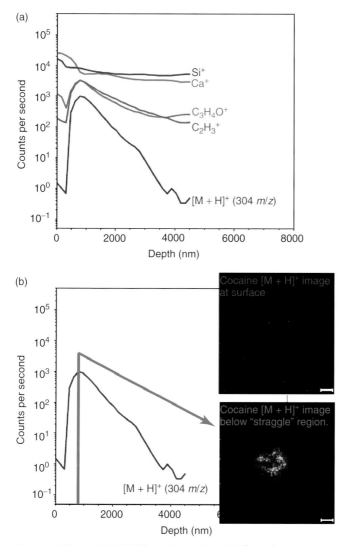

Figure 6.15 (a) ToF-SIMS depth profiles of Si^+, Ca^+, the nonspecific $C_2H_3^+$ organic fragment, the characteristic $C_3H_4O^+$ fragment of PLGA, and the cocaine $[M + H]^+$ molecular ion. The depth scale is based on the steady-state sputter rate of PLGA and similar polymers. (b) ToF-SIMS depth profile of the cocaine $[M + H]^+$ molecular ion with corresponding molecular images: The top image shows the secondary ion image of cocaine $[M + H]^+$ at the surface of the as-received FIB-milled polymer sphere, and the bottom image shows the secondary ion image of cocaine $[M + H]^+$ extracted from the raw data file at the depth of maximum cocaine signal. In each image, the scale bar is 10 μm. Data courtesy of Physical Electronics (2010).[45]

determined by the lateral resolution of the imaging technique rather than by typical sputter depth profiling characteristics of the cutting beam. For cluster beams such as C_{60}^+ or Ar_n^+, this limits the "depth" resolution to about 1 μm. If higher image resolution is required, the system must be equipped with a third ion gun providing, for instance, an LMIG for image acquisition. In this case, "depth" resolution of the order of 100 nm might be achievable. Note that this value is much larger than that obtained by ordinary sputter depth profiling (~10 nm). Moreover, it should be noted that the spatial resolution in the stacking direction (perpendicular to the cutting beam) is *not* equivalent to the spot size of the cutting beam, but rather is determined by the thickness of the laterally damaged layer. Experience reveals that this will limit the achievable resolution to values of the order of 10–30 nm.

6.3   IMPORTANT CONSIDERATIONS FOR ACCURATE 3D REPRESENTATION OF DATA

6.3.1   Beam Rastering Techniques

Imaging surface analysis can, in principle, be done in two different ways. In the *microscope* mode, a large surface area is being excited at the same time and the resulting signal is detected using a stigmatic imaging technique. Instrumentation for the operation of SIMS in this mode is available using either electromagnetic mass filtering or ToF mass spectrometry, delivering a lateral resolution on the micrometer scale. In the *microprobe* mode, the ion beam used for image acquisition is focused and digitally rastered across the desired analysis field-of-view where the resulting signal is detected as a function of the momentary beam position. The lateral resolution achievable in this mode is largely defined by the spot size of the excitation beam.

When beam rastering is used, the ion beam used for sample erosion must be focused (or defocused) and rastered such that there is a laterally homogenous ion flux density across the desired erosion area. In principle, rastering can be performed either by scanning the beam in analog mode or by stepping the beam across a digital raster pattern. While the former has often been used in conventional sputter depth profiling where the erosion area is generally chosen to be much larger than the "gated" analysis area, it is of limited use for 3D depth profiling experiments, as it is extremely hard to achieve a laterally homogenous current density profile across the entire eroded crater. Therefore, digital rastering schemes should be employed where the beam is stepped across a freely selectable pixel sequence within the desired erosion area. The beam itself can, in principle, be operated in a continuous mode, but in this case extreme care must be taken that the computer controlling the experiment does not perform any other task except stepping the beam and acquiring the data.[‡] Otherwise, the beam might spend ill-defined times on subsequent pixels, possibly resulting in an extremely inhomogeneous erosion profile. It is therefore

---

[‡]Note that this is impossible to assure if the experiment is controlled by software running under a multitasking operation system.

better to operate the beam in a pulsed mode in order to fix the dwell time on each pixel. As commonly known from FIB technology,[51−51] this time should be chosen as small as possible to minimize material redeposition across the eroded volume (Fig. 6.12).

Moreover, the raster scanned area during erosion should not be much larger than the analyzed area when applying a certain projectile ion fluence (as is commonly done in conventional sputter depth profiling), but rather be matched to the image acquisition area when acquiring a molecular 3D profile. Multiple raster frames using short pixel dwell time should be preferred as compared to a single raster with long dwell time when applying the selected ion fluence during an erosion cycle in order to minimize material redeposition effects. Note that the conditions described above for 3D analysis of organic samples, specifically in reference to equal erosion and image areas, is different from what is typically done during 2D sputter depth profiling. When acquiring sputter depth profiles, the selected or gated analysis area is usually much smaller than the erosion area so as to avoid crater edge effects. In 3D profiling this is not necessary because the gating area for depth profiles extracted from the acquired data set can always be selected and optimized in retrospect (see following paragraphs).

Finally, the focal width of the eroding beam must be matched to the lateral pixel size. If the beam spot size is much larger than the pixel dimension, the applied fluence will not match the value calculated from the selected rastering area, and large edge effects will distort the profile at the boundary of the eroded crater. If, on the other hand, the spot size is small compared to the pixel dimension, a corrugated fluence profile is generated which results in the erosion of an egg carton-shaped crater bottom. It is evident that such an erosion profile will spoil the depth resolution and must therefore be avoided. Optimum conditions are established when the beam spot size (full width at half maximum, FWHM) matches the pixel dimension. In cases where it cannot be avoided to work with a too tightly focused beam (for instance, if an extremely large erosion area is needed), the raster pattern should be dithered between subsequent frames. For that purpose, a pixel is divided into subpixels, and the entire raster pattern is shifted across these subpixels between subsequent multiple raster frames.

6.3.2   Geometry Effects

As a first step in 3D depth profiling, it is necessary to define an appropriate coordinate system. For the case of an ideally flat surface, this is a straightforward task leading to lateral coordinates (x, y) in the direction parallel to the surface and a vertical (depth or height) coordinate z along the surface normal. In cases where the analyzed surface exhibits any significant topography, however, this choice is not trivial and one would define these coordinates relative to the *average* surface. Sputter erosion, on the other hand, always proceeds along the *local* surface normal that does not necessarily coincide with the overall depth coordinate, z. Moreover, the sputter yield (i.e., the average number of atoms or molecule equivalents removed per projectile impact) is known to depend on the projectile impact angle with respect to the surface normal, which may vary locally for a heavily corrugated surface. For the description of erosion phenomena, it is convenient to define the sputter yield in terms of the *yield volume*, Y_v, that is, the average sample volume removed per projectile impact. The (local) erosion rate is then given by

$$\frac{dz'}{df} = Y_v \text{ or } \frac{dz'}{dt} = j_p \cdot Y_v \qquad (6.1)$$

where z' denotes the depth eroded along the local surface normal, f is the projectile ion fluence (or dose), and $j_p = df/dt$ is the projectile ion flux density. Note that j_p can vary locally, for instance, because of an inhomogeneous current density profile of the rastered ion beam. The erosion rate along the depth coordinate z is then given as

$$\frac{dz}{df} = \frac{dz'}{df} \cdot \cos \vartheta \qquad (6.2)$$

with ϑ describing the local surface tilt, that is, the angle between the local surface normal and the z-axis.

In many cases of practical interest, the tilt angles ϑ are very small and can be neglected. In this context, it should be noted that surface topography images usually show a largely magnified vertical scale. For instance, in the example depicted in Figure 6.8, the maximum slope within the trench walls is of the order of 75 nm/μm, which translates to a maximum ϑ of $4°$. In these cases, the erosion rate can safely be approximated by Equation 6.1. There are, however, other cases one could imagine where this approximation fails. An example of such a scenario would be the erosion of a polymer bead as depicted in Figure 6.14, which might, for instance, serve as a prototype of a single cell. If such an object is to be analyzed by 3D sputter depth profiling, one must take the surface angle into account both via the sputter yield and the correction of Equation 6.2. Using the known impact angle dependence of the sputter yield, one can simulate the erosion profile for such an object by numerical integration of Equation 6.1. An example of this is shown in Figure 6.16, where the shape of an originally hemi-elliptical object on a flat surface is calculated for equidistant ion fluence steps using the angular sputter yield variation as measured by Kozole et al.[54] for a cholesterol film bombarded with a 40 keV C_{60}^+ ion beam. As displayed in Figure 6.16a, the experimental yield data were fitted to a suitable functional form and cut off by a quadratic function for impact angles above $75°$.[§] The bottom panel shows the resulting height profile of the object at different projectile ion fluences, where the length units of the axes are arbitrary as long as both are the same. The direction of projectile impact is indicated by the arrow. The evolution of the shape of the bombarded object at various stages of the erosion is evident. At points where the incident ion beam strikes tangential to the surface (or, which are shadowed from the impinging ion beam), erosion is not possible and the erosion rate is zero. As a consequence, topographical features develop which may even give rise to instabilities, leading to a macroscopic roughening of the surface. It is obvious that under such circumstances the assumption of a constant and homogeneous erosion rate (which implicitly forms the basis of pseudo-3D

[§]The sputter yield must necessarily go to zero for impact angles approaching $90°$.

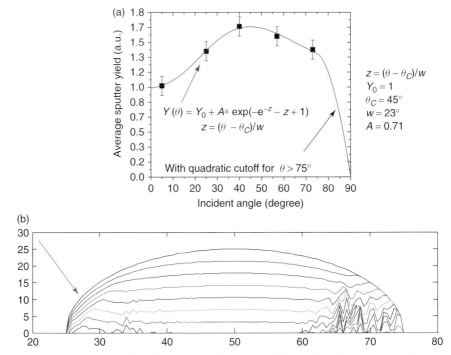

Figure 6.16 Cross-sectional erosion profile of an elliptical object bombarded with a parallel cluster ion beam under $45°$ impact with respect to the substrate surface as indicated by the blue arrow in panel (b). The different colors refer to equidistant steps of the projectile ion fluence, with the blue line depicting the original shape of the object at zero fluence. The numbers at the axes can be interpreted in any length unit as long as both are identical. The data were calculated using the impact angle dependence of the sputter yield depicted in panel (a) (data points measured for cholesterol under C_{60}^+ bombardment[54] and fitted by the indicated functional form).

image stacking) must be largely misleading. It should also be noted that sample rotation during ion bombardment may reduce but not completely mitigate these effects.

6.3.3 Depth Scale Calibration

Although the pseudo-3D representation depicted in Figures 6.3 through 6.6 provides some qualitative insight, more steps are required to achieve a true 3D representation of the analyzed sample volume. The crucial goal is to evaluate the z-coordinate at which a particular voxel (i,j,k) corresponding to a pixel (j,k) in an acquired image number (i) was actually located in the sample. Three factors determine the height (depth) of a voxel. First, the topography of the original surface before sputter depth profiling must be known. In the following examples described, the topography was determined from AFM images of the surface measured before the depth profile analysis. Second, the ion fluence applied during the depth profiling experiment must be converted into eroded depth. In general, this depth scale calibration will be nonlinear because the erosion rate can vary significantly between different vertical layers. Moreover, as outlined earlier, erosion rates can, in principle, also vary between different lateral areas of the sample, thus making the depth progression between subsequently acquired images pixel-dependent.

As a starting point to address the depth scale calibration, it is necessary to register topography images obtained before and after (and ideally also during) the depth profile analysis in a pixel-by-pixel manner with the acquired chemical images. Since orientation and displacement can be different for all these data sets, at least two marker points must exist at the surface that can be unambiguously identified in all images. Note that these markers must be located outside the eroded crater area because otherwise they will be destroyed during the depth profile acquisition. If the sample does not *a priori* contain suitable features, they can be imprinted using a focused ion beam. As an important step, the topography images must be cropped and converted to the same pixel resolution as the chemical images. If image area and erosion area were chosen to be the same during the depth profile analysis, this task is straightforward as the erosion area is clearly visible as the sputter crater in the topography data. Otherwise, one has to rely on a known relation between image and erosion area thereby adding ambiguity, or at least more complexity, to the data analysis. The success of the registration process must be carefully examined by overlaying chemical and (registered) topography images in order to find a perfect match. This step is important as only slight registration errors may translate into huge depth scale calibration errors in cases where the sample exhibits strong topographical features (such as the trenches in the example of Figure 6.8).

Finally, as topography images typically provide only a relative change in surface height, the respective zero of all topography images must be shifted such that the average height of all pixels located outside the eroded crater area is the same. Then, subtraction of the images directly reveals the total depth that has been eroded during the entire depth profile as a function of the lateral pixel coordinate.

For the example of Figure 6.8, the resulting topography data are shown in Figure 6.17. If the erosion rates were identical at each pixel, the contour plot in Figure 6.17 would consist of only a single color. Clearly, the data show that this assumption is incorrect for the example presented here. Depending on the lateral position inside the crater, the total eroded depth spans a range of +300 to −50 nm, where negative values indicate the *buildup* instead of removal of material during ion bombardment. This finding is of central importance in converting a pseudo-3D image to a true three-dimensional representation as it indicates that the depth scale calibration must be performed *individually* on each lateral pixel of the analyzed area.

For the example presented here (Figure 6.17), analysis shows that the large lateral variation of the erosion rate is related to the chemical damage and gallium implantation produced by the FIB beam in connection with the large topography of the sample before the depth profile analysis. In particular, the dark blue or black colored patches inside the letters "PSU" indicate that the bottom of the FIB-eroded trenches *rises up* in the course of the depth profile, indicating an apparent filling

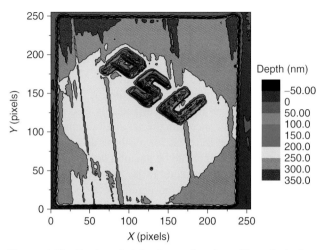

Figure 6.17 Total eroded depth as a function of lateral pixel coordinate for the image depth profile depicted in Figure 6.8. Reproduced from Wucher et al.,[58] with permission from ACS Publications.

of the trench holes under C_{60}^+ bombardment which is caused by redeposition of material sputtered from the trench walls. It is interesting to note that this phenomenon occurs although the trench walls are actually not very steep,[¶] forming a maximum slope of less than $5°$ with respect to the original film surface. It can therefore be expected that effects similar to this will often play a role on samples featuring aspect ratios of that order or larger. In the regions showing a net buildup of material instead of removal, the depth scale is obviously ill-defined. In the remaining part of the crater, total erosion depths between 0 and 300 nm are observed.

As discussed earlier, the most direct approach to calculate the pixel-dependent erosion rate is to divide the total eroded depth at each pixel by the total projectile ion fluence. Assuming this average erosion rate to be constant throughout the depth profile, this would result in a constant (but pixel-dependent) height step between subsequent images (provided the chemical images are separated by equivalent fluence erosion cycles). The assumption of a constant erosion rate, on the other hand, is questionable when profiling across a vertical sequence of different layers and therefore also needs to be examined. For that purpose, it is advantageous to extract "regional" depth profiles from the acquired data set. These are obtained by summing the intensity of a particular measured mass spectrometric signal over all pixels within a selected area (region) of interest and plotting the result as a function of erosion time or projectile ion fluence. The selection of suitable regions depends on the nature of the investigated sample and therefore cannot be generalized. For the example of Figure 6.8, two different regions can be identified, namely (1) the area where the virgin molecular film is still present and (2) the area where the original

[¶]The steepness generally appears very exaggerated in AFM images because of the largely different vertical and lateral length scales.

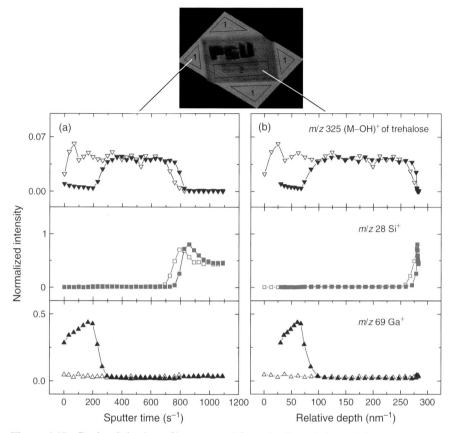

Figure 6.18 Regional depth profiles extracted from the data of Figure 6.8. The plotted signals (arbitrary units) represent the intact molecular film (blue), the implanted gallium (red), and the silicon substrate (green). Open symbols represent the depth profile from the pristine undamaged region, while the closed symbols represent the depth profiles from the Ga-damaged regions. Recreated from Wucher et al.,[58] with permission from ACS Publications.

surface was subjected to Ga^+ ion prebombardment. For illustration purposes, the respective profiles are shown in Figure 6.18.

The data obtained for region (1) depicted in Figure 6.18a represent a typical molecular depth profile as described in detail in Chapter 5. From the fluence needed to completely remove the entire molecular film in connection with the measured total eroded depth, one can calculate the erosion rate of the undisturbed film. In the calculation, it must be taken into account that the erosion rate varies strongly between the molecular overlayer and the substrate. As a first order approximation,[55] this can be done by interpolation in the interface region according to

$$\dot{z} = \chi_{film} \cdot \dot{z}_{film} + \chi_{substrate} \cdot \dot{z}_{substrate} \qquad (6.3)$$

where $\dot{z}_{\text{film}}$ and $\dot{z}_{\text{substrate}}$ denote the erosion rates of the molecular film and the substrate, respectively. The weighing factors χ_X must, in some way, be derived from the measured mass spectrometric signals. Various recipes have been published in the literature.[55−61] The most straightforward approach is to use a linear interpolation according to

$$\chi_X = \frac{I_X/I_X^{\max}}{\sum_Y I_Y/I_Y^{\max}} \quad (X, Y = \text{film, substrate}) \tag{6.4}$$

where $I_{\text{film}}^{\max}$ and $I_{\text{substrate}}^{\max}$ denote the respective maximum signal detected throughout the depth profile. The advantage of Equation 6.4 is that it automatically ensures the normalization $\sum_i \chi_i = 1$ and is easily extended to more than one interface (i.e., more than two significant signals). Using a known (measured) substrate erosion rate, the value of $\dot{z}_{\text{film}}$ can then be determined by integrating Equation 6.3 and equating the result with the measured total eroded depth.

In region (2) depicted in Figure 6.18b, the depth scale calibration is more complicated. At the beginning of the profile, erosion proceeds through the FIB-prebombarded material that is apparently heavily doped with gallium. In this regime, the molecular ion signals from the film are very low, indicating severe chemical damage induced by the Ga^+ impact. Both factors lead to an altered erosion rate when compared to the undisturbed film. After application of a certain ion fluence, on the other hand, the Ga signal disappears and the molecular ion signals rise to the same level as in the undisturbed film. In this regime, the undamaged film underneath the altered layer has been uncovered and is being removed with the same erosion rate as in region (1), until the substrate is reached. An essential, but so far unknown, quantity is the erosion rate within the Ga-doped/damaged layer. Using the same interpolation scheme as depicted in Equations 6.3 and 6.4, we can now write the erosion rate as

$$\dot{z} = \chi_{\text{Ga}} \cdot \dot{z}_{\text{altered layer}} + \chi_{\text{film}} \cdot \dot{z}_{\text{film}} + \chi_{\text{substrate}} \cdot \dot{z}_{\text{substrate}} \tag{6.5}$$

Inserting $\dot{z}_{\text{film}}$ and $\dot{z}_{\text{substrate}}$ from Equation 6.3, it is now possible to integrate Equation 6.5, set the result equal to the measured total eroded depth in this area, and solve for $\dot{z}_{\text{altered layer}}$. For the specific example of Figure 6.8, one finds a substantially reduced erosion rate across the damaged layer, which is only about 1/3 of that measured in the undisturbed film. In the silicon substrate, the erosion rate is by a factor of 37 smaller than in the intact molecular film. These findings illustrate the importance of an appropriate depth scale calibration that is determined by integrating Equation 6.5 as a function of sputter time or fluence. In the example of Figure 6.8, the thickness of the altered layer can be calculated to be about 50 nm.

We reiterate that the linear interpolation of erosion rates according to Equation 6.4 naturally represents only a first order approximation that may not be accurate in all cases, particularly for interfaces between layers with widely different sputter yields (e.g., a molecular film and an inorganic substrate). This approach is further questionable[62] in cases where intermediate layers (for instance oxide or hydroxide layers) are present at such an interface.[61] Moreover, it has been demonstrated that erosion rates induced by, for instance, a fullerene cluster ion beam may drop with increasing eroded depth even if a homogeneous sample is being bombarded. The exact reason for this effect, which appears to be significantly reduced at low temperature[36,59,63] or if the erosion is performed with a rare gas cluster ion beam,[64] is currently not well understood. Certainly, the damage mechanisms described in Chapter 5 play a role. Therefore, it is of utmost importance to measure the erosion rate as a function of eroded depth. In lieu of an *in situ* measurement, this has to be done in an extremely time- and sample-intensive way by applying many different ion fluences and measuring the resulting erosion craters *ex situ*. Experiments of this kind have been conducted,[54,65,66] but it appears difficult to resolve the erosion rate variation across the shallow interface in this manner.[54]

An elegant strategy to obtain more detailed information is to erode a wedge-shaped crater by applying a linearly increasing fluence between both sides of the erosion area as was discussed earlier. In this way, one single topography measurement of the resulting crater is sufficient to determine (i) the nonlinear relation between fluence and erosion rates (from the slope of the wedge) and (ii) the fluence-dependent evolution of surface roughness under prolonged ion bombardment (from the high frequency "noise" of the wedge surface).[35] In principle, the wedge erosion can be interrupted at regular intervals and images can be taken at each step. After the topography measurement of the wedged crater, the sample can be reinserted and the same procedure can be applied in the opposite direction, thereby now eroding the wedge down to the substrate. The resulting image stack represents a set of tomographic slices in different directions, which can afterwards be converted into a true 3D representation of the eroded sample volume.

6.4   THREE-DIMENSIONAL IMAGE RECONSTRUCTION

In order to reconstruct the true three-dimensional structure of the analyzed sample, the location (x, y, z) of each voxel (i, j, k) of the data set needs to be calculated. Determination of x and y from the pixel indices (j) and (k) is straightforward, as the pixel steps Δx and Δy are easily calculated from the total number of pixels and the measured lateral dimensions of the sputter crater. Since the correlation between (x, y) and (j, k) remains constant—at least to first order—throughout the entire depth profile, this mapping only needs to be done once. The problematic part is the correlation between the vertical coordinate (z) and the image index (i) (denoting the first, second, third, etc. acquired images per sputter cycle), which will be different for each pixel (j, k). In order to calculate $z(i, j, k)$, the nonlinear depth scale calibration described in Section 6.3 must be employed for each pixel of the lateral image area separately. In cases where different substructures are crossed in the vertical direction, signals representative for the specific chemistry of these structures must be identified and utilized for the erosion rate interpolation procedure. It is important to note that this process is highly sample specific and must be optimized separately for each individual system. It is this fact that renders true 3D imaging so hard to generalize, and it is also the reason why practically all published data on 3D sputter depth profiling have only been visualized in the pseudo-3D mode (in some cases with substrate level correction as described in the following paragraphs).

We will illustrate the process again using the example of Figure 6.8. Starting from the original topography measurement, we define $z(i = 1, j, k)$ for the first image taken before sputter erosion begins. In principle, the depth interval Δz eroded in each subsequent sputtering cycle must be calculated by interpolating the erosion rate between the values appropriate for the undisturbed molecular film, the altered layer, and the substrate, respectively, based on the respective mass spectral signals recorded in the previous image. Here, a problem arises because of the small number of secondary ions detected on each pixel, which makes a proper interpolation difficult. Hence, it will, in most cases, be necessary to sum (and average) the signal over a few neighboring pixels. The averaging process, however, results in a loss of lateral resolution which might lead to erroneous results in cases where erosion rates change drastically as a function of lateral image position. Such a situation may, for instance, occur if the substrate is reached in specific areas, while other neighboring areas still consist of molecular film material. In the specific example discussed here, we summed the signal over an area of 5 × 5 pixels and still found the resulting averaged signal too noisy for an interpolation scheme. In these cases, it might be better to *switch* instead of *interpolate* the erosion rate based on a threshold value of the averaged signals. In the particular example of the data presented in Figure 6.8, the erosion rates were therefore assumed to switch between the values of the altered layer and the intact molecular film if the averaged Ga signal exceeded a threshold value of 0.2 counts/pixel, and between the values of the molecular film and the substrate if the averaged Si signal exceeded a threshold value of 0.2 counts/pixel, respectively. The choice of the switching threshold is critical and needs to be optimized for each sample. The resulting three-dimensional representation of the data is depicted in Figure 6.19. In order to generate the displayed isosurface rendering, the irregularly spaced (x, y, z, I) [‖] data need to be converted into an isospaced grid. For that purpose, the data set is first scanned to find the limiting values of the coordinates x, y, and z. While this is straightforward for x and y (with x_{min}, y_{min} and x_{max}, y_{max} corresponding to the lateral image area), the limits z_{min} and z_{max} depend on the measured surface topography. Then, a three-dimensional cell grid is spanned between these limits with a predefined cell dimension $\Delta x \times \Delta y \times \Delta z$. While Δx and Δy are naturally defined by the image pixel size, the value of Δz can be arbitrarily set. As the next step, the (x, y, z, I) data are binned into the cell grid by summing the intensity of all data points whose coordinates (x, y, z) fall within a particular cell. Owing to the irregular nature of the measured data set (induced by the fact that the sample surface and therefore the image slices are not flat), there are many cells that contain no data point at all. In other words, these cells do not belong to the analyzed sample, as no point of the sample volume probed by the depth profile has coordinates (x, y, z) that fall into such a cell. These cells therefore need to be distinguished from those that do contain data points (in other words, belong to the sample), albeit

---

[‖] Note that I denotes the intensity of a particular mass spectrometric signal, and therefore there are as many such data sets as there are mass peaks of interest.

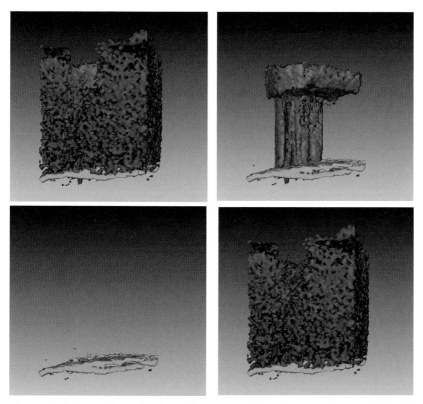

Figure 6.19 Three-dimensional sample reconstruction with corrections, calculated from the data presented in Figure 6.8. Reproduced from Wucher et al.,[75] with permission from Springer.

with zero signal intensity, I. For that purpose, these cells were given a negative intensity value in order to be able to exclude them from the rendering process by means of a filter. For the visualization displayed in Figure 6.19, the depth sampling interval was set as $\Delta z = 1$ nm and the resulting cell array was downsampled to combine $5 \times 5 \times 5$ cells into one supercell. Isosurface rendering was performed using threshold intensity values of 25 (Ga), 20 (Si), and 1 (peptide) counts per supercell. It is seen that the substrate is now reached at approximately the correct depth across the entire crater. Moreover, the thickness of the altered layer in the Ga^+ prebombarded area is consistent with the depth profile data of Figure 6.18b.

At this point, a few words are in order regarding the interpretation of isosurface images. It should be noted that the threshold values used in such a visualization play an extremely important role as the displayed features represent contour surfaces connecting points where the respective signal closely approximates these threshold values. Changing these values may therefore drastically change the visualization result. From a fundamental perspective, the rendering process only makes sense if there is a clear separation between regions with signal that is higher or lower than the selected threshold value. If there is a connected volume of high signal surrounded by a volume with low signal, then the result of the rendering process will be a closed surface, showing the contour of the high signal region. This is visible for the red and green surfaces displayed in Figure 6.19, which separate the gallium-implanted sample volume (red) and the silicon substrate (green) from the intact molecular film volume. In this case, the spatial extension of the displayed feature depends on the threshold value and will shrink and expand with increasing or decreasing threshold, respectively. If the threshold is lowered down to the noise level of the signal, the rendered isosurface will disintegrate into many little features. This is visible for the blue surface that basically separates volumes with any peptide intensity from those with zero intensity. In this case, the rendered image displays a volumetrically filled region, where the definition of an isosurface becomes meaningless. Nevertheless, the resulting image is still usable here to visualize the volume of the intact molecular film.

In order to better visualize the data, each secondary ion image can be superimposed with the corresponding surface topography and displayed separately. In this way, the entire depth profile can be viewed as a series of sequential three-dimensional images illustrating the development of the topography along with the chemical composition at the momentary surface. For the example of Figure 6.19, snapshots at particularly interesting points of the profile are shown in Figure 6.20. It is seen that the first image displays the original surface topography locating the gallium signal (red) inside the FIB-eroded crater and within the deep trenches, respectively. During the evolution of the profile, it is evident that the virgin trehalose (blue) overlayer is more rapidly removed than the Ga^+-altered layer, resulting in a rather flat surface after the altered layer has been removed. While the remaining overlayer is quickly eroded in the regions away from the trenches, no change in height is observed at some points inside the walls of the original trenches. At the bottom of the deep trenches, the surface is seen to move upward as depth profiling proceeds.

The three-dimensional data set produced using the above procedures is still not free from artifacts. This is due to the fact that a surface topography measurement using techniques such as AFM employs correction procedures that level the average height across an acquired image. As a consequence, any large-scale gradient of the surface height will automatically be leveled. In order to illustrate the consequence of this problem, a cross-sectional view of the data set presented in Figure 6.19 is shown in Figure 6.21. It is seen that the data indicate a slope of the film–silicon interface, which must clearly be wrong as it is known that the substrate surface is flat. In principle, such information must be utilized by applying another correction to the data set, which renders the film–substrate interface horizontal. For that purpose, the index i_{switch} is identified for each pixel stack (j, k) where the erosion rate is switched from the film to the substrate value. Then, all z values of this pixel stack are shifted in parallel in order to render $z(i_{switch})$ at the same height. Note that this "leveling correction" (which is the same as that published by Breitenstein et al.[67–69] and applied to the data presented in Figure 6.5[70] and Figure 6.26[71]) requires prior knowledge about the sample and is therefore not easily generalized.

What can be deduced from Figure 6.21 is that the thickness of the molecular film deposited on the substrate must vary across the analyzed lateral area by as

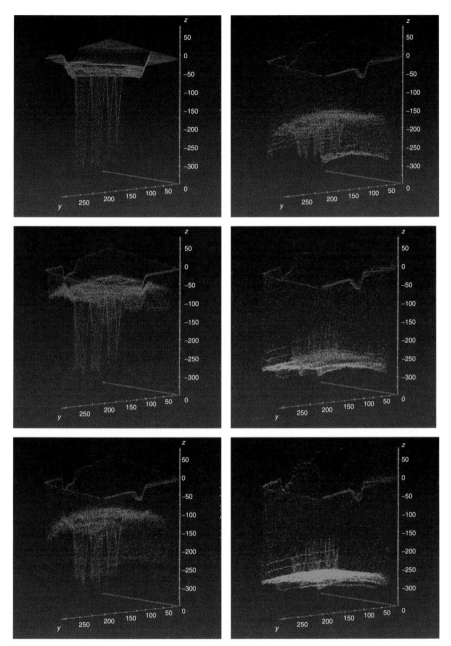

Figure 6.20   Three-dimensional visualization of selected SIMS images from the data set presented in Figure 6.8 overlayed with topography.   Reproduced from Wucher et al.,[58] with permission from ACS Publications.

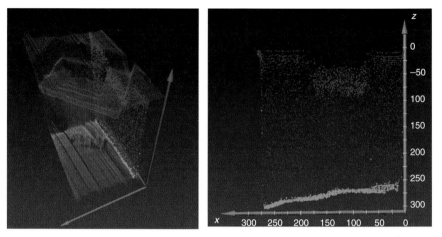

Figure 6.21 Vertical cross-sectional view of the data set presented in Figure 6.8.[58] Figure reprinted from Wucher et al.,[75] with permission from Springer.

much as 10% of the entire film thickness ($\sim$30 nm in this case). This finding has important consequences with respect to the apparent depth resolution in sputter depth profiling experiments. From Figure 6.21, it is entirely clear that a sputter depth profile acquired with a gating area of about half the sputter crater dimension must necessarily exhibit an artificial interface width contribution of about 20 nm which arises exclusively from large-scale film thickness fluctuations. In order to investigate and eliminate such effects, three-dimensional depth profiling data is extremely helpful, as regional depth profiles can be extracted with different gating area from one single measured data set. For the particular case discussed here, the apparent depth resolution was found to improve significantly with decreasing gating area as shown in Figure 6.22.[72] By extrapolating the resulting data to zero area, the "intrinsic" depth resolution can be determined as an important characteristic of the depth profiling method ($\sim$7 nm in the present case[72]).

6.5   DAMAGE AND ALTERED LAYER DEPTH

On the basis of the concepts described above, one can investigate the influence of a FIB ion beam on a molecular film in more detail. Of particular interest are the following questions:

1. How does the altered (damaged) layer evolve as a function of accumulated ion fluence?

2. Does a polyatomic ion, for instance, Au_3^+, generate less damage than a monatomic (Au^+ or Ga^+) beam?

In order to address the first question, we refer to an experiment similar to that described in the previous section. In this case, a Langmuir–Blodgett (LB) multilayer film of 400 nm thickness deposited on a silicon substrate was

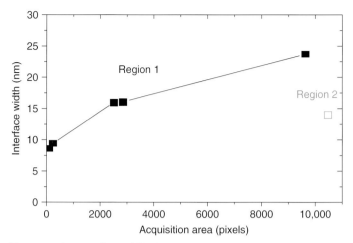

Figure 6.22   Interface width determined from selected pixels within the imaged area of the data presented in Figure 6.8 versus "gating area" selected for the analysis. Reproduced from Wucher et al.,[72] with permission from Elsevier.

prebombarded with 15 keV Au^+, Au_2^+, and Au_3^+ ion beams to varying fluences and subsequently analyzed in a 3D depth profile using a 40 keV C_{60}^+ ion beam for sputtering and image acquisition.[73] Interestingly, even under prolonged bombardment with the gold beam (up to fluences of several 10^{15} ions/cm^2) the sample could not be eroded to more than a depth of about 15 nm. It is presently unclear what causes this effect; possible explanations could include charging or a beam chemistry-induced quenching of the erosion rate similar to that sometimes observed under C_{60} bombardment. Subsequent sputter depth profiling with the C_{60}^+ beam revealed a severely damaged subsurface region at the gold-bombarded part of the surface. The surface topography within the image area before and after acquisition of the 3D depth profile is shown in Figure 6.23. As a technical side note, the figure shows that the data had to be subjected to a special smoothing algorithm in order to remove spikes originating from dust particles which would otherwise corrupt the depth scale calibration. The data show that significant surface topography evolves during erosion of the gold prebombarded area, while this is not the case in regions where the virgin molecular film is being removed. It is also observed that the erosion rate of the damaged subsurface layer must be smaller than that of the intact film. The three-dimensional reconstruction of the sample along with a 3D visualization of selected image frames is shown in Figures 6.24 and 6.25, respectively. A strongly enhanced signal of C^+ and C_2^+ secondary fragment ions can be found in the altered subsurface layer, which is visualized by the red isosurface in Figure 6.24. At the same time, the molecular ion signal of the intact LB film (blue) is practically absent in this volume, indicating almost complete fragmentation of the molecular film within the altered layer influenced by the gold prebombardment. Again, one finds a significantly reduced erosion rate when profiling across the altered layer as compared to the intact molecular film, indicating that this might be a general feature of cluster beam molecular depth profiling.

240

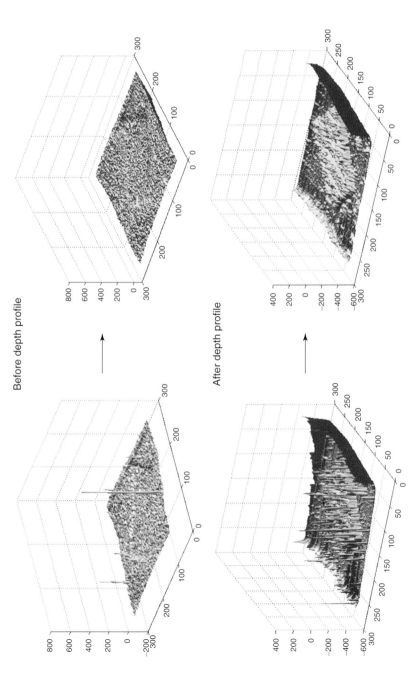

Before depth profile

After depth profile

Figure 6.23 AFM images (vertical axis in nm, horizontal axes in µm) of the sputtered crater bottom taken before and after a 3D sputter depth profile of a 300 nm Langmuir–Blodgett molecular film on silicon. The sample was prebombarded with an LMIG Au^+ ion beam in the central area visible in the images. Left panels: data before smoothing; right panels: data after smoothing. Data courtesy of Lu et al.[73]

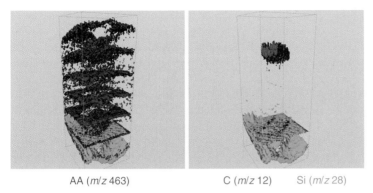

AA (*m/z* 463)   C (*m/z* 12)   Si (*m/z* 28)

Figure 6.24   Three-dimensional reconstruction of a Langmuir–Blodgett molecular film composed of 400 nm lipid film deposited on a silicon substrate which was prebombarded with a 15 keV Au$^+$ ion beam. The profile was acquired using a 40 keV C$_{60}{}^+$ cluster ion beam for sputtering and image acquisition. Data courtesy of Lu et al.[73] Green = Si substrate, blue = arachidic acid (AA) molecular signal, and red = carbon fragments.

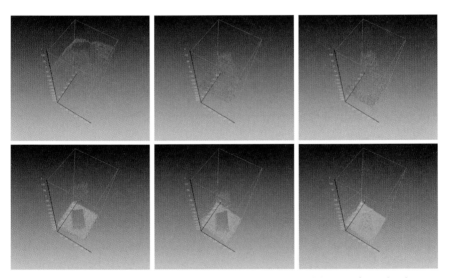

Figure 6.25   Three-dimensional visualization of selected SIMS images from the data set presented in Figure 6.24 overlayed with topography. Data courtesy of Lu et al.[73]

Using the same procedures as described above, one can determine the thickness of the altered layer as a function of the gold projectile ion fluence (C. Lu, A. Wucher and N. Winograd, unpublished data). The altered layer thickness is found to increase with increasing gold ion fluence even when sputter erosion has already stopped, reaching a steady-state value of about 100 nm after impact of about 6×10^{14} Au$^+$ ions/cm^2. This value is even larger than the one determined for a 15 keV Ga$^+$ bombardment of trehalose (50 nm), a finding that might be due to a larger penetration depth of the heavier gold projectile and/or the ordered structure of the LB multilayer film. In any case, the data confirm that damaged layer thicknesses of several ten nanometers coming out as fit parameters of the erosion dynamics model (see Chapter 5) are probably realistic. From the perspective of the FIB-ToF method, the data indicate that the best ion beam milling conditions, in terms of altered layer thickness (i.e., straggle) and the cleanup/polishing requirements, may be achieved with the use of the lower mass Ga ion.

In regards to the second question, while the altered layer thickness decreases with increasing nuclearity, this may not necessarily translate to more ideal FIB-SIMS etching. Simulation data for Au_3^+ and Au^+ sputtering, for example, shows that the lateral component of damage is greater for Au_3^+ than for Au^+. This is because Au has only a single trajectory, while the Au_3^+ breaks up into three divergent trajectories. Thus, in the case of FIB-ToF, in which vertical slicing is performed, it is likely that more polishing may be required to remove the lateral component of the beam-induced damage caused by Au_3^+ than for Au^+. Thus far there have not been any studies to verify this hypothesis.

6.6 BIOLOGICAL SAMPLES

In many biological applications, samples need to be kept at cryogenic temperatures during the entire analysis. While this is no problem with regard to the sputter depth profile (commercially available ToF-SIMS instrumentation is capable of handling and analyzing frozen samples), it poses a significant problem for the determination of surface topography. The protocol described is therefore likely to be impractical on samples such as, for instance, frozen-hydrated cells. Although there is commercially available instrumentation for low temperature AFM, these instruments usually have a very restricted maximum field-of-view (of the order of 10 μm) which is too small to capture an image of the entire sputter crater. Since we are not aware of commercially available optical interferometry instrumentation with nanometer height resolution that would be compatible with cryogenic sample cooling, measurement of the surface topography appears problematic and the depth scale calibration described in Section 6.3 cannot easily be performed for this type of sample. Therefore, the data treatment in most of the work published in this field is restricted to the assumption of a constant erosion rate across the entire eroded volume. Although this assumption is prone to fail when profiling across largely different matrices, one might argue that frozen-hydrated samples mostly consist of water ice as the main matrix material, thus rendering the assumption of a constant erosion rate not too far-fetched. Note, however, that this argument fails if the sample exhibits steep topographical features with large angle slopes. In order to still arrive at a meaningful three-dimensional reconstruction of the analyzed volume, one can determine the erosion rate from the ion fluence needed to remove a water ice film of known thickness (deposited, for instance, by dosing water vapor into the vacuum system). Then, utilizing the condition that the substrate is known to be flat, it is possible to use the leveling correction by rescaling the height coordinate based on the appearance of substrate related signals in the mass spectrum. An example of such an analysis is shown in Figure 6.26, where single HeLa cells deposited

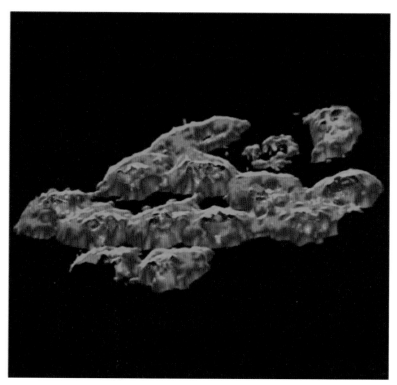

Figure 6.26 Three-dimensional visualization of the membrane (green, m/z 184.1, phosphocholine) and nucleus (red, m/z 136.1, adenine) chemistry in frozen-hydrated HeLa-M cells. Reproduced from Fletcher et al.[71] with permission from Wiley.

on a stainless steel substrate were subjected to a 3D depth profile using a 40 keV C_{60}^+ ion beam. It is seen that the analysis of specific molecular signals allows one to reconstruct the shape of the cell membrane as well as the location of the cell nuclei fairly well. It should be noted again, however, that the chosen isosurface rendering does not easily permit a quantitative estimate of the dimension of, for instance, the cell nucleus without a thorough investigation of the role of the used threshold value.

6.7 CONCLUSIONS

High resolution three-dimensional chemical characterization at the molecular level still poses a significant challenge in surface and thin-film analysis. The advent of focusable cluster ion beams has greatly enhanced the capabilities of surface mass spectrometry in this field. Using cluster ion beams such as C_{60}^+ or Ar_n^+, it is possible to erode molecular surfaces by ion sputtering without destroying the molecular integrity of the sample. As a consequence, techniques such as imaging sputter depth profiling or FIB-ToF tomography can now be extended to molecular structures as well, thereby opening up new application fields such as organic semiconductors, polymer chemistry, and biology. Using LMIGs providing cluster ion beams such as Au_n^+ and Bi_n^+, the lateral resolution of such an analysis can be pushed to the submicrometer level allowing, in principle, for a spatial resolution on the order of 100 nm. However, one must keep in mind that the useful resolution is often limited by detection sensitivity. Moreover, care needs to be taken when converting measured data into true 3D information about the analyzed structure. While simple image stacking may in some cases be a reasonable approximation, effects such as differential sputtering may in other cases cause severe distortions of the measured profiles. These effects must therefore be examined for every projectile/sample combination and corrected, in order to arrive at a reasonable 3D representation of the analyzed sample volume.

REFERENCES

1. Chabala, J. M.; Levi-Setti, R.; Wang, Y. L. *Appl. Surf. Sci.* **1988**, 32 (1–2), 10–32.
2. Levi-Setti, R.; Wang, Y. L.; Crow, G. *Appl. Surf. Sci.* **1986**, 26 (3), 249–264.
3. Levi-Setti, R.; Chabala, J. M.; Hallegot, P.; Yuh, L. W. *Microelectron. Reliab.* **1989**, 9 (1–4), 391–399.
4. Levi-Setti, R., Hallegot, P., Girod, C., Chabala, J. M., Li, J., Sodonis, A., and Wolbach, W. *Surf. Sci.* **1991**, 246 (1–3), 94.
5. Chabala, J. M.; Levi-Setti, R.; Wang, Y. L. *J. Vac. Sci. Technol.* **1988**, 6 (3), 910–914.
6. Levi-Setti, R.; Chabala, J. M.; Wang, Y. L. *Ultramicroscopy* **1988**, 24 (2–3), 97–113.
7. Levi-Setti, R., Chabala, J., and Wang, Y. L. *Scanning Microsc.* **1987**, 1 (1), 13–22.
8. Hallegot, P., Girod, C., Le Beau, M. M.; Levi-Setti, R. *Int. Soc. Opt. Eng.* **1991**, 1396, 311–315.
9. Hindie, E., Hallegot, P., Chabala, J. M., Thorne, N. A., Coulomb, B., Levi-Setti, R.; Galle, P. *Scanning Microsc.* **1988**, 2 (4), 1821–1829.
10. Pumphrey, G. M.; Hanson, B. T.; Chandra, S.; Madsen, E. L. *Environ. Microbiol.* **2009**, 11 (1), 220–229.
11. Chandra, S.; Pumphrey, G.; Abraham, J. M.; Madsen, E. L. *Appl. Surf. Sci.* **2008**, 255 (4), 847–851.
12. Chandra, S. *Appl. Surf. Sci.* **2004**, 231, 467–469
13. Chandra, S. *Appl. Surf. Sci.* **2003**, 203, 679–683.
14. Chandra, S.; Smith, D. R.; Morrison, G. H. *Anal. Chem.* **2000**, 72 (3), 104A–114A.
15. Fahey, A., Gillen, G., Chi, P.; Mahoney, C. *Appl. Surf. Sci.* **2006**, 252, 7312–7314.
16. Gillen, G.; Batteas, J.; Michaels, C. A.; Chi, P.; Small, J.; Windsor, E.; Fahey, A.; Verkouteren, J.; Kim, K. J. *Appl. Surf. Sci.* **2006**, 252 (19), 6521–6525.
17. Gillen, G.; Fahey, A. *Appl. Surf. Sci.* **2003**, 203–204, 209–213.
18. Gillen, G.; Roberson, S.; Fahey, A.; Walker, M.; Bennett, J.; Lareau, R. T. *AIP. Conf. Proc.* **2001**, 550, 687–691.
19. Gillen, G., Walker, M., Thompson, P., and Bennett, J. *J. Vac. Sci. Technol.* **2000**, 18 (1), 503–508.
20. Gillen, G.. Secondary Ion Mass Spectrometry Using Polyatomic and Cluster Primary Ion Beams. In Microbeam Analysis 2000; IOP publishing: Bristol, **2000**, pp. 339–340.
21. Gillen, G., Roberson, S., Ng, C.; Stranick, M. *Scanning* **1999**, 21 (3), 173–181.
22. Niehuis, E.; Grehl, T.; Kollmer, F.; Moellers, R.; Rading, D.; Kersting, R.; Hagenhoff, B. *Surf. Interface Anal.* **2011**, 43 (1–2), 204–206.
23. Gillen, G.; Fahey, A.; Wagner, M.; Mahoney, C. *Appl. Surf. Sci.* **2006**, 252 (19), 6537–6541.
24. Fisher, G. L.; Belu, A. M.; Mahoney, C. M.; Wormuth, K.; Sanada, N. *Anal. Chem.* **2009**, 81 (24), 9930–9940.
25. D. Rading et al., Data presented by IonToF GmbH at the Annual SIMS Workshop, Baltimore and the International SIMS conference, Riva del Garda, **2011**.

26. Fletcher, J. S.; Lockyer, N. P.; Vaidyanathan, S.; Vickerman, J. C. *Anal. Chem.* **2007**, 79 (6), 2199–2206.
27. Nygren, H.; Borner, K.; Malmberg, P.; Hagenhoff, B. *Appl. Surf. Sci.* **2006**, 252 (19), 6975–6981.
28. Carado, A.; Passarelli, M. K.; Kozole, J.; Wingate, J. E.; Winograd, N.; Loboda, A. V. *Anal. Chem.* **2008**, 80 (21), 7921–7929.
29. Hill, R.; Blenkinsopp, P.; Thompson, S.; Vickerman, J.; Fletcher, J. S. *Surf. Interface Anal.* **2011**, 43 (1–2), 506–509.
30. Fletcher, J. S.; Rabbani, S.; Henderson, A.; Blenkinsopp, P.; Thompson, S. P.; Lockyer, N. P.; Vickerman, J. C. *Anal. Chem.* **2008**, 80 (23), 9058–9064.
31. Thompson, D. A.; Johar, S. S. *Appl. Phys. Lett.* **1979**, 34 (5), 342–345.
32. Hsu, C. M.; McPhail D. S. *Nucl. Instrum. Meth.* **1995**, 101 (4), 427–434.
33. Skinner, D. K. *Surf. Interface Anal.* **1989**, 14 (9), 567–571.
34. Voigtmann, R.; Moldenhauer, W. *Surf. Interface Anal.* **1988**, 13 (2–3), 167–172.
35. Mao, D.; Wucher, A.; Winograd, N. *Anal. Chem.* **2010**, 82 (1), 57–60.
36. Mao, D.; Lu, C.; Winograd, N.; Wucher, A. *Anal. Chem.* **2011**, 83, 6410–6417.
37. Crecelius, A. C.; Cornett, D. S.; Caprioli, R. M.; Williams, B.; Dawant, B. M.; Bodenheimer, B. *J. Am. Soc. Mass Spectrom.* **2005**, 16 (7), 1093–1099.
38. Eberlin, L. S.; Ifa, D. R.; Wu, C.; Cooks, R. G. *Angew. Chem. Int. Ed.* **2010**, 49 (5), 873–876.
39. Hinder, S. J.; Lowe, C.; Maxted, J. T.; Watts, J. F. *Surf. Interface Anal.* **2004**, 36 (12), 1575–1581.
40. Holzer, L.; Indutnyi, F.; Gasser, P. H.; Munch, B.; Wegmann, M. *J. Microsc.* **2004**, 216 (1), 84–95.
41. Stevie, F. A.; Vartuli, C. B.; Giannuzzi, L. A.; Shofner, T. L.; Brown, S. R.; Rossie, B.; Hillion, F.; Mills, R. H.; Antonell, M.; Irwin, R. B. *Surf. Interface Anal.* **2001**, 31 (5), 345–351.
42. Heymann, J. A. W.; Hayles, M.; Gestmann, I.; Giannuzzi, L. A.; Lich, B.; Subramaniam, S. *J. Struct. Biol.* **2006**, 155 (1), 63–73.
43. Stevie, F. A.; Downey, S. W.; Brown, S. R.; Shofner, T. L.; Decker, M. A.; Dingle, T.; Christman, L. *J. Vac. Sci. Technol. B* **1999**, 17, 2476.
44. Giannuzzi, L. A.; Utlaut M. *Surf. Interface Anal.* **2011**, 43 (1–2), 475–478.
45. G. L., Fisher et al., Data presented at the Annual SIMS workshop, Norfolk, VA, 2010 and the International SIMS XVIII conference, Riva del Gorda, Italy, 2011.
46. Volkert, C. A.; Minor, A. M. *MRS Bull.* **2007**, 32 (05), 389–399.
47. Fisher, G. L.; Belu, A. M.; Mahoney, C. M.; Wormuth, K.; Sanada, N. *Anal. Chem.* **2009**, 81 (24), 9930–9940.
48. Volkert, C. A.; Minor, A. M. *MRS Bull.* **2007**, 32 (05), 389–399.
49. Prenitzer, B. I.; Urbanik-Shannon, C. A.; Giannuzzi, L. A.; Brown, S. R.; Irwin, R. B.; Shofner, T. L.; Stevie, F. A. *Microsc. Microanal.* **2003**, 9 (03), 216–236.
50. Mahoney, C. M.; Fahey, A. J.; Belu, A. M. *Anal. Chem.* **2008**, 80 (3), 624–632.
51. de Winter, D. A. M.; Mulders, J. J. L.. *Redeposition Characteristics of Focused Ion Beam Milling for Nanofabrication*; AVS, **2007**, pp. 2215–2218.
52. Bhavsar, S. N.; Aravindan, S.; Rao, P. V. *Precis. Eng.* **2012**, 36 (3), 408–413.
53. Fu, Y. Q.; Bryan, N. K. A.; Shing, O. N.; Hung, N. P. *Int. J. Adv. Manuf. Technol.* **2000**, 16 (12), 877–880.
54. Kozole, J., Wucher, A.; Winograd, N. *Anal. Chem.* **2008**, 80, 5293–5301.
55. Cheng, J.; Wucher, A.; Winograd, N. *J. Phys. Chem. B.* **2006**, 110 (16), 8329–8336.
56. Wagner, M. S. *Anal. Chem.* **2005**, 77 (3), 911–922.
57. Cheng, J.; Winograd, N. *Anal. Chem.* **2005**, 77 (11), 3651–3659.
58. Wucher, A., Cheng, J.; Winograd, N. *Anal. Chem.* **2007**, 79 (15), 5529–5539.
59. Zheng, L.; Wucher, A.; Winograd, N. *J. Am. Soc. Mass Spectrom.* **2008**, 19, 96–102.
60. Wucher, A.; Cheng, J.; Winograd, N. *Appl. Surf. Sci.* **2008**, 255 (4), 959–961.
61. Green, F. M.; Shard, A. G.; Gilmore, I. S.; Seah, M. P. *Anal. Chem.* **2009**, 81 (1), 75–79.
62. Mahoney, C. M. *Mass Spectrom. Rev.* **2010**, 29 (2), 247–293.
63. Lu, C.; Wucher, A.; Winograd, N. *Anal. Chem.* **2010**, 83 (1), 351–358.
64. Lee, J. L. S.; Ninomiya, S.; Matsuo, J.; Gilmore, I. S.; Seah, M. P.; Shard, A. G. *Anal. Chem.* **2010**, 82 (1), 98–105.
65. Shard, A. G.; Green, F. M.; Brewer, P. J.; Seah, M. P.; Gilmore, I. S *J. Phys. Chem. B.* **2008**, 112 (9), 2596–2605.

66. Shard, A. G.; Brewer, P. J.; Green, F. M.; Gilmore, I. S. *Surf. Interface Anal.* **2007**, 39 (4), 294–298.
67. Breitenstein, D.; Rommel, C. E.; Stolwijk, J.; Wegener, J.; Hagenhoff, B. *Appl. Surf. Sci.* **2008**, 255 (4), 1249–1256.
68. Breitenstein, D.; Rommel, C. E.; Mollers, R.; Wegener, J.; Hagenhoff, B. *Angew. Chem. Int. Ed.* **2007**, 46 (28), 5332–5335.
69. Breitenstein, D.; Batenburg, J. J.; Hagenhoff, B.; Galla, H. J. *Biophys. J.* **2006**, 91 (4), 1347–1356.
70. Fletcher, J. S.; Vickerman, J. C. *Anal. Bioanal. Chem.* **2010**, 396 (1), 85–104.
71. Fletcher, J. S.; Rabbani, S.; Henderson, A.; Lockyer, N. P.; Vickerman, J. C. *Rapid Commun. Mass Spectrom.* **2011**, 25 (7), 925–932.
72. Wucher, A.; Cheng, J.; Zheng, L. L.; Willingham, D.; Winograd, N. *Appl. Surf. Sci.* **2008**, 255 (4), 984–986.
73. Lu, C.; Wucher, A.; Winograd N. *Surf. Interface Anal.* **2012**, doi:10.1002/sia.4838.
74. Nygren, H.; Hagenhoff, B.; Malmberg, P.; Nilsson, M.; Richter, K. *Microsc. Res. Tech.* **2007**, 70 (11), 969–974.
75. Wucher, A.; Cheng, J.; Zheng, L.; Winograd N. *Anal. Bioanal. Chem.* **2009**, 393 (8), 1835–1842.

CHAPTER 7

CLUSTER SECONDARY ION MASS SPECTROMETRY (SIMS) FOR SEMICONDUCTOR AND METALS DEPTH PROFILING

Greg Gillen and Joe Bennett

7.1 INTRODUCTION

Sputter depth profiling has long been used to investigate the distribution of subsurface elemental species. Detection and identification of secondary ions formed during the sputtering process can provide exceptional sensitivity for many elements in a wide range of materials and forms the basis for secondary ion mass spectrometry (SIMS) depth profiling. Under properly chosen primary ion beam conditions, it is possible to combine the technique's excellent sensitivity with superior depth resolution to create an unparalleled method for accurately determining the distribution of elements as a function of depth in a sample. These strengths have led to the successful application of SIMS depth profiling to study inorganic materials in areas such as metallurgy,[1] geology,[2] and microelectronics.[3] Current limitations in depth profiling of inorganic materials by SIMS most often result from the physics of the sputtering process, which can introduce multiple artifacts into a SIMS depth profile. These include degraded depth resolution due to collisional mixing,[4,5] ion and sputter yield transients at surfaces,[6,7] and sputter-induced roughening.[8–10] Fortunately, by proper selection of primary ion beam conditions it has so far proven possible to minimize or at least mitigate many of these artifacts. However, in the area of microelectronics, the evolution of advanced microfabrication capabilities has led to a continued reduction in the size of device components as reflected by the recent introduction of 22 nm node semiconductor device fabrication technologies. The very shallow dopant distributions, junction depths, and abrupt compositional variations in these next generation microelectronic devices place severe demands on analytical tools

Cluster Secondary Ion Mass Spectrometry: Principles and Applications, First Edition.
Edited by Christine M. Mahoney.
© 2013 John Wiley & Sons, Inc. Published 2013 by John Wiley & Sons, Inc.

such as SIMS, requiring further improvements in capabilities for ultra shallow depth profiling. This chapter will describe the origin of some of the artifacts that currently limit SIMS depth profiling performance as well as the evolution of ion beam technology. Specific emphasis will be placed on the application of cluster primary ion beams for ultra high resolution depth profiling. Examples will be drawn primarily from depth profiling investigations of various semiconductor materials and metal films.

7.2   PRIMARY PARTICLE–SUBSTRATE INTERACTIONS

7.2.1   Collisional Mixing and Depth Resolution

The interaction between the incoming energetic primary particle (atom, ion, or cluster) and the atoms in the substrate leads to collisional mixing—the term coined to describe the repositioning of atoms in the substrate that occurs during this interaction. To a first order, collisional mixing results in a repositioning of substrate and dopant atoms biased in the forward direction, deeper into the substrate.[4] Therefore, as the magnitude of collisional mixing increases, the depth resolution will degrade. Collisional mixing is strongly dependent on the primary ion beam penetration depth[11] which in turn is dependent on the primary particle(s) mass, impact energy, and incidence angle.[12−14] Investigation of collisional mixing can be facilitated by profiling samples that contain single or multiple delta layer structures. Delta structures are created using sophisticated vacuum deposition techniques to grow or deposit ultrathin layers of pure materials interspaced by matrix material. The idealized delta layer would be one atomic layer thick with atomically abrupt interfaces. In practice, these characteristics are seldom achievable, but today's deposition methods are capable of creating very thin layers with minimal intermixing. In a depth profile, any deviation from the idealized square wave approximation of a delta layer will provide a quantifiable assessment of the collisional mixing process. Profiles from a delta structure-containing sample were used to obtain the data represented in Figure 7.1. Profiles of a B delta layer in Si were obtained under O_2^+ bombardment at $0°$ (with respect to the surface normal) at various energies to highlight the energy dependence of collisional mixing. As the O_2^+ energy increases, the slope of the decaying tail of the profile increases.[13] This indicates that there is increased collisional mixing occurring at higher beam energies. Similar effects are observed in some organic species (see Chapter 5).

Historically, mono- or di-atomic primary ion beams (e.g., Ar^+, Cs^+, O^-, O_2^+) have been employed for SIMS depth profiling. Early experiments of sputtering semiconductor materials with these beams indicated that, in general, profiling with low energy primary ions (≤ 1 keV) with oblique incidence angles ($\geq 50°$) minimizes the penetration depth and reduces collisional mixing thereby improving depth resolution.[15−17] There exists, however, a special case where excellent depth resolution can be obtained when profiling Si with low energy (500 eV), normal incidence ($0°$) O_2^+.[18] In SIMS, the abruptness of the decrease (or increase) in secondary ion intensities as a function of depth is often used as a means to quantify depth resolution. The abruptness can be determined by calculating the decay

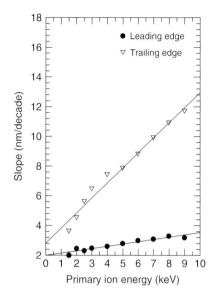

length, $\lambda_{(1/e)}$,

$$\lambda = \frac{(x_1 - x_2)}{\ln\left(\dfrac{I_{x_2}}{I_{x_1}}\right)} \tag{7.1}$$

where x_1 and x_2 are depths (in nanometers) and I_{x_1} and I_{x_2} are the secondary ion intensities at those depths. In the special case where $I_{x_2}/I_{x_1} = 10$, then $\lambda =$ nm/decade. Dowsett and Chu[19] have demonstrated that under normal incidence O_2^+ bombardment the decay length for B sputtered from a delta layer in Si could be approximated by $1.39 E_p^{0.56}$ where E_p is the primary beam energy. This leads to a prediction of $\lambda_{1/e} < 1$ nm when profiling with 500 eV O_2^+ at $0°$. A successful demonstration of that prediction is shown in Figure 7.2.

Profiling a series of B delta structures in Si with low energy O_2^+ at $0°$ produces $\lambda_{1/e} = 0.73$ nm ($\lambda = 1.65$ nm/decade). Figure 7.2 also shows that similar results can be obtained when profiling with 500 eV O_2^+ at $45°$ in combination with a high partial pressure of O_2 in the analysis chamber, that is, oxygen flooding. These results were obtained as a part of a SEMATECH-sponsored B depth profiling round-robin study aimed at investigating experimental conditions for optimizing depth resolution.[20] One of the best decay lengths reported for B profiling under oxygen bombardment is $\lambda = 0.83$ nm/decade for 150 eV O_2^+ at $45°$ with no O_2 flooding.[17]

7.2.2 Transient Effects

The profiles in Figure 7.2 also illustrate an artifact that is common to many profiles—transient behavior that occurs at the beginning of a profile and can

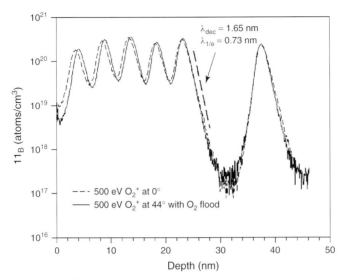

Figure 7.2 ^{11}B depth profiles from a B delta-doped sample under 500 eV O_2^+ bombardment at different incidence angles and chamber pressures. Both profiles show excellent depth resolution. λ_{dec} refers to the special case of calculating the decay length where the intensity decreases by 10x.

affect ion yields and sputter rates. In the case of the B profiles in Figure 7.2, the different sputtering conditions give rise to sputter rate variations at the beginning of the profiles that lead to differences in the apparent depth of the first four B delta layers. The origin of this phenomenon has been the subject of several studies[21-23] and algorithms have been created to correct for the artifact under specific sputtering conditions.[22] The extent (i.e., width) of the transient region and its effects can be minimized through the use of appropriate primary ion beam conditions, that is, low energy ions incident at normal or oblique angles.[7] The incorporation of the primary ion into the substrate can also influence the probability of secondary ion formation, that is, the ion yield. Until steady-state equilibrium sputtering conditions are achieved, there exists the possibility for ion yield variations at the beginning of the profile.[24-26] This variation can lead to errors in the experimentally calculated concentrations of elements near the surface. Again, proper selection of sputter conditions can minimize this artifact. In Figure 7.3, the width (in nanometer) of the transient region, where the ion yield of Si^+ is changing, can be seen to vary with primary ion beam angle. In this case, O_2^+ incident at $0°$ gives the narrowest transient region.[24]

Magee et al.[27] have developed an analysis protocol that seeks to correct profiles on a point-by-point basis by taking into account differences in relative ion yields and sputter rates that occur in the transient region of profiles in Si. This protocol is able to produce reasonably accurate profiles from the top few atomic layers of the sample. Recent studies with Cs^+ ions at energies of 150–200 eV show the potential for removal of the ion yield and sputter rate variations without data

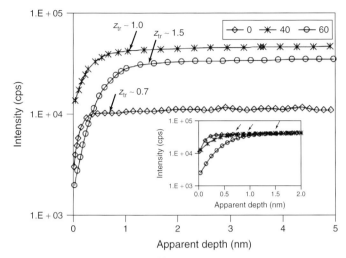

Figure 7.3   Depth profiles of $^{30}Si^+$ under O_2^+ bombardment at approximately $0°$, $40°$, and $60°$ incidence angles. The arrows show the end of the transient width. The insert shows the $^{30}Si^+$ profiles normalized to the profile at $40°$. Reproduced from Chanbasha and Wee,[24] with permission from Elsevier.

correction.[28] The general trend appears to be toward the use of ever decreasing beam energies to provide the most accurate profiles.

7.2.3 Sputter-Induced Roughening

Another artifact associated with ion bombardment of solids (metals and semiconductors) is sputter-induced roughening. Silicon, for example, is prone to roughening under O_2^+ and Cs^+ bombardment at lower impact energies and oblique incidence angles.[9,29–31] Roughening has also been observed while profiling III–V materials.[8,10] The onset of roughness leads to changes in sputter rate and ion yield. The onset of roughness is dependent on several factors including the primary ion beam energy and angle and can occur at such shallow depths that the accuracy of the profile is affected. An example of ion yield variations due to roughness is shown in Figure 7.4. Profiles from Si under 2 keV O_2^+ bombardment show significant change in the SiO^+ ion yield at different depths as a function of incidence angle. An atomic force microscopy (AFM) image (Fig. 7.5a) obtained from the bottom of a sputtered crater formed at an incidence angle of $45°$ (with respect to the surface normal) at a depth of 880 nm shows visible roughness on the 5–10 nm scale. However, at $55°$ minimal roughness was observed even out as far as 9.5 μm (Fig. 7.5b). In recent work with ultra low energy primary ion beams, roughening has been shown to be negligible at energies ≤200 eV.[17,32] The roughening of Si can be minimized or eliminated entirely when the proper primary ion beam conditions are employed.

Si and III–V substrates are not the only materials plagued by sputter-induced roughening. Many metals, polycrystalline materials, and organic materials are also

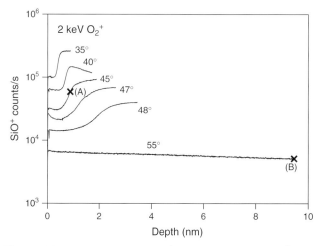

Figure 7.4 Depth profiles of SiO^+ taken under 2 keV O_2^+ bombardment at different angles. Locations (A) and (B) indicate depths where AFM images (shown in Figure 7.5) were collected.

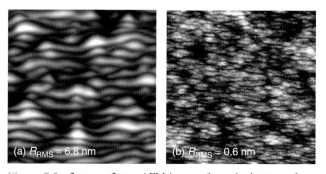

Figure 7.5 2 μm × 2 μm AFM images from the bottom of craters sputtered in Si using 2 keV O_2^+ at (A) 45° and (B) 55°. Image (A) taken at depth of 880 nm, image (B) from 9.5 μm (Fig. 7.4)

prone to roughening. The presence of multiple grain structures and boundaries, all oriented in different directions to the primary beam, can lead to nonuniform sputter yields between grains. As a result the depth resolution can be greatly degraded. One approach to reducing this artifact is to rotate the sample while sputtering.[33,34] Sample rotation averages out the angular dependence of the sputter yield for the various grain orientations with respect to the ion beam direction. Sample rotation has been successfully used for SIMS depth profiling of various metal films employed in the semiconductor industry (e.g., Al, Ta, Cu).[35,36] Rotation can also be used to reduce the previously discussed roughening of Si or GaAs.[10,37] However, sample rotation is not a straightforward approach, requiring special sample stages and well aligned optics.

7.3 POSSIBLE IMPROVEMENTS IN SIMS DEPTH PROFILING — THE USE OF CLUSTER PRIMARY ION BEAMS

Historically, one of the most widespread uses of SIMS for inorganic depth profiling applications has been to support semiconductor materials characterization. Microelectronic devices often rely on the intentional doping of trace elements into micrometer or submicrometer regions of silicon or related materials to provide the required electrical behavior. Also, many devices incorporate metal films or contact lines to provide electrical contact to these active areas. In both instances, the analytical characterization of semiconductor materials benefits greatly from the unique ability of SIMS depth profiling to reproduce accurately—with extremely high sensitivity—the distribution of elemental species as a function of depth. While there has been much debate concerning the optimal analysis conditions required to mitigate some of the beam-induced artifacts discussed in the earlier sections, it is generally agreed that the most promising approach for depth profiling of ultra shallow structures is the use of the lowest possible primary ion beam energy. Given the impressive results demonstrated earlier, with decay lengths approaching 0.8 nm/decade, it seems reasonable to ask how low of a primary ion energy can be used? From an engineering standpoint, the primary instrumental concern is designing low impact energy primary ion beam systems that provide adequate beam current, current uniformity, and spot size. Through continued evolution of ion source and primary ion column design it is now possible to produce such primary ion beams with energies of a few hundred electronvolts.[38] In practice, the use of low energy primary ion beams for SIMS has been limited not as much by ion source or optical design issues but rather by the well-known rapid decrease in sputter yield as a function of decreasing primary ion impact energy. In addition, the reduction in sputter yield using conventional ion beam species can make depth profiling at ultra low energies prohibitively long. Clegg reported that the sputter yield for Si under O_2^+ bombardment quickly falls to impractical values at energies <250 eV.[39] Ultimately, the threshold for sputtering of Si will limit the lowest possible primary beam energy that can be used, and is on the order of 50–100 eV for O_2^+ and Cs^+.[40] This fundamental limitation in the SIMS sputtering process and its influence on obtainable depth resolution was recently discussed by Vandervorst[41] and is described in Figure 7.6.

To explore continued improvements in SIMS depth resolution while addressing the issue of the sputtering threshold and the reduction in sputter rate at lower energies, a number of researchers have turned to the use of cluster primary ion projectiles. The use of cluster primary ion beams for SIMS depth profiling of inorganic materials has several potential advantages for semiconductor characterization. First, compared to commonly used monoatomic primary ions, a cluster ion dissociates upon impact with the surface, with each constituent atom of the cluster retaining a fraction of the initial beam energy. The reduction in cluster-constituent atom energy results in a significantly reduced penetration depth for a cluster impact.

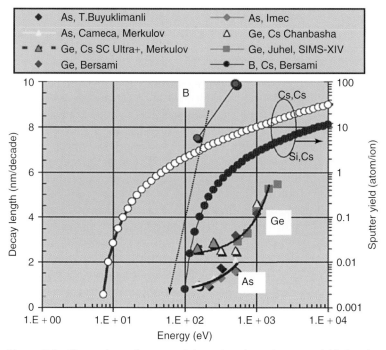

Figure 7.6 Comparison of measured decay lengths and sputter yield showing reduction in sputter yield as a function of primary ion energy. Reproduced from Vandervorst,[41] with permission from Elsevier.

This has important implications for inorganic depth profiling, as a reduced primary ion penetration depth would generally be expected to improve the measured depth resolution. Furthermore, a decreased primary ion penetration depth should also reduce surface ion yield transient effects, as the depth over which the primary ion beam concentration stabilizes will also be decreased. Recall from Chapter 1 that the reduction in primary beam impact energy is given by $E_c = \frac{E_0 \times M_c}{M_t}$ where E_c is the energy of an individual constituent atom after impact of the cluster ion with the surface, E_0 is the impact energy of the intact cluster beam, M_c is the mass of a cluster atom constituent, and M_t is the total mass of the intact cluster ion.

This cluster effect is illustrated in Figure 7.7, which shows Monte Carlo simulations of primary ion trajectories [SRIM (stopping and range of ions in matter) (www.srim.org)] and mean projected ranges of various primary ions in silicon under normal incidence bombardment. The partitioning of energy for the cluster projectile impact results in lower net individual atom impact energies with the effect being more pronounced as the number of atoms in the cluster is increased. In this example, a C_{60} projectile on silicon with an impact energy of 3 keV would have an individual carbon impact energy of 50 eV and a projected range of less than 1 nm. In addition to the reduction in impact energy, a cluster ion impact may also exhibit nonlinear sputtering behavior, thus allowing for low bombardment energies without the commensurate loss in sputter yield. First observed by

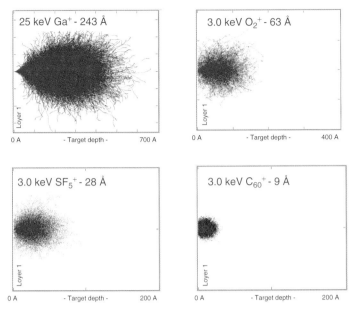

25 keV Ga⁺ - 243 Å

Figure 7.7   Comparison of primary ion trajectories for various SIMS ion beams in silicon at normal incidence. The mean projected range is also given.

Anderson and Bay in 1974,[42] the sputter yield from the impact of a cluster ion is often significantly enhanced compared to monatomic ions of the same mass and nominal impact energy. In fact, nonlinear sputtering phenomena can be even more dramatic for larger cluster projectiles allowing sputtering of materials at impact energies that would be considered subthreshold for monatomic ions. For example, massive gaseous cluster ion bombardment (GCIB) using Ar or SF_6 clusters containing hundreds to thousands of individual constituents have sputtering thresholds for Si that can approach a few electronvolts per atom (see Chapter 2).[43]

7.4   DEVELOPMENT OF CLUSTER SIMS FOR DEPTH PROFILING ANALYSIS

7.4.1 CF_3^+ Primary Ion Beams

One of the first demonstrations of a polyatomic primary ion beam for practical SIMS depth profiling analysis was presented by Reuter and Scilla in 1988.[44] They compared 5 keV impact CF_3^+ and O_2^+ primary ion beams generated in a cold cathode discharge ion source running on a quadrupole-based SIMS instrument. The performance of each beam was evaluated by depth profiling of B and Cu implants in silicon, metal multilayers, and an AlGaAs marker layer in GaAs. Qualitative improvements in depth resolution were observed for all three systems with the observed improvements thought to be related to both a lower penetration depth of the primary ion beam as well as a reduction in sputter-induced topography.

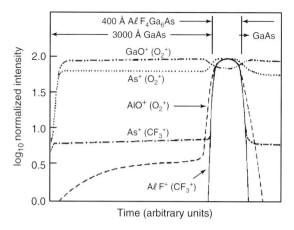

Figure 7.8 Depth profile of the respective secondary ion signals for an AlGaAs spike in GaAs depth profiled with 5 keV CF_3^+ and O_2^+ primary ion beams at normal incidence. Reproduced from Reuter and Scilla,[44] with permission from ACS Publications.

Additional benefits resulting from the use of the CF_3^+ beam was a reduction in oxygen enhanced diffusion of copper and enhanced secondary ion yields (for some species) relative to oxygen bombardment. This work suggested that the ability to use cluster primary ions containing reactive species such as fluorine also offers the possibility to "tune" the chemistry of the primary ion beam for different samples, providing greater flexibility in enhancing secondary ion yields for maximum sensitivity. Figure 7.8 shows a representative result from this work demonstrating sharper interfaces and greater dynamic range in the profile of the AlGaAs/GaAs sample for CF_3^+ bombardment as compared to O_2^+. It is also notable that this work utilized preexisting and commercially available ion sources making the approach available to a variety of SIMS users without a requirement for unique or specialized ion source development.

7.4.2 NO_2^+ and O_3^+ Primary Ion Beams

Another early example of using small cluster ions for depth profiling in semiconductors may be found in the work of Ronsheim and Fitzgibbon in 1992[45] who demonstrated that the use of the NO_2^+ trimer generated in a conventional duoplasmatron ion source (with N_2O feed gas) on a commercial magnetic sector instrument gave superior depth resolution for depth profiling of bipolar junction transistors as compared to O_2^+ bombardment under the same conditions. Later work also demonstrated that they could achieve a 30% decrease in trailing edge broadening for profiling of low energy As implants in Si.[46] An example of this approach for depth profiling of a 5 keV As implant in silicon is shown in Figure 7.9, which compares the profiles obtained with both O_2^+ and NO_2^+ primary ion beams. It is also important to note from this work that the approach of using cluster projectiles to improve depth resolution is particularly attractive on commonly used magnetic sector SIMS instruments which utilize a high sample voltage bias to provide high secondary ion collection efficiency. In this type of instrument, the use of a primary ion beam of the same polarity as the sample bias results in deflection of the primary beam reducing net impact energy and increasing beam impact angle

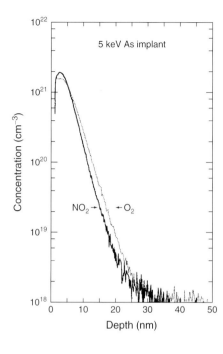

Figure 7.9   Comparison of NO_2^+ and O_2^+ depth profiles of a 5 keV As implant in silicon. Reproduced from Ronsheim and Lee,[46] with permission from the American Vacuum Society.

relative to the surface normal. Both effects tend to reduce the mixing depth and can improve depth resolution with the caveat that the deflection effect generally precludes (without specific instrument redesign) the use of ultra low energy primary ion beams that are more routinely used on quadrupole SIMS instruments with the sample stage at ground potential.

In a similar manner, Yamazaki and Mitani demonstrated in 1997[47] that O_3^+ primary ion beams, also produced in a standard duoplasmatron ion source operating on a magnetic sector SIMS instrument, could provide improved decay lengths for B delta-doped samples as compared to O_2^+ under the same conditions.

7.4.3 SF_5^+ Polyatomic Primary Ion Beams

An important milestone in the development of cluster SIMS for depth profiling of semiconductors was the early work of Iltgen et al. from the University of Muenster in 1997–1998[48,49] who compared a variety of primary ion beams (Ar, Xe, O_2^+, SF_5^+) for dual beam depth profiling on a time-of-flight (ToF)-SIMS instrument. In this mode of operation, a low energy sputter gun (0.3–2 keV) was used for sputtering while a high energy (10–25 keV Ar^+ or a Ga^+ liquid metal ion source) was used for analysis. With this mode of operation the $1/e$ decay length for a single B delta layer in Si was found to be 0.53 nm for SF_5^+ bombardment at 600 eV impact energy, as shown in Figure 7.10. This early result remains one of the very best decay lengths ever published for SIMS depth profiling and points out the potential of cluster SIMS for depth profiling in semiconductors. Figure 7.11 shows the effect on decay length of various experimental conditions including primary

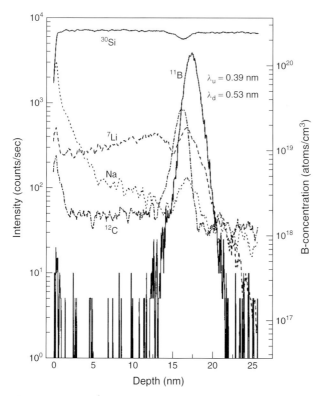

Figure 7.10 SF_5^+ depth profile of a B delta layer in Si. λ_u and λ_d refer to the decay lengths for the upslope and downslope of the B peak. Reproduced from Iltgen et al.,[48] with permission from the American Vacuum Society.

beam species, beam energy, and incidence angle. In all cases, the use of the SF_5^+ primary ion beam was found to provide improved depth resolution. Note also that oxygen flooding was used in all examples.

Following up on the work at the University of Muenster, Gillen and coworkers at NIST developed an SF_5^+ ion source for use on a magnetic sector SIMS instrument.[50–52] This work demonstrated the potential advantages of SF_5^+ bombardment for reduction of sputter-induced topography in metals with significant improvements in layer resolution during depth profiling of a Ni/Cr multilayer depth profiling reference material with O_2^+ and SF_5^+ at 3 keV impact energy, as shown in Figure 7.12. Similar improvements were noted in SF_5^+ depth profiling of B delta-doped layers although the ultimate depth resolution was limited by sputter-induced topography formation and was only found to be better than O_2^+ when using oxygen backfilling as shown in Figure 7.13.

7.4.4 CSC_6^- and C_8^- Depth Profiling

Much of the early work for cluster depth profiling focused on the analysis of electropositive elements in semiconductor materials using positively charged primary

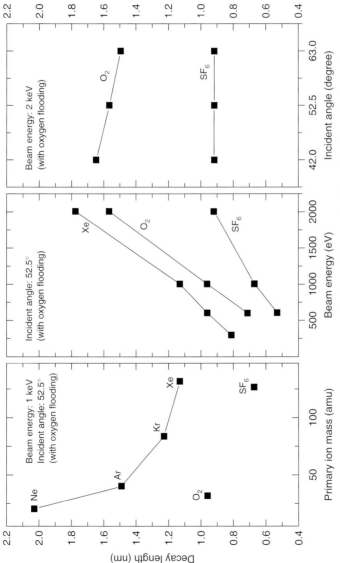

Figure 7.11 Comparison of measured decay lengths for B delta layer in Si as a function of primary ion mass, energy, and incidence angle showing the improvement resulting from using a polyatomic primary ion projectile. Reproduced from Iltgen et al.,[48] with permission from the American Vacuum Society.

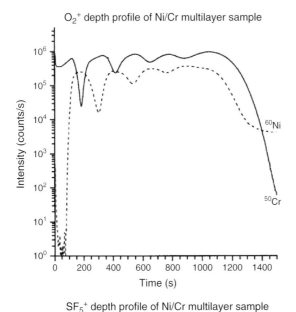

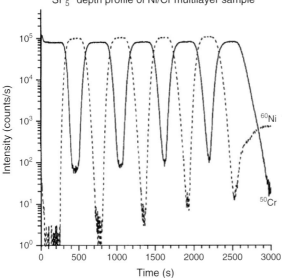

Figure 7.12 Depth profiles of a NIST Ni/Cr metal multilayer sample comparing 3.0 keV impact O_2^+ and SF_5^+ primary ion beams. Reproduced from Gillen et al.,[50] with permission from the American Vacuum Society.

ion cluster beams (SF_5^+, CF_3^+, NO_2^+). For analysis of electronegative elements that would benefit from Cs^+ bombardment, achieving low bombardment energies is problematic on magnetic sector SIMS instruments. In this case, the positive Cs^+ ion beam is accelerated to the sample surface increasing the impact energy, with a more normal incidence angle and a commensurate increase in mixing depth. To address this issue, in 2001 Gillen et al.[53] utilized a negative Cs sputter ion source on a magnetic sector SIMS instrument to generate negative cluster ions of C_x^-

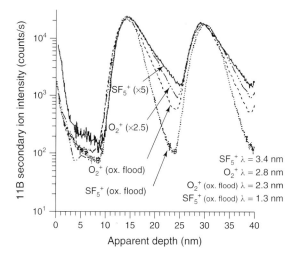

Figure 7.13 SIMS depth profiles of the first two layers of a 15 nm spacing B delta-doped structure profile with SF_5^+ and O_2^+ at 3.0 keV and $52°$ incidence angle with and without oxygen gas flooding at $7–9 \times 10^{-4}$ Pa. Reproduced from Gillen et al.,[51] with permission from the American Vacuum Society.

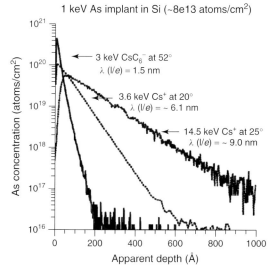

Figure 7.14 Comparison of three SIMS depth profiles of a 1 keV arsenic implant in silicon acquired with either Cs^+ or CsC_6^- as the bombarding species while monitoring $AsSi^-$ negative secondary ions. Reproduced from Gillen et al.,[53] with permission from the American Vacuum Society.

($x = 4–10$) and CsC_x ($x = 2–8$) or metal clusters such as Al_x where ($x = 1–8$), for various SIMS depth profiling applications. The use of a negatively charged, cesium-containing ion beam offers two advantages. First, the primary ion is decelerated (and deflected) as it nears the sample surface leading to a reduction in impact energy and a more oblique incidence angle. Second, the cluster ion dissociates upon impact with the surface further reducing the net impact energy of the primary ion beam constituents. Figure 7.14 shows SIMS depth profiles of a 1 keV As implant in Si comparing analysis using Cs^+ and CsC_6^- primary ions. In this example, the Cs^+ profiling was conducted with an unmodified 14.5 keV impact energy Cs^+ source as well as a modified 3.6 keV low energy Cs^+ source. The CsC_6^- ion beam gave a factor of 6 improvement in apparent $1/e$ decay length in comparison to the

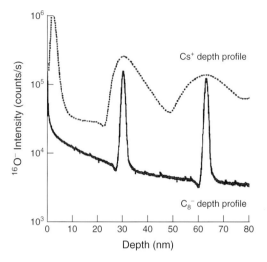

Figure 7.15 Comparison of C_8^- and Cs^+ depth profiles of a Cr metal sample containing CrO_2 monolayers spaced 30 nm apart. Reproduced from Gillen et al.,[52] with permission from the American Institute of Physics.

standard Cs^+ profiling conditions and a factor of 4 in comparison to the lower energy Cs^+ bombardment. It was also found that bombardment with CsC_6^- can lead to a formation of ripple topography in the SIMS crater bottom, which may result in a shift in the apparent depth scale. Follow up experiments by Loesing in 2002 demonstrated that sample rotation in conjunction with the use of CsC_6^- is an optimal combination for improved depth resolution analysis.[54]

The negative carbon containing primary ion cluster beams produced by the sputter source were also found to be advantageous for reducing topography for metals analysis. Figure 7.15 shows a comparison of two depth profiles of a Cr/CrO^- depth profiling standard using a C_8^- primary ion beam as compared to a Cs^+ primary ion beam.

The approach of using a cesium sputter ion source for production of various small cluster ion beams including Si_n^- and Cu_n^- for SIMS depth profiling applications was also demonstrated by Belykh et al. in 2007 (see Chapter 3).[55]

7.4.5 $Os_3(CO)_{12}$ and $Ir_4(CO)_{12}$ Primary Ion Beams

Another interesting type of primary ion beam developed for SIMS depth profiling analysis are the metal cluster ion sources that utilize triosmium dodecarbonyl $Os_3(CO)_{12}$ or tetrairidium dodecarbonyl $Ir_4(CO)_{12}$ produced by sublimation of the appropriate compound and subsequent electron impact ionization. Fujiwara et al.[56,57] demonstrated the use of these beams to examine various B delta-doped structures and obtained sub 1 nm $1/e$ decay lengths in favorable cases when using oxygen backfill. An example of a SIMS depth profile of a boron delta-doped structure using the $Ir_4(CO)_{12}$ primary ion beam at 5 keV impact energy is shown in Figure 7.16. The impact energy of these metal-containing ion beams is ultimately limited by beam deposition effects that will be discussed in more detail in the next section.

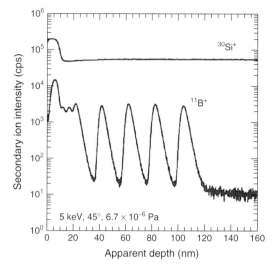

Figure 7.16   Example of SIMS depth profiles under 5 keV $Ir_4(CO)_7^+$ primary ion bombardment with oxygen flooding. Reproduced from Fujiwara et al.,[57] with permission from the Japanese Journal of Applied Physics.

7.4.6 C_{60}^+ Primary Ion Beams

Early results with cluster ion beams for depth profiling suggested that bigger cluster ion beams with large numbers of constituents would be desirable for further reduction in mixed depth and subsequent improvements in SIMS depth resolution. First introduced as a commercial ion source by the Vickerman group from the University of Manchester in 2003,[58] the use of C_{60} primary ion sources on commercial SIMS instruments has developed into an area of active investigation. While the initial focus for these studies was for organic surface analysis and depth profiling, the potential benefits for semiconductor materials depth profiling were soon explored. In 2004, Sun et al. from the group at Penn State demonstrated C_{60} depth profiling of a Ni/Cr NIST depth profiling test material.[59] The use of the C_{60} ion beam appeared advantageous for reducing sputter-induced topography in metal systems in comparison to other ion beams available on their system at that time. In Chapter 1 (Fig. 1.3), an unpublished C_{60} depth profile of this Ni/Cr material taken on a magnetic sector SIMS instrument at NIST was shown. As compared to analysis of the same sample, on the same instrument, using O_2^+ and SF_5^+ (Figure 7.12) relative improvements in depth resolution are noted. The decay lengths obtained for such metals systems using C_{60} were subsequently found to not be as good as those obtained using ultra low energy Ar^+ bombardment under similar conditions.[41] In 2006, Gillen et al. studied depth profiling of silicon substrates and As delta-doped structures using C_{60}.[60] Unexpectedly, the $1/e$ decay lengths for As profiling using C_{60} was almost five times worse than the decay length obtained from the analysis of the same sample using 500 eV Cs^+ (5.0 vs 1.1 nm). The severe degradation in depth resolution was thought to result from the development of sputter-induced topography possibly enhanced by localized carbon deposition.[60] The observation of carbon deposition and sputter-induced topography formation effects during C_{60} bombardment of silicon has subsequently been confirmed by several groups.[61,62]

The use of the C_{60} ion beam for depth profiling of metal films and alloys analysis was further explored by Kim et al. in 2007 and the potential for major element quantification using C_{60} depth profiling were noted.[63,64] Goacher, in 2010, examined depth profiling of gold/silicon and GaAs multilayers comparing Cs^+ and C_{60} depth profiling on a ToF-SIMS instrument. While some advantages for C_{60} analysis were noted, the primary conclusion was that Cs^+ bombardment provided better depth resolution than C_{60}^+ for all samples examined.[64,65]

The early studies on C_{60} SIMS depth profiling[60] demonstrated that the use of this beam for silicon depth profiling can result in several undesirable effects including (i) deposition of a uniform amorphous carbon layer on the substrate at lower impact energies, (ii) sputter-induced topography limiting depth resolution, and (iii) localized deposition effects at higher primary ion dose. The deposition effect was found to be energy- and incident angle dependent and precluded the use of the C_{60} beam for sputtering experiments below about 12 keV impact energy (depending on impact angle and use of oxygen backfill). The requirement for using higher impact energies partially negates the advantage of using such a cluster projectile. Figure 7.17 shows a plot of sputtering rate versus C_{60}^+ impact energy with deposition dominating at lower beam energies.[61] Even at higher energies where amorphous deposition was not observed, the formation of surface features resulting from localized carbon deposition in the sputtered crater bottom were observed. These features referred to as "*Hershey Kisses*" are shown in Figure 7.18. The deposition phenomenon, particularly in silicon, has been modeled in detail by Krantzman in 2011 and is partially related to the strong Si–C bonds that are

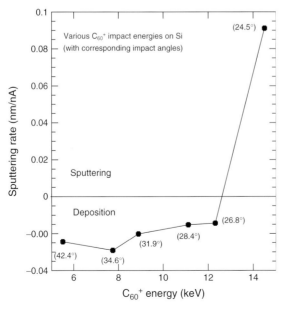

Figure 7.17 Sputter rate for C_{60} bombardment of silicon as a function of beam energy (incident angle). Sputtering of silicon was observed above 12 keV. Reproduced from Gillen et al.,[60] with permission from Elsevier.

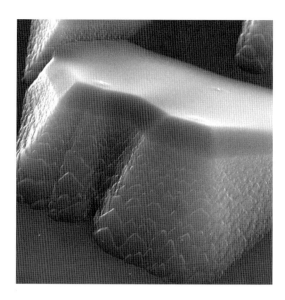

Figure 7.18   Scanning electron micrograph of a sputter-induced "Hershey Kiss" in the bottom of a C_{60} SIMS crater in silicon. The structure is composed of an amorphous carbon cap which appears to protect a localized region of the underlying substrate. Reproduced from Gillen et al.,[60] with permission from Elsevier.

formed during primary ion beam implantation of carbon which helps to mitigate the resputtering of carbon from the surface.[66]

In addition to the deposition and topography development during C_{60} bombardment of silicon, there is recent evidence that additional factors may complicate the interpretation of SIMS depth profiles of shallow layers in silicon. Cross-sectional transmission electron microscopy (TEM) analysis of C_{60} bombarded silicon suggested that the resulting altered layer was significantly deeper than would be expected on the basis of a simple energy partitioning model (Bennett and Gillen, Unpublished Work). This observation was recently confirmed in 2011 by Morita who found that the range of C_{60}^+ was 50% greater than the nominal range of an individual carbon atom.[67] Clearly, additional study is required to elucidate some of these factors and their ultimate influence of the quality of SIMS depth profiles.

7.4.7   Massive Gaseous Cluster Ion Beams

Gaseous cluster ion beam sources (GCIB) have been available for many years primarily for semiconductor processing applications. While not yet fully exploited for inorganic depth profiling, their use is promising because of subthreshold sputtering and the potential for use of reactive primary ion beam species to produce ultra low energy chemical sputtering. They may be particularly applicable for dual beam depth profiling applications where the requirements for focusing of the sputtering beam are reduced. Of particular note would be reactive cluster ion beams containing oxygen or SF_x clusters.[68,69] To date, we could not find any published work specifically addressing the use of such sources for ultra shallow depth profiling applications.

7.5 CONCLUSIONS AND FUTURE PROSPECTS

Cluster bombardment for SIMS depth profiling of inorganic materials and semiconductors has been explored for a number of primary ion beam species and sample types. Generally speaking, the process of energy partitioning during sputtering with a cluster projectile leads to a low energy per constituent atom and provides a commensurate improvement in depth resolution compared to monatomic or diatomic sputtering under comparable conditions of total beam energy and incident angle. For a given instrumental configuration, this can provide improved analytical performance over what would be possible with a more conventional beam on the same instrument. However, it has not generally been observed that cluster ions beams can provide demonstrable advantages for depth resolution when evaluated against conventional SIMS depth profiling approaches using state of the art, ultra low energy ion beams. Whether minimization of transient effects and higher sputter yields provide sufficient justification for using cluster beams has to be evaluated on a case-by-case basis. The recent development of commercially available GCIB may offer substantial improvements in this situation.

REFERENCES

1. Degreve, F.; Thorne, N. A.; Lang, J. M. *J. Mater. Sci.* **1988**, 23 (12), 4181–4208.
2. Ganguly, J.; Tirone, M.; Hervig, R. L. *Science* **1998**, 281 (5378), 805.
3. Chu, P. K. *Mater. Chem. Phys.* **1994**, 38 (3), 203–223.
4. Littmark, U.; Hofer, W. O. *Nucl. Instrum. Methods Phys. Res. B* **1980**, 168 (1–3), 329–342.
5. Wittmaack, K. *Vacuum* **1984**, 34 (1–2), 119–137.
6. Dowsett, M. G.; Ormsby, T. J.; Cooke, G. A.; Chu, D. P. *J. Vac. Sci. Technol. B* **1998**, 16, 302.
7. Wittmaack, K. *Surf. Interface Anal.* **1996**, 24 (6), 389–398.
8. Stevie, F. A.; Kahora, P. M.; Simons, D. S.; Chi, P. *J. Vac. Sci. Technol. A* **1988**, 6 (1), 76–80.
9. Jiang, Z. X.; Alkemade, P. F. A. *Appl. Phys. Lett.* **1998**, 73, 315.
10. Cirlin, E. H.; Vajo, J. J.; Hasenberg, T. C.; Hauenstein, R. J. *J. Vac. Sci. Technol. A* **1990**, 8 (6), 4101–4103.
11. Ronsheim, P. A.; Tejwani, M. *J. Vac. Sci. Technol. B* **1994**, 12 (1), 254–257.
12. Iltgen, A.; Benninghoven, A.; Niehuis, E. *Proceedings of the 11th International Conference of Secondary ion mass spectrometry (SIMS XI);* Wiley-VCH, Chichester; **1998**; p 367.
13. Dowsett, M. G.; Barlow, R. D.; Fox, H. S.; Kubiak, R. A. A.; Collins, R. *J. Vac. Sci. Technol. A* **1992**, 10 (1), 336.
14. Wittmaack, K. *J. Vac. Sci. Technol. A* **1985**, 3 (3), 1350–1354.
15. Magee, C. W.; Frost, M. R. *Int. J. Mass Spectrom.* **1995**, 143, 29–41.
16. Bennett, J. *Surf. Interface Anal.* **1997**, 25 (6), 454–457.
17. Juhel, M.; Laugier, F.; Delille, D.; Wyon, C.; Kwakman, L. F.; Hopstaken, M. *Appl. Surf. Sci.* **2006**, 252 (19), 7211–7213.
18. Smith, N. S.; Dowsett, M. G.; McGregor, B.; Phillips, P. *Proceedings of the 10th International Conference on Secondary Ion Mass Spectrometry (SIMS X);* Wiley-VCH, New York; **1997**; p 363.
19. Dowsett, M. G.; Chu, D. P. *J. Vac. Sci. Technol. B* **1998**, 16 (1), 377.
20. Bennett, J.; Diebold, A. *Proceedings of the 12th International Conference on Secondary Ion Mass Spectrometry (SIMS XII);* Wiley-VCH, Amsterdam; **2000**; p 541.
21. Wittmaack, K. *J. Vac. Sci. Technol. B* **2000**, 18 (1), 1–6.
22. Tomita, M.; Tanaka, H.; Koike, M.; Kinno, T.; Hori, Y.; Yoshida, N.; Sasaki, T.; Takeno, S. *Appl. Surf. Sci.* **2008**, 255 (4), 1311–1315.

23. van der Heide, P. A. W.; Lim, M. S.; Perry, S. S.; Bennett, J. *Nucl. Instrum. Methods Phys. Res. B* **2003**, 201 (2), 413–425.

24. Chanbasha, A. R.; Wee, A. T. S. *Appl. Surf. Sci.* **2006**, 252 (19), 7243–7246.

25. Vandervorst, W.; Shepherd, F. R. *Appl. Surf. Sci.* **1985**, 21 (1), 230–242.

26. Tomita, M.; Hongo, C.; Murakoshi, A. *Proceedings of the 12th International Conference on Secondary Ion Mass Spectrometry (SIMS XII);* Wiley-VCH, Amsterdam; **2000**; p 489.

27. Magee, C. W.; Hockett, T. H.; Buyuklimanli, T. H.; Abdelrehim, I.; Marino, W. *AIP Conf. Proc.* **2012**, 142–145.

28. Vandervorst, W.; Janssens, T.; Huyghebaert, C.; Berghmans, B. *Appl. Surf. Sci.* **2008**, 255 (4), 1206–1214.

29. van der Heide, P. A. W.; Lim, M. S.; Perry, S. S.; Bennett, J. *Appl. Surf. Sci.* **2003**, 203, 156–159.

30. Kataoka, Y.; Yamazaki, K.; Shigeno, M.; Tada, Y.; Wittmaack, K. *Appl. Surf. Sci.* **2003**, 203, 43–47.

31. Alkemade, P. F. A.; Jiang, Z. X. *J. Vac. Sci. Technol. B* **2001**, 19, 1699.

32. Jiang, Z. X.; Lerma, J.; Sieloff, D.; Lee, J. J.; Backer, S.; Bagchi, S.; Conner, J. *J. Vac. Sci. Technol. B* **2004**, 22 (2), 630–635.

33. Zalar, A. *Surf. Interface Anal.* **1986**, 9 (1), 41–46.

34. Sykes, D. E. *Surf. Interface Anal.* **1999**, 28 (1), 49–55.

35. Stevie, F. A.; Moore, J. *Surf. Interface Anal.* **1992**, 18 (2), 147–152.

36. Liu, R.; Wee, A. T. S.; Liu, L.; Hao, G. *Use of Sample Rotation in SIMS Profiling of Ta Barrier Layers to Cu Diffusion*; **2000**; p 98.

37. Liu, R.; Wee, A. T. S. *Appl. Surf. Sci.* **2004**, 231, 653–657.

38. Merkulov, A.; Peres, P.; Choi, S.; Horreard, F.; Ehrke, H. U.; Loibl, N.; Schuhmacher, M. *J. Vac. Sci. Technol. B Microelectron. Nanometer Struct.* **2010**, 28, C1–C48.

39. Clegg, J. B. *J. Vac. Sci. Technol. A.* **1995**, 13 (1), 143–146.

40. Cooke, G. A.; Ormsby, T. J.; Dowsett, M. G.; Parry, C.; Murrell, A.; Collart, E. J. H. *J. Vac. Sci. Technol. B* **2000**, 18 (1), 493–495.

41. Vandervorst, W. *Appl. Surf. Sci.* **2008**, 255 (4), 805–812.

42. Andersen, H. H.; Bay, H. L. *J. Appl. Phys.* **1974**, 45 (2), 953–954.

43. Toyoda, N.; Yamada, I. *Nucl. Instrum. Methods Phys. Res. B* **2010**, 268 (19), 3291–3294.

44. Reuter, W.; Scilla, G. *J. Anal. Chem.* **1988**, 60 (14), 1401–1404.

45. Ronsheim, P. A.; Fitzgibbon, G. *J. Vac. Sci. Technol. B* **1992**, 10 (1), 329–332.

46. Ronsheim, P. A.; Lee, K. L. *J. Vac. Sci. Technol. B* **1998**, 16 (1), 382–385.

47. Yamazaki, H.; Mitani, Y. *Nucl. Instrum. Methods Phys. Res. B* **1997**, 124, 91–94.

48. Iltgen, K.; Bendel, C.; Benninghoven, A.; Niehuis, E. *J. Vac. Sci. Technol. A* **1997**, 15 (3), 460–464.

49. Kruger, D.; Iltgen, K.; Heinemann, B.; Kurps, R.; Benninghoven, A. *J. Vac. Sci. Technol. B* **1998**, 16 (1), 292–297.

50. Gillen, G.; King, R. L.; Chmara, F. *J. Vac. Sci. Technol. A* **1999**, 17 (3), 845–852.

51. Gillen, G., Walker, M., Thompson, P., Bennett, J. *J. Vac. Sci. Technol. B* **2000**, 18 (1), 503–508.

52. Gillen, G., Roberson, S., Fahey, A., Walker, M., Bennett, J., and Lareau, R. T. *AIP. Conf. Proc.* **2001**, 550, 687–691.

53. Gillen, G.; King, L.; Freibaum, B.; Lareau, R.; Bennett, J.; Chmara, F. *J. Vac. Sci. Technol. A* **2001**, 19, 568.

54. Loesing, R.; Guryanov, G. M.; Phillips, M. S.; Griffis, D. P. *J. Vac. Sci. Technol. B* **2002**, 20, 507.

55. Belykh, S. F.; Palitsin, V. V.; Veryovkin, I. V.; Kovarsky, A. P.; Chang, R. J. H.; Adriaens, A.; Dowsett, M. G.; Adams, F. *Rev. Sci. Instrum.* **2007**, 78, 085101.

56. Fujiwara, Y.; Kondou, K.; Teranishi, Y.; Nonaka, H.; Fujimoto, T.; Kurokawa, A.; Ichimura, S.; Tomita, M. *J. Appl. Phys.* **2006**, 100, 043305.

57. Fujiwara, Y.; Kondou, K.; Watanabe, K.; Nonaka, H.; Saito, N.; Fujimoto, T.; Kurokawa, A.; Ichimura, S.; Tomita, M. *Jpn. J. Appl. Phys.* **2007**, 46 (11), 7599.

58. Weibel, D.; Wong, S.; Lockyer, N.; Blenkinsopp, P.; Hill, R.; Vickerman, J. C. *Anal. Chem.* **2003**, 75 (7), 1754–1764.

59. Sun, S.; Wucher, A.; Szakal, C.; Winograd, N. *Appl. Phys. Lett.* **2004**, 84, 5177.

60. Gillen, G.; Batteas, J.; Michaels, C. A.; Chi, P.; Small, J.; Windsor, E.; Fahey, A.; Verkouteren, J.; Kim, K. J. *Appl. Surf. Sci.* **2006**, 252 (19), 6521–6525.

61. Kozole, J.; Winograd, N. *Appl. Surf. Sci.* **2008**, 255 (4), 886–889.
62. Yu, B. Y.; Lin, W. C.; Chen, Y. Y.; Lin, Y. C.; Wong, K. T.; Shyue, J. J. *Appl. Surf. Sci.* **2008**, 255 (5), 2490–2493.
63. Kim, K. J.; Moon, D. W.; Park, C. J.; Simons, D.; Gillen, G.; Jin, H.; Kang, H. *J. Surf. Interface Anal.* **2007**, 39 (8), 665–673.
64. Kim, K. J.; Simons, D.; Gillen, G. *Appl. Surf. Sci.* **2007**, 253 (14), 6000–6005.
65. Goacher, R. E.; Gardella, J. A. *Appl. Surf. Sci.* **2010**, 256 (7), 2044–2051.
66. Krantzman, K. D.; Garrison, B. J. *Surf. Interface Anal.* **2011**, 43, 123–125.
67. Morita, Y.; Nakajima, K.; Suzuki, M.; Narumi, K.; Saitoh, Y.; Vandervorst, W.; Kimura, K. *Nucl. Instrum. Methods Phys. Res. B.* **2011**, 269 (19), 2080–2083.
68. Ninomiya, S.; Ichiki, K.; Yamada, H.; Nakata, Y.; Seki, T.; Aoki, T.; Matsuo, J. *Rapid Commun. Mass Spectrom.* **2009**, 23 (20), 3264.
69. Fenner, D. B.; Shao, Y. *J. Vac. Sci. Technol. A* **2003**, 21, 47.

CHAPTER 8

CLUSTER ToF-SIMS IMAGING AND THE CHARACTERIZATION OF BIOLOGICAL MATERIALS

John Vickerman and Nick Winograd

8.1   INTRODUCTION

It is now clear that the use of metal cluster and polyatomic primary ions can greatly enhance the capability of time-of-flight secondary ion mass spectrometry (ToF-SIMS) to tackle the analysis of complex chemical systems. Biological materials present the greatest challenge, but also probably the greatest opportunity to apply the full capabilities of SIMS in label-free analysis.[1] This chapter will seek to provide a brief overview of the present capabilities and the future possibilities of ToF-SIMS using cluster primary ions in the study of biological cells and tissue.

There are basically two types of ToF-SIMS instrumentation available at present. By far, the majority of instruments utilize a pulsed primary beam and a reflectron time of flight mass spectrometer. Very recently a new concept of instrument has appeared that uses a DC primary beam and a hybrid two-stage mass spectrometer arrangement. First, we summarize the benefits common to both instruments and then outline those specific to the new ToF-SIMS platforms. The challenges facing ToF-SIMS in biological studies will then be discussed, followed by some illustrative examples of the present status of ToF-SIMS in biological analyses.

Cluster Secondary Ion Mass Spectrometry: Principles and Applications, First Edition.
Edited by Christine M. Mahoney.
© 2013 John Wiley & Sons, Inc. Published 2013 by John Wiley & Sons, Inc.

8.2 THE CAPABILITIES OF ToF-SIMS FOR BIOLOGICAL ANALYSIS

Analysis without sample pretreatment. As a mass spectrometric-based method, ToF-SIMS enables molecular analysis without the requirement to use tags or labels. Furthermore, in contrast to matrix-assisted laser desorption/ionization (MALDI), SIMS is performed without any sample pretreatment, although matrix-assisted SIMS is a variant that is being explored to extend its capability. With mass resolution in the region of 5000–10,000, and good mass accuracy, ToF-SIMS is, in theory, able to identify most major components of biological systems of mass below m/z ~2000, although other factors to be discussed later influence this capability.

Upper layer sensitivity and molecular depth profiling. ToF-SIMS is a surface-sensitive technique, probing information from only the top few monolayers of a sample. However, because of the low levels of chemical damage usually generated when polyatomic cluster primary ions such as SF_5^+, C_{60}^+, and large argon clusters are used to sputter molecular systems, the chemistry of subsurface layers may also be exposed (via sputter removal) so that their composition can be distinguished from the surface. Hence, molecular depth profiling and 3D imaging becomes possible.

Excellent spatial resolution. Liquid metal cluster ion beams can be focused to sub-100 nm beam diameter at the surface. In principle, submicrometer spatial resolution is feasible if the ion yield is sufficient under the static conditions that are required for these beams. Although argon cluster beams are more difficult to focus, C_{60}^+ beams can be focused to at least 1 μm so voxel analysis at this level is possible.

Analysis of samples in the frozen state. Appropriately equipped instruments can now handle and analyze samples in the frozen hydrated state. Fast freezing of biological samples ensures that there is as little deviation from the natural state as possible. Facilities are now available to fracture cells and tissue while frozen so that the internal chemical structure can be probed without the sample having to be warmed up or exposed to the atmosphere.

8.3 NEW HYBRID ToF-SIMS INSTRUMENTS

8.3.1 Introduction

Because polyatomic ion beams generate much less chemical damage in most molecular analyte systems, it is now possible to use a DC primary beam. The application of a DC beam allows for rapid 3D analysis, with increased sensitivies and higher spatial resolutions than can be offered using the more conventional systems. In conventional ToF-SIMS instruments, a pulsed primary beam has to be used. The pulse length determines the mass resolution, but the movement of the beam during the pulsing can limit the attainable spatial resolution. When a DC beam is used, the mass spectrometry is separated from the ion formation process, so mass resolution and spatial resolution can be separated.

Two new types of ToF-SIMS instruments emerged in the early 2000s that are designed to exploit the capabilities of polyatomic cluster ions with a DC beam configuration. The new SIMS instruments are basically variants of the *ortho*-ToF designs used for MALDI and electrospray ionization (ESI) mass spectrometry. A sample of a stream of secondary ions generated by the DC beam are either pushed out into a ToFMS by a pusher plate as in the Q-Star qToF instrument, or are collected in a long buncher and accelerated into a ToFMS as in the J105 instrument from Ionoptika, Figure 8.1.[2,3] While an analysis is carried out, a further sample of the secondary ion stream is collected. Some loss of the secondary ion stream occurs because of the sampling process, although this can be minimized for significant parts of the spectrum. In general, however, the transmission efficiency of these hybrid instruments is rather lower than the simple pulsed-beam reflectron ToF designs.

8.3.2   Benefits of New DC Beam Technologies

The DC-beam systems potentially offer enhancements to the ToF-SIMS capability as follows:

Analysis beyond the static limit in 2D and 3D analysis. Unlike conventional ToF-SIMS designs that alternate sputter and analysis cycles, the new designs allow for simultaneous collection of ions for analysis while the beam is sputtering the sample. This means that all sputtered material is used for analysis. Because C_{60} polyatomic ions do not generate as much chemical damage as liquid metal cluster beams, analysis is not restricted by the static regime to 1% of the sample surface. The potential yield from any 2D pixel is increased by 100 times; furthermore, analysis is not limited to the surface layer, so multiple layers can be acquired enabling 3D analysis and imaging without switching beams.

Spatial resolution and mass resolution are independent of one another. A DC beam can be focused to its optimum capability depending on beam energy and current. The properties of the beam raster can be defined relative to the pixel size and sputter rates required. The quality of the mass spectrometry is dependent only on capabilities of the mass spectrometer and independent of how the secondary ions are formed. In principle, the best spatial resolution and mass resolution can be obtained at the same time.

Spectral and image acquisition times are greatly reduced. The primary beam of conventional ToF-SIMS instruments is on for only 10^{-4} of the time. This greatly extends the time for acquisition of the large data sets involved in 2D and 3D images that could take many hours or even days. The use of the DC beam arrangement has reduced these times to a few hours or even minutes. For example, to analyze each of the 10^8 molecules in a

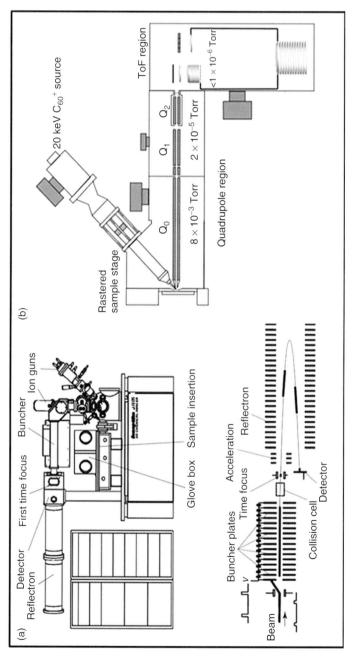

Figure 8.1 Schematics of the J105 3D Chemical Imager (a) and the C_{60} equipped Q-Star (b) instruments. Both instruments can use continuous beams. The J105 bunches the secondary ion stream to a focus at the entrance to the ToF analyzer while the Q-Star orthogonally pulses sections of the secondary ion stream in to a ToF analyzer.

1 μm × 1 μm × 10 nm voxel using a 10 pA ion beam with a 100 ns pulse width and a 10 kHz repetition rate, 10 s of instrumental time is required (assuming a sputter yield of 100). Therefore, it would take 182 h to acquire a 256 voxel × 256 voxel image. However, by eliminating the pulsed nature of the ion beam, the same image can be acquired in only 11 min or 10^3 less time.

MS/MS analysis is now possible. Despite the capability for accurate mass measurement, one of the great drawbacks of ToF-SIMS for the analyses of biological samples has been the inability to carry out tandem MS analyses on large ions to determine their chemical structure. These new instruments basically have two mass spectrometers in tandem, so precursor ions can be selected by the first MS (a quadrupole in the case of the qToF, and the buncher in the case of the J105) and the fragments formed in a collision cell can be analyzed by the ToFMS to enable in many cases the precise chemical structure to be determined. This potentially brings ToF-SIMS into the mainstream of biological mass spectrometry.

Combining this new instrumentation with the high quality accessible with conventional instruments, ToF-SIMS is now potentially in a position to contribute to a very significant degree to understanding the 2D and 3D chemistry of biosystems. However, there are a number of challenging issues that need to be factored into the design of experiments and in the assessment of data.

8.4 CHALLENGES IN THE USE OF ToF-SIMS FOR BIOLOGICAL ANALYSIS

8.4.1 Sample Handling of Biological Samples for Analysis in Vacuum

Sample preparation is essential when acquiring meaningful SIMS images of single cells or tissue.[4−6] In order to prepare cells for analysis *in vacuo*, the 3D integrity of the sample must be preserved in the solid state with micrometer precision. Common strategies for cell preservation include freeze-drying, freeze-etching, freeze-fracturing, chemical fixation, and sugar vitrification.[7,8] These methods can be very successful in maintaining the physical state of the biological materials. However, when it is also required that the chemistry is not disturbed, the situation can be significantly more complicated and there are few objective studies that seek to assess the usefulness of each preparation protocol. There seems to be a great deal of dogmatic statement, or hope, expressed that such-and-such-a-process does not, or is not "believed," to affect the chemical and physical localization of the compounds under study. Chemical fixation clearly has to change the chemistry of the samples being studied. It seems to be recognized that a particular fixative does not work on all the biological molecules in a cell or tissue.[9] When the chemistry of the material under study is complex, adding further to the chemistry of the system may not be the best way forward.

It is clear that fast freezing in liquid propane or isopentane is effective in ensuring that cell or tissue structure is not damaged by water crystallizing.[10] This process also would seem to have a good chance of maintaining the chemical localization. However, it is what the analyst does with the material thereafter that is probably crucial for the chemical analysis. Frequently, after storage in liquid nitrogen or at -80 $°$C, the material is freeze-dried, which entails the material warming up to room temperature. Many biologists seem to be comfortable that the procedure does not redistribute the chemistry significantly, but it is difficult to find any definitive studies that confirm this conclusion. The problem is that imaging mass spectrometry might be one of the only "discovery" techniques that could assess the validity of this view. A "gold standard" technique to cross-reference against is difficult to identify. However, such studies are needed if freeze-drying is to be relied upon. Studies have demonstrated conclusively that cholesterol can be lost completely from samples being freeze-dried if the process is prolonged.[11,12] In contrast to freeze-drying, maintaining the sample in the frozen hydrated state followed by freeze-fracturing cells or tissue to expose regions of interest would seem to have some advantages. The fracturing process is best carried out inside the vacuum system to prevent condensation of water ice on the sample. A number of approaches have been successful.[13] Most recently, the development of a reverse "mousetrap" device has been described.[14,15] The frozen material, cells, or tissue are sandwiched between two metal plates, and the whole device is cooled. The top plate is spring loaded. When everything is in place at the analysis position, the trigger is pushed in to release the top plate that flips backward to fracture the material and to expose two surfaces of material. In the case of cells, this procedure may expose the inside of the cell for analysis. Where an overlayer of water ice is formed during the fracturing process it can be removed using polyatomic ion bombardment with no damage to the underlying molecules.[8] Therefore, a good strategy for maintaining the 3D structure of biological samples while exposing an intact, meaningful surface for SIMS imaging involves combining freeze-fracture technology and cluster ion beams. Despite freeze-fracturing probably being the best approach, because it is quicker and easier, much sample preparation is carried out using freeze-drying. Sample washing is also frequently involved. Both washing and drying can result in the movement or loss of molecules from the cells or tissues. One of the exemplar studies below further investigates the preparation of cells for analysis.

8.4.2 Analysis is Limited to Small to Medium Size Molecules

The SIMS process seems to be able to lift off and ionize molecules of a restricted size range, that is, up to m/z of a few thousands. Proteins and large peptides seem to be inaccessible, although the use of matrix-assisted SIMS does seem to enable some large peptides to be accessed.[16] The highest yields are obtained for molecules such as lipids and small metabolites.[17] Such measurements, as have been made, suggest that sputter yields of molecules in the size range between m/z 300 and 1500 are much the same at about 200 per C_{60} impact[18]; however, it is clear that ionization probabilities, although low, vary widely and for many molecules are influenced by their basicity.[19] Thus it seems that ToF-SIMS can be used to monitor small molecules in bio-systems, but the all-important proteins will need a matrix to be added in order to be probed by MALDI and possibly matrix-assisted SIMS.

8.4.3   Ion Yields Limit Useful Spatial Resolution for Molecular Analysis to not Much Better than 1 μm

Despite the fact that liquid metal ion beams can deliver a probe size at the sample of less than 100 nm, because of the static SIMS limitation (10^{-2} molecules for analysis) and the very small ionization probability (frequently less than 10^{-4}),[20,21] even with 50% transmission fewer than five secondary ions could be expected from a 100% component of a 2D 1 μm pixel. If the static limit is relaxed in the case of polyatomic primary ions, a 2D pixel would yield 10–100 ions from a 100% component, such that analysis at the 10% level might become feasible. If the whole of the 3D 1 μm voxel is used for analysis, 10^3–10^4 ions would become available even with 10% instrument transmission. Thus, the ability to use a DC beam and collect all the ions sputtered might enable analysis of composition down to the 1% level, which is beginning to become useful. However, it is clear that it will be difficult to obtain useful information on molecular distributions below 1 μm 2D pixel dimensions unless the ionization probability of the species studied is significantly increased. Small fragment ions and salt cationized ions sometimes fall into this category. Nevertheless, to be able to analyze with submicrometer resolution on a routine basis will require a significant increase in molecular ionization probability.

8.4.4   Matrix Effects Inhibit Application in Discovery Mode and Greatly Complicates Quantification

The secondary ion formation mechanism of a particular ion is profoundly influenced by the chemical and electric environment from which it is sputtered. The complications arising from this phenomenon have been known for almost as long as SIMS has been used as an analytical technique.[22] It is the principal reason why it is frequently inaccurately affirmed that SIMS is not a quantitative technique. It certainly can be very difficult, sometimes impossible; however, in many cases it is possible to quantify SIMS data. SIMS is not alone in being influenced by the matrix effect. MALDI in fact exploits the matrix effect to preferentially emit and ionize some molecules rather than others. Yields from ESI are greatly influenced by the components in the electrospray jet and salts affect all ion yields in mass spectrometry.[23–25] The issue for SIMS is that there is now ample data to demonstrate that in any chemical system the other chemicals can enhance the ion yield of a molecule such as, for example cholesterol, while in a system of differing composition the ion yield might be suppressed.[12,26–28] This has the consequence that it is impossible to conclude that the absence of an ion in a spectrum or image implies the absence of the related molecule from a bio-system. Similarly, without much background calibration work, it is not possible to conclude much about relative amounts of a molecule from the observed ion yield. These issues can also complicate our understanding of an ion image. Relative yields of a single component in different regions of a sample may be influenced by the differing chemistry in each region. This can go as far as total suppression of signal or gross enhancement (an example will be shown later). Some researchers suggest that SIMS and MALDI can be used in what they term a *discovery mode*, by which we understand that one can find out the composition of an unknown sample. The ion yield and matrix effect issues mean that this is entirely impossible and could be a dangerous assumption.

The matrix effect has to be considered in experiment design and in data assessment. Its mode of operation is not fully understood. A number of studies have shown that gas phase basicity has a role where protonation is important to generate $[M + H]^+$ ions.[19] Molecules with a high basicity scavenge protons suppressing $[M + H]^+$ ions of lower basicity components. Experiments have shown that this suppression can be lifted to some degree by the supply of protons to the system.[29,30]

8.4.5 The Complexity of Biological Systems can Result in Data Sets that Need Multivariate Analysis (MVA) to Unravel

Biological systems are the most complex that can be imagined. While the composition of synthesized chemical systems are known, the detailed composition of even well-studied biological systems are not fully known. Add to this the increased complexity introduced by the matrix effect, understanding and interpreting the SIMS spectrum can be a daunting task. Where the task is to try to understand the difference between samples or sets of samples, a "stare and compare" approach is impossible and it is essential to use computational multivariate analysis (MVA) techniques.[31] However, as many of these techniques are based on an assumption of a linear variance between some chemical difference and the spectra, for these to give useful results it is important to ensure that some irrelevant effect such as differences of sample preparation or analysis procedure does not interfere. In the end it is not too interesting to know that two samples or a series of samples are confirmed to be different; the whole point of the application of MVA techniques is to provide insights into the chemical reasons for the differences. It is noticeable that many studies stop at demonstrating a difference rather than probing the chemistry that seems to be the principal reason for using mass spectrometry in biological research.

8.5 EXAMPLES OF BIOLOGICAL STUDIES USING CLUSTER-ToF-SIMS

When considering the use of ToF-SIMS to research a biological problem, it is obviously important to take into account the capabilities and challenges presented by the equipment available. A crucial question is, "Is it possible to estimate whether the molecules to be studied can be detected at the concentration and spatial resolution desired?" Frequently it is suggested that ToF-SIMS using metal cluster primary ions is capable of sub-100 nm spatial resolution. However, we have seen that it is highly unlikely that any useful yield or distribution of molecular ions will be detected at submicrometer resolution under static conditions. There are just too few molecules in a pixel area below 1 µm taken together with the generally very low ion yields.

The ionization probability is a crucial factor that is little thought about. It can vary between 10^{-2} down to $<10^{-6}$. This obviously has an enormous influence on the detectability of molecules. It is essential to know something about the yields of molecules to be studied and if nothing is known, to carry out background experiments to investigate the issue. As mentioned earlier, lipids appear to be of a suitable molecular size. In addition, their ion yield is high enough for detection at physiological concentrations. As a consequence, one of the most successful application areas for ToF-SIMS in biology has been the study of the role of lipids. There is an extensive ToF-SIMS literature reporting their detection in a range of tissues and cells. For example Table 8.1 provides a systematic list of lipid-related ions identified in various mammalian tissue sections using ToF-SIMS and organized using the lipid classification system established by the Lipid MAPS consortium.[17] In multicomponent biological systems the matrix effect will inevitably be a significant parameter affecting ion yields. It is vital to be aware of its possible influence. Many researchers just close their eyes to the possibility that it might have a role. Although it can be difficult to investigate with complex systems, its influence must be factored into our interpretation of results, as failure to do this could have catastrophic consequences, particularly in medically related studies.

If 2D and 3D studies are required by a research study, the relative benefits of single beam polyatomic sputtering with simultaneous analysis, compared to inter-leaving analysis and sputtering using conventional pulsed beam ToF-SIMS needs to be assessed. In principle, the former should provide more rapid analysis and higher yield with a considerable reduction in the stability of the sample, particularly when analysis in the frozen state is required.

8.5.1   Analysis of Tissue

In the characterization of tissue, ToF-SIMS offers dual functionality: for macroscale studies where the whole tissue is of interest, the beam can be defocused to cover a large field of view. For microscale analyses selected regions of interest can be probed with a highly focused ion beam for a more detailed view and the generation of 2D images. Rat brain sections have become a well-established model system for tissue-based ToF-SIMS studies using both approaches for lipid-based investigations. Sjovall and coworkers were the first to report a number of sulfogly-cosphingolipids (sulfatides) and cholesterol in the white matter of a rat brain as well as glycerophospholipids molecules, specifically glycerphosphocholines (GPCho) and glycerophosphoinositols (GPIns), in the gray matter of a rat brain section using a Bi_3^+ metal cluster source.[11,32] More recently, Benabdellah and coworkers in a fascinating study comparing the use of MALDI and ToF-SIMS, confirmed these findings while imaging a sagittally sliced rat brain section, Figure 8.2.[33] In addition to the variety of lipids identified in the sample, their analysis of this model system demonstrated the connection between spatial and chemical information allowing inferences to be made between anatomical features and physiological functions. For example, the coronal brain sections are distinguished by the large

TABLE 8.1 Lipid species-protonated ions, adducts and pseudomolecular ions-identified in various mammalian tissue sections using ToF-SIMS and organized using the lipid classification system established by the Lipid MAPS consortium.

	LM_ID	Sub-class	label(C:DB)	Mass	Formula	Species	Tissue
Glycerophospholipid	GP1001	Glycerophophates/ Diacylglycerophosphates/	PA(34:0)	675.5	$C_{37}H_{72}O_8P$	$[M-H]^-$	Muscle [79]
			PA(34:1)	673.5	$C_{37}H_{70}O_8P$	$[M-H]^-$	Brain [60]
			PA(36:1)	701.5	$C_{29}H_{74}O_8P$	$[M-H]^-$	Muscle [79]
	GP0601	Glycerophosphoinositols/ Diacylglycerophoshoinositols/	PI(36:4)	857.5	$C_{45}H_{78}O_{13}P$	$[M-H]^-$	Brain [59,60]
			PI(38:4)	885.6	$C_{47}H_{82}O_{13}P$	$[M-H]^-$	Brain [59,60], Adipose [74], Liver [70]
			PI(38:3)	887.6	$C_{47}H_{84}O_{13}P$	$[M-H]^-$	Liver [70]
	GP0701	Glycerophosphoinositolmono- phosphates/ Diacylglycerophos- phoinositolmonophosphates/	PIP(38:4)	965.6	$C_{47}H_{83}O_{16}P2$	$[M-H]^-$	Brain [59]
	GP0101	Glycerophosphocholines/ Diacylglycerophosphocholines/	PC(34:2)	758.6	$C_{42}H_{81}NO_8P$	$[M+H]^+$	Liver [70]
			PC(34:1)	760.6	$C_{42}H_{83}NO_8P$	$[M+H]^+$	Muscle [79], Brain [59,60], Liver [70]
			PC(34:1)	699.6	$C_{39}H_{72}O_8P$	$[M-H-TMA]^-$	Brain [60]
			PC(36:1)	788.6	$C_{44}H_{87}NO_8P$	$[M+H]^+$	Brain [59] [60]
			PC(32:0)	734.6	$C_{40}H_{81}NO_8P$	$[M+H]^+$	Muscle [79], Brain [59,60,71]
Glycerolipids	GL0101	Monoradylglycerols/ Monoacylglycerols/	MAG(16:1)	311.3	$C_{19}H_{35}O_3$	$[M+H-OH]^+$	Liver [70]
			MAG(16:0)	313.3	$C_{19}H_{37}O_3$	$[M+H-OH]^+$	Liver [70]
			MAG(18:1)	339.3	$C_{21}H_{39}O_3$	$[M+H-OH]^+$	Liver [70]
			MAG(18:0)	341.3	$C_{21}H_{41}O_3$	$[M+H-OH]^+$	Liver [70]

		m/z	Formula	Ion	Tissue [Ref]
GL0201	Diradylglycerols/ Diacylglycerols/				
	DAG(30:2)	519	$C_{33}H_{59}O_4$	$[M+H-OH]^+$	Liver [70]
	DAG(30:1)	521	$C_{33}H_{61}O_4$	$[M+H-OH]^+$	Liver [70]
	DAG(30:0)	523	$C_{33}H_{63}O_4$	$[M+H-OH]^+$	Liver [70], Adipose [74]
	DAG(32:2)	547	$C_{35}H_{63}O_4$	$[M+H-OH]^+$	Liver [70]
	DAG(32:1)	549	$C_{35}H_{65}O_4$	$[M+H-OH]^+$	Liver [70]
	DAG(32:0)	551	$C_{35}H_{67}O_4$	$[M+H-OH]^+$	Adipose [73,74], Liver [70], Muscle [75]
	DAG(34:3)	573	$C_{37}H_{65}O_4$	$[M+H-OH]^+$	Liver [70]
	DAG(34:2)	575	$C_{37}H_{67}O_4$	$[M+H-OH]^+$	Liver [70]
	DAG(34:1)	577	$C_{37}H_{69}O_4$	$[M+H-OH]^+$	Adipose [73], Liver [70], Muscle [75]
	DAG(34:0)	579	$C_{37}H_{71}O_4$	$[M+H-OH]^+$	Adipose [74]
	DAG(36:4)	599	$C_{39}H_{67}O_4$	$[M+H-OH]^+$	Liver [70]
	DAG(36:3)	601	$C_{39}H_{69}O_4$	$[M+H-OH]^+$	Liver [70]
	DAG(36:2)	603	$C_{39}H_{71}O_4$	$[M+H-OH]^+$	Adipose [73], Liver [70], Muscle [75]
	DAG(36:0)	607	$C_{39}H_{75}O_4$	$[M+H-OH]^+$	Adipose [74]
GL0301	Triradylglycerols/ Triacylglycerols/				
	TAG(48:0)	805	$C_{51}H_{97}O_6$	$[M-H]^-$	Adipose [74]
	TAG(50:3)	851	$C_{53}H_{97}O_6Na$	$[M+Na]^+$	Liver [70]
	TAG(50:2)	829	$C_{53}H_{97}O_6$	$[M-H]^-$	Muscle [79]
	TAG(50:2)	853	$C_{53}H_{98}O_6Na$	$[M+Na]^+$	Liver [70]
	TAG(50:1)	855	$C_{53}H_{100}O_6Na$	$[M+Na]^+$	Liver [70]
	TAG(50:0)	833	$C_{53}H_{101}O_6$	$[M-H]^-$	Adipose [74]
	TAG(50:0)	857	$C_{53}H_{102}O_6Na$	$[M+Na]^+$	Liver [70]
	TAG(52:4)	877	$C_{55}H_{98}O_6Na$	$[M+Na]^+$	Liver [70]

(Continued)

TABLE 8.1 (Continued)

LM_ID	Sub-class	label(C:DB)	Mass	Formula	Species	Tissue
		TAG(52:3)	855	$C_{55}H_{99}O_6$	$[M-H]^-$	Adipose [74]
		TAG(52:3)	879	$C_{55}H_{100}O_6Na$	$[M+Na]^+$	Liver [70]
		TAG(52:2)	857	$C_{55}H_{101}O_6$	$[M-H]^-$	Adipose [73], Muscle [79]
		TAG(52:2)	881	$C_{55}H_{102}O_6Na$	$[M+Na]^+$	Liver [70]
		TAG(52:1)	883	$C_{55}H_{104}O_6Na$	$[M+Na]^+$	Liver [70]
		TAG(52:0)	861	$C_{55}H_{105}O_6$	$[M-H]^-$	Adipose [74]
		TAG(52:0)	885	$C_{55}H_{106}O_6Na$	$[M+Na]^+$	Liver [70]
		TAG(54:4)	881	$C_{57}H_{101}O_6$	$[M-H]^-$	Muscle [75]
		TAG(54:3)	883	$C_{57}H_{99}O_6$	$[M-H]^-$	Adipose [73]
Fatty Acyls	FA0101 Fatty Acids and Conjugates/ Straight chain fatty acid/	FA(14:0)	227.2	$C_{14}H_{27}O_2$	$[M-H]^-$	Liver [70]
		FA(16:0)	255.2	$C_{16}H_{31}O_2$	$[M-H]^-$	Adipose [73,74], Muscle [75,79], Liver [70]
		FA(18:0)	283.2	$C_{18}H_{35}O_2$	$[M-H]^-$	Muscle [75,79], Adipose [74], Liver [70]
	FA0103 Fatty Acids and Conjugates/ Unsaturated fatty acid/	FA(16:1)	253.2	$C_{16}H_{29}O_2$	$[M-H]^-$	Adipose [73], Muscle [75,79], Liver [70]
		FA(16:2)	251.2	$C_{16}H_{27}O_2$	$[M-H]^-$	Muscle [79]
		FA(18:3)	277.2	$C_{18}H_{29}O_2$	$[M-H]^-$	Muscle [79]
		FA(18:2)	279.2	$C_{18}H_{31}O_2$	$[M-H]^-$	Adipose [73], Muscle [79], Liver [70]

Category	ID	Common name	m/z	Formula	Ion	Tissue [Ref]
		FA(18:1)	281.2	$C_{18}H_{33}O_2$	[M-H]⁻	Adipose [73], Muscle [79], Liver [70]
Sterol Lipids	ST01010001	Cholesterol and break derivatives/				
		FA(20:4)	303.2	$C_{20}H_{31}O_2$	[M-H]⁻	Muscle [79]
		CH	369.3	$C_{27}H_{45}$	[M+H-H$_2$O]⁺	Adipose [73], Liver [70], Brain [59,63,71]
		CH	385.3	$C_{27}H_{46}O$	[M-H]⁻	Muscle [79], Brain [59]
		CH	385.3	$C_{27}H_{46}O$	[M-H]⁺	Brain [59,63], Liver [70]
		7-ketocholesterol	399.3	$C_{27}H_{43}O_2$	[M+H]⁺	Aorta [77]
Prenol lipids	PR02020001	Quinones and hydroquinones/Vitamin E/				
		α-tocopherol	429.3	$C_{29}H_{49}O_2$	[M-H]⁻	Muscle [79], Liver [70]
	PR02010004	Quinones and hydroquinones/ubiquinones/	430.3	$C_{29}H_{50}O_2$	[M]⁺	Retina [78], Liver [70]
		coenzyme Q9	795.6	$C_{54}H_{83}O_4$	[M-H]⁻	Muscle [79]
Sphingolipids	SP0602	Acidic glycosphingolipids/ Sulfoglycosphingolipids (sulfitides)/	778.5	$C_{40}H_{76}SNO_{11}$	[M-H]⁻	Brain [66]
		C16[a]				
		C16-OHb	794.6	$C_{40}H_{76}SNO_{12}$	[M-H]⁻	Brain [66]
		C18	806.6	$C_{42}H_{80}SNO_{11}$	[M-H]⁻	Brain [59,60,66]
		C18-OH	822.5	$C_{42}H_{80}SNO_{12}$	[M-H]⁻	Brain [59,60,66]
		C20	834.6	$C_{44}H_{84}SNO_{11}$	[M-H]⁻	Brain [59,60,66]
		C22-OH	850.6	$C_{44}H_{84}SNO_{12}$	[M-H]⁻	Brain [59,60,66]
		C22	862.6	$C_{46}H_{88}SNO_{11}$	[M-H]⁻	Brain [59,60,66]
		C23	876.7	$C_{47}H_{90}SNO_{11}$	[M-H]⁻	Brain [59,66]
		C22-OH	878.6	$C_{46}H_{88}SNO_{12}$	[M-H]⁻	Brain [59,60,66]
		C24:1	888.6	$C_{48}H_{90}SNO_{11}$	[M-H]⁻	Brain [59,60,66]
		C24	890.6	$C_{48}H_{92}SNO_{11}$	[M-H]⁻	Brain [59,60,66]
		C25:1	902.6	$C_{49}H_{92}SNO_{11}$	[M-H]⁻	Brain [66]

(Continued)

TABLE 8.1 (Continued)

LM_ID	Sub-class	label(C:DB)	Mass	Formula	Species	Tissue
		C24:1-OH or C25	904.6	$C_{48}H_{90}SNO_{12}$	$[M-H]^-$	Brain [59,60,66]
		C24-OH	906.6	$C_{48}H_{92}SNO_{12}$	$[M-H]^-$	Brain [59,60,66]
		C26:1	916.6	$C_{50}H_{94}SNO_{11}$	$[M-H]^-$	Brain [66]
		C25:1-OH or C26	918.6	$C_{49}H_{92}SNO_{12}$	$[M-H]^-$	Brain [66]
		C26:1-OH	932.7	$C_{50}H_{94}SNO_{12}$	$[M-H]^-$	Brain [66]
		C26-OH	934.6	$C_{50}H_{96}SNO_{12}$	$[M-H]^-$	Brain [66]
SP0501	Neutral glycosphingolipids/Simple Glc series/	C18:0	750.6	$C_{42}H_{81}NO_8Na$	$[M+Na]^+$	Brain [63], Aorta [77]
		C24:0c	834.6	$C_{48}H_{93}NO_8Na$	$[M+Na]^+$	Brain [63,64], Aorta [77]
		C24:1	832.6	$C_{48}H_{91}NO_8Na$	$[M+Na]^+$	Brain [63,64]
		Ch24:0d	850.6	$C_{48}H_{93}NO_9Na$	$[M+Na]^+$	Brain [63,64], Aorta [77]
		Ch24:1	848.6	$C_{48}H_{91}NO_9Na$	$[M+Na]^+$	Brain [63,64]
		Ch23:0	836.6	$C_{47}H_{91}NO_9Na$	$[M+Na]^+$	Brain [64]
		Ch22:0	822.6	$C_{46}H_{89}NO_9Na$	$[M+Na]^+$	Brain [64]
SP0301	Phosphosphingolipids/Ceramide phosphocholinesSphingomylin	SM(34:1)	616.5	$C_{34}H_{67}NO_6P$	$[M-(C_2H_2(N(CH_3)_3))]^-$	Liver [70]
		SM(34:1) SM(34:1)	642.6	$C_{36}H_{69}NO_6P$	$[M-(N(CH_3)_3)]^-$	Liver [70]
		SM(34:1)	687.6	$C_{38}H_{76}N_2O_6P$	$[M-CH_3]^-$	Liver [70]

[a] C16 = (3′-Sulf)Galβ-Cer(d18:1/16:0).
[b] C16-OH = (3′-Sulf)Galβ-Cer(d18:1/2-OH-16:0).
[c] C24:0 = GalCer(d18:1 / 2-OH-24:0).
[d] Ch24:0 = GalCer(d18:1 / 2-OH-24:0).

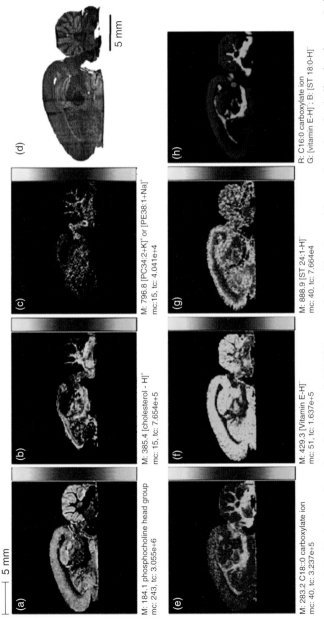

Figure 8.2 Sagittal section of rat brain: (a–d) SIMS images obtained in the positive mode—phosphocholine headgroup (*m/z* 184), cholesterol (*m/z* 385 and 796.8)—and optical image of the tissue; (e–h) SIMS images obtained in the negative mode—stearic (18 : 0) fatty acid fragment (*m/z* 283), vitamin E (*m/z* 429.3), and sulfoglycosphingolipid (sulfitide, d18 : 1/24 : 1)—and the overlay of these ions. [fatty acid (red), vitamin E (green), and sulfitide (blue).] Reproduced from Benabdellah et al.,[33] with permission from Springer Verlag.

283

region of cholesterol (m/z 369.3 and 385.3) that correlates to the corpus callosum. The negative ion at m/z 429, vitamin E is perfectly colocalized with cholesterol. Cholesterol and vitamin E are well known to be mostly located in the white matter, especially in the corpus callosum, where they also play a role in the axonal myelinization. The corpus callosum is a bundle of nerve fibers that bridges the right and left hemispheres of the brain. To ensure efficient electrical signal conduct across the corpus callosum, the neural fibers are coated with myelin sheath. Overall, the corpus callosum is easily distinguished from the cerebral cortex and other regions of the brain in SIMS images based on its distinct chemical composition. This aspect of the study does not challenge the spatial resolution of ToF-SIMS under static SIMS conditions. The images were acquired by scanning a Bi_3^{2+} beam over 87.5×87.5 μm^2 pixels and moving the stage to deliver 256×256 pixel images over an area of 22.4×22.4 mm^2. The molecules detected and displayed are in high concentration per pixel. Under these circumstances, the ion beam is operated such that high mass resolution is possible. Nevertheless, the ion at m/z 796.8 could have one of three structures: protonated phosphatidylethanolamine C40 : 4, or sodium cationized phosphatidylethanolamine C38 : 1, or potassium cationized phosphatidylcholine C34 : 2. This illustrates the need for MS–MS facilities to distinguish the structure of these ions effectively. Overall, the study demonstrated that MALDI generally delivered higher yields than ToF-SIMS at pixel sizes above 50 μm. In the main the ion distributions for ions detected by both techniques were well correlated in the 2D images. However, the ion m/z 796.8 referred to earlier was perfectly anticorrelated in SIMS and MALDI images! This was attributed to the matrix effect operating rather differently in the two systems and this illustrates the importance of understanding how this can influence results.

The effects on ion yield of decreasing the pixel size being interrogated are also very nicely illustrated (see Figure 8.3). When the pixel size was decreased by a factor 44 from 87.5 to 2 μm, reducing the number of molecules per pixel by $\sim$1900, the ion fluence could be increased from 8.4×10^8 to 2.5×10^{11} ions/cm^2, an increase of $\times$300 to compensate to some extent while still staying below the static limit. The image contrast was maintained, the spatial resolution and detail visible was much greater and the range of molecules detectable was about the same, although maximum counts per pixel were rarely above 10. Increasing the spatial resolution to 390 nm was only possible by moving to unit mass resolution, the ion dose was increased to 2×10^{12} ions/cm^2 which is beyond the static limit for many molecules.[34,35] Even so, molecules above m/z 450 were undetectable. Other than fragment ions with high ion yield, the maximum count per pixel was less than 2 or 3 ions. Cholesterol and vitamin E are detectable and in both ion modes show similar localization, complementary to sulfatides, phosphocholine (PC) head group, or fatty acid fragments.

The capability of cluster primary ion ToF-SIMS illustrated here has been exploited particularly by the group of Touboul et al. in a range of studies of lipid-related diseases, such as Duchenne muscular dystrophy,[36,37] Fabry disease,[38] nonalcoholic fatty liver disease,[39] atherosclerosis,[40] and cystic fibrosis,[41] as well as cancers. These diseases stem from dysfunctional metabolic processes and result in abnormal concentrations of biomolecules. Chemical images across diseased

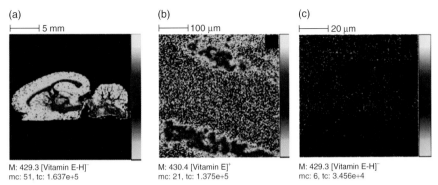

(a)
⊢———⊣5 mm

(b)
⊢———⊣100 μm

(c)
⊢———⊣20 μm

M: 429.3 [Vitamin E-H]⁻
mc: 51, tc: 1.637e+5

M: 430.4 [Vitamin E]⁺
mc: 21, tc: 1.375e+5

M: 429.3 [Vitamin E-H]⁻
mc: 6, tc: 3.456e+4

Figure 8.3   256 × 256 Ion images of a sagittal rat brain section at three different levels of spatial resolution. From left to right the pixel size decreases from 87.5 μm to 2 μm to 390 nm. The ion dose increases from 8.4×10^8 ions/cm² to 2.5×10^{11} ions/cm² to 2×10^{12} ions/cm². The ion images originate from vitamin E. (a) and (c) are the negative ion at m/z 429.3, while (b) of the positive ion at m/z 430.4. The ion yields counts per pixel are indicated by the color coding, yellow representing the maximum. The range is 0–51 for the 87.5 μm pixels in (a); 0–21 for the 2 μm pixels in (b), and 0–6 for the 390 nm pixels in (c). Adapted from Benabdellah et al.,[33] with permission from Springer Verlag.

tissue reveal areas of abnormal chemistry; such scarcity or overabundance of a particular biomarker can link cellular dysfunction with anatomical specificity. Le Naour et al. found a higher concentration of unsaturated diacylglycerides (DAGS) and triacylglycerides (TAGS), as well as increased cholesterol signals in steatotic vesicles taken from an individual with fatty liver disease compared to normal tissue.[39] While these studies are very impressive they do emphasize that even with the benefits of cluster primary beams, ToF-SIMS is almost exclusively sensitive to lipids and similar smallish molecules that (i) are present in quite significant concentrations, (ii) can be lifted off the surface efficiently, and (iii) have relatively high ion yields. To overcome these restrictions will require means to increase the ionization probability. In this regard, the use of matrix-assisted SIMS is being explored using MALDI type matrices. As in the case of MALDI, these matrices are able to extract molecules of interest to the surface and using proton donors, ionization to M + H ions can be encouraged. As already mentioned, this approach has enabled large peptides from tryptic digests to be analyzed using $C_{60}{}^+$ ToF-SIMS enabling the source proteins to be identified.[16] Matrix assistance can be seen as a possible way forward for SIMS; however, the fact that a matrix has to be used means that this variant of SIMS cannot be said to be sample pretreatment free.

8.5.2   Drug Location in Tissue

One area within which imaging ToF-SIMS promises exciting new possibilities is drug discovery, especially as a replacement or a complementary technique to autoradiography and fluorescence microscopy for mapping the location of drugs within tissue sections. However, these pharmaceutical compounds will be at a much lower surface abundance than the native compounds that have been imaged in the past with SIMS. It is also likely that the drug will metabolize. It may be the case that there are too few drug molecules within the upper monolayers to yield a detectable signal from the sample. It is in cases such as this that the low damage accumulation seen with the polyatomic ion beams will be significant because analysis will not be restricted by the static limit, allowing full exploitation of all available molecules in a pixel or voxel.

To illustrate the possibilities and some of the challenges, Figure 8.4 shows the mapping of the drug raclopride within a section of rat brain containing the striatum from an animal dosed with the drug *in vivo*.[35] The estimated concentration of the drug in the bulk organ is in the region of 2 ppm as determined by ESI-MS. Raclopride is a molecule that specifically binds to dopamine-D2 receptors within the brain, and a ^{11}C-containing variant is commonly used in positron emission tomography (PET) to map the location of these receptors. The selected ion images shown in Figure 8.4a–c relate to the shaded area of the optical image in Figure 8.4d. In order to identify the different domains within the section, the distribution of the PC head group ion at m/z 184 (Fig. 8.4a) and cholesterol ion, m/z 369 (Fig. 8.4b), are shown. From the accompanying spectrum, peaks relating to the $[M + H]^+$ of the drug raclopride are clearly visible to the left of a cluster of peaks relating to cholesterol fragments within the drug containing section and absent from a representative control section spectrum. The distribution of this peak was then imaged using the C_{60}^+ ion beam and presented in Figure 8.4c. However, the distribution demonstrated by the SIMS analysis does not accurately corroborate with the known locations of D2 receptor sites.[42] These data, while encouraging with respect to detection levels, offer two possible explanations for the result obtained, both of which involve common challenges encountered with SIMS and other mass spectrometry analysis. One possible explanation is that there is redistribution of the drug within the sample on entering the vacuum. It has been shown that cholesterol migration at room temperature can yield different results from the same tissue section depending on the temperature the sample is held at during analysis,[11] although there is much more to be understood of the mechanisms behind this molecular movement. The other possible cause is *the matrix effect*. It is to be noted from Figure 8.4c that the $[M + H]^+$ signal from the drug closely resembles the distribution of cholesterol (Fig. 8.4b) which is indicative of the myelin-rich white matter of the brain. It has previously been shown that cholesterol can provide a chemical environment that promotes the protonation of a species, while the phosphatidylcholine containing lipids act to suppress the $[M + H]^+$ signal from a typical drug compound.[26]

In order to investigate whether a matrix effect could explain the results in Figure 8.4, a control section of rat brain was spin coated with 5 µl of a 1 × 10^{-3} M concentration solution of a related drug molecule, haloperidol, which should provide a submonolayer covering of the drug over the surface. When the tissue section was imaged, there were significant similarities to the raclopride result in Figure 8.4. In Figure 8.5, the distribution of drug $[M + H]^+$ signal (m/z 376) across two different domains of the tissue is shown. The signal from the PC head group (m/z 184) indicates the distribution of gray matter, while the peak from the

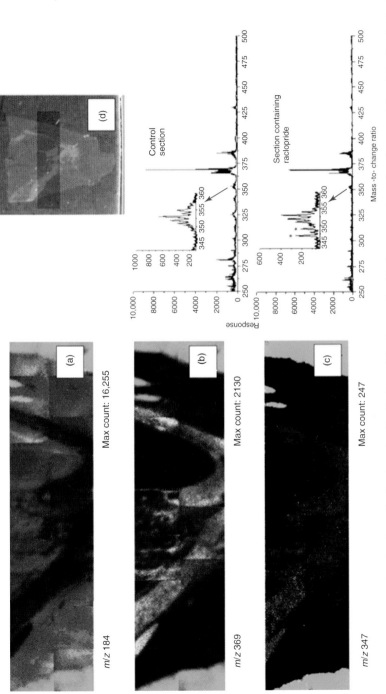

Figure 8.4 A SIMS imaging investigation into the distribution of the drug raclopride from a section taken from a rat dosed *in vivo*. In the inset mass spectra, the lower spectrum demonstrates the presence of the $[M + H]^+$ of the raclopride at m/z 347 with its isotope peak at m/z 349, which are absent from the control spectrum. The images are shown on a thermal scale and illustrate the distribution of (a) m/z 184 from phosphatidylcholine head group, (b) m/z 369 from cholesterol, and (c) m/z 347 from the drug raclopride, the maximum counts per pixel of the ion of interest are quoted. In all of the images, the green pixels represent the substrate. The optical image (d) shows the area that was analyzed. The total imaged area was 1.6×8.0 mm, with an ion dose of 8.8×10^{11} ion/cm². Reproduced from Jones et al.,[35] with permission from Elsevier.

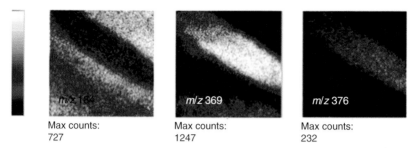

Max counts: 727   Max counts: 1247   Max counts: 232

Figure 8.5 Distribution of the molecular signal of the drug haloperidol ([M + H]$^+$ signal at m/z 376) spun cast onto a section of brain with respect to the chemical domains of the tissue. The signal from cholesterol (m/z 369) and phosphatidylcholine (m/z 184) are shown to indicate the different chemical domains within the tissue surface; the analyzed area is 800 × 800 μm with a dose of 8 × 10^{10} ion/cm^2. Although the drug species covered the whole area visible in the image, the molecular signal is only detected from the cholesterol-rich areas. Reproduced from Jones et al.,[35] with permission from Elsevier.

cholesterol (m/z 369) is again used to demonstrate the localization of white matter. The haloperidol signal has been suppressed in the gray matter region, whereas it is clearly evident from the white matter. This model system demonstrates the severity of the suppression/enhancement effects that can be encountered across a two-domain system such as brain tissue sections. Without prior knowledge of the system in question it would be easy to assume that the peak at m/z 376 was linked to the constituents of the white matter along with the cholesterol.

As we have already discussed, matrix effects due to the salt content of biological systems are found to interfere quite severely with all mass spectrometries.[23–25] With SIMS and MALDI, salts can be removed by washing samples in ammonium formate solutions that convert the salts to volatile formates. However, great care has to be exercised because this procedure can remove other compounds and also cause redistribution of molecules in tissue and cells. It would be advantageous if matrix effects could be overcome. Where the effects are due to competition for protons, there may be procedures to provide excess protons at the sample surface to reduce competition.[29,30] Another approach is to use laser post-ionization to probe the high yield of neutrals above the sputtered surface where the compound of interest has formed radical cations.[43] To demonstrate the possible benefits of this approach, the authors prepared a model sample consisting of the drug atropine in 1 : 1 mixtures with both cholesterol and dipalmitoyl phosphatidylcholine (DPPC). The samples were analyzed using C$_{60}$$^+$ SIMS in the conventional manner and laser postionization of the sputtered neutrals. The results presented in Figure 8.6 clearly show a great difference in the intensity of the drug molecule within the SIMS experiment, with the intensity of the [M + H]$^+$ at m/z 290 and major fragment at m/z 124 differing by an order of magnitude between the two lipid matrices. When the same samples are analyzed using laser postionization, the most abundant representative ion seen from the drug is the fragment peak at m/z 124, the [M + H]$^+$ ion of course is not generated by laser postionization. When the intensity of this peak is

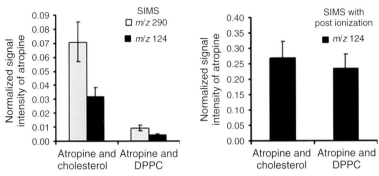

Figure 8.6   The analysis of a drug molecule mixed with two abundant biological lipids: cholesterol and DPPC. The SIMS analysis, reliant upon ionization at or just above the surface demonstrates strong differences between the ionization probability of the drug atropine to its $[M + H]^+$ ion (m/z 290) and major fragment (m/z 124), whereas when the molecule is laser postionized the difference is within experimental error. The ion used to identify the atropine in the laser postionization was a characteristic fragment at m/z 124. Reproduced from Jones et al.,[35] with permission from Elsevier.

compared across the two samples, the difference is within experimental error. This suggests that the same amount of the drug molecule is present at the surface to be sampled, and the same number are being sputtered into the vacuum; however, the nature of the sample has a great effect on the percentage of these molecules that enter the vacuum in a charged state. Clearly, there are potential benefits of separating the desorption and ionization steps within surface mass spectrometry.

8.5.3   Microbial Mat — Surface and Subsurface Analysis in Streptomyces

Polyatomic primary ions not only offer the possibility of good sensitivity in 2D imaging at micrometer spatial resolution, their use also offers the possibility of delving into the subsurface chemistry. An early biological investigation that illustrated these capabilities was a study by Vaidyanathan et al. of the antibiotic production of the mycelial, soil-dwelling microorganism *Streptomyces coelicolor*. *S. coelicolor* is a member of the actinomycetes family of gram-positive bacteria, which are prolific in the production of secondary metabolites of therapeutic importance. The microorganism produces at least four antibiotics, including methylenomycin, calcium-dependent-antibiotic (CDA), actinorhodin (a blue pigment), and undecylprodigiosin (a red pigment). Normally, actinorhodin is produced in excess of undecylprodigiosin. Under salt stress growth conditions (2.5% NaCl), the trend is reversed and the red pigment is observed at the surface in excess of the blue. However, little was known about the distribution of these antibiotics within the bacteria. A detailed ToF-SIMS study of the bacteria under normal and salt-stressed culture conditions was recently performed where depth profiling of the bacteria was carried out using C_{60}^+.[44] A bacterial suspension was transferred to a piece of a silicon wafer and the bacteria were heat fixed for analysis in the ToF-SIMS

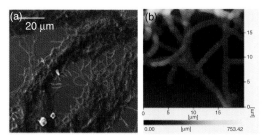

Figure 8.7   Images of the bacterial sample on silicon as presented for analysis on ToF-SIMS, acquired with a scanning electron microscope (a), and an atomic force microscope (b). The samples were exposed to the vacuum conditions of the ToF-SIMS before analysis by SEM and AFM. The dimensions of the bacteria and its mycelial nature can be inferred from the images. Reproduced from Vaidyanathan et al.,[44] with permission from the ACS publications.

instrument. SEM and atomic force microscopy (AFM) analyses were carried out after ToF-SIMS analysis to check the integrity of the material (Figure 8.7).

Surface and subsurface ToF-SIMS imaging of the antibiotic distribution is demonstrated on a control and a salt-stressed bacterial population. The mass spectra of the two main antibiotics showed that actinorhodin was characterized by peaks at m/z 368 and 483 and undecylprodigiosin by a peak at m/z 392; a minor contributor to undecylprodigiosin, butylcyclohexylprodiginin displayed a peak at m/z 394. Figure 8.8 shows a set of images from the control sample grown under normal conditions. The top row of figures show the general organic distribution (characterized by m/z 57, $C_4H_9^+$) in a section of the bacterial mat, the distribution of protein material indicated by the sum of several immonium ion signals from the amino acids, and finally the distribution of some lipid tail group ions. On the bottom row we have images of the distributions of the two antibiotics. It can be seen that actinorhodin, the blue compound, appears to present a greater yield than the red compounds, which is in accord with the general observation that the blue antibiotic is more commonly observed at the surface. There is some difference in localization, but it is not great.

Turning to the salt-stressed sample, Figure 8.9, an SEM image of a sample area analyzed (bar ∼100 μm), is shown together with images of the distribution of the two antibiotics before and after sputter etching. In the top row, the spectrum of the surface of the salt-stressed sample is shown. The two antibiotics are clearly evident. In this case the red antibiotic now appears to show a higher yield. In the upper inset spectrum, in the region of m/z 500, spectral features from the tail groups of membrane surface lipids can also be seen. After sputter etching the surface with a primary ion fluence of 6×10^{13} ions/cm^2 the actinorhodin (m/z 368) in the lower spectrum has disappeared while undecylprodigiosin (m/z 392) remains. AFM measurements suggest that approximately 10–50 nm of surface has been removed. In the inset of the lower spectrum the remnant of peaks in the m/z 500 region suggests that some cell membrane remains. The disappearance of m/z 368 suggests that while the microorganism secretes actinorhodin at the surface as a first line of defense,

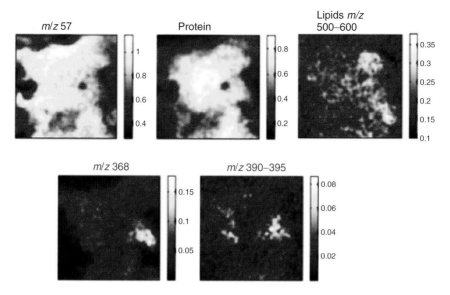

Figure 8.8 Images acquired using 40 keV $C_{60}{}^+$ of the control sample of *S. coelicolor* grown under normal conditions. The images are 128×128 pixels from a 200×250 μm field of view, total ion fluence $<10^{13}$ cm^{-2}. The images are normalized to total ion count at each pixel. Protein is the sum of ion counts for the immonium ions at *m/z* 30, 44, 70, 84, 86, and 110. Reproduced from Vaidyanathan et al.,[44] with permission from the ACS publications.

when under stress the red pigments are generated in excess in the subsurface as a second line of defense. This ability to depth profile molecules in a biological matrix opens up exciting possibilities. In this example, a heterogeneous bacterial mat has been studied. There was no attempt to determine accurately the depth removed in uncovering the surface; however, useful and biologically interesting information was extracted. The issues involved in more quantitative depth profiling and the prospect for three dimensional imaging will be explored in the next section.

8.5.4 Cells

Analysis and imaging of single cells using ToF-SIMS challenges the capability of securing adequate ion yield from small areas. The dimensions of most biological cells lie between 10 and 50 μm in diameter; this means that to locate and distinguish the spatial distribution of chemistry requires spatial resolution of around 1 μm, preferably submicrometer. There are some cell systems with very large dimensions that have allowed studies to be carried out which demonstrate the potential offered by ToF-SIMS analysis. A recent study that combined MALDI and C_{60}—ToF-SIMS using the Q-Star *ortho*-ToF instrument referred to earlier, investigated the lipid composition of a single neuron from an *Aplysia californica*

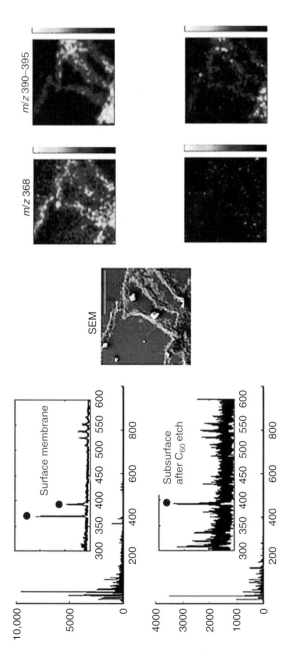

Figure 8.9 Spectra (signal intensity (counts) vs m/z) and images acquired using 40 keV C_{60}^+ of the salt-stressed sample of *S. coelicolor*. The upper spectrum and images show the yield and distribution of the two antibiotics at the surface of a $200 \times 250 \ \mu m^2$ area of the bacterial mat. The lower spectrum and images show the yield and antibiotic distribution of the subsurface uncovered followed sputtering with a C_{60}^+ fluence of 10^{13} cm^{-2}. Reproduced from Vaidyanathan et al.,[44] with permission from the American Chemical Society.

sea slug.[45] The neuron dimensions are in the millimeter range and despite the limited imaging capability of the Q-star, a clear total ion image could be acquired, Figure 8.10d. SIMS and MALDI spectra were compared. The major peaks in the SIMS spectra are found at m/z 709.5, 719.5, 725.5, 768.5, and 786.5, Figure 8.10a, whereas the major peaks in the MALDI spectrum were m/z 746.5, 768.5, and 796.5, Figure 8.10b. The lipid molecular ion at m/z 746.5 produces a peak at m/z 768.5 when it attaches to a sodium ion, and this species is observed in both spectra. Loss of $-N(CH_3)_3$ yields the expected fragment at m/z 709.5. In order to confirm this idea, tandem MS was needed and was available by exploiting the qToF configuration of the Q-star. The precursor ions can be selected by the scanning the quadrupole and these are fragmented in a quadrupole collision cell. The resulting fragments are analyzed using the ToFMS. Unfortunately, the yield of precursor ions using SIMS was not sufficient to carry out MS–MS analysis so the ions at m/z 746.5 and 768.5 were studied in the MALDI mode. The tandem MS spectra produced 184.07 fragments, Figure 8.10c, typically associated with glycerophosphocholine and sphingomyelin. Although sphingomyelin lipids also produce a characteristic

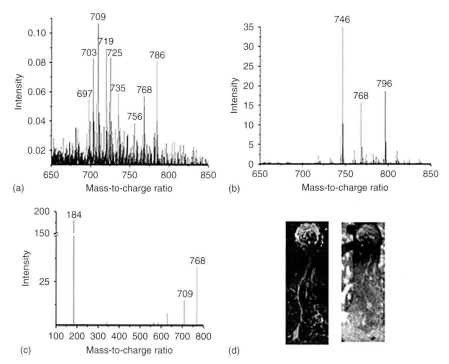

Figure 8.10   Lipid profile obtained from a single neuron with SIMS (a) and from a compilation of neurons with MALDI (b). The tandem MS spectrum shows that m/z 709 and 184 are major fragments of m/z 768.5, the sodiated adduct of major lipid component m/z 746.5 (c). Optical image (d, left) and black and white SIMS total ion image (d, right) of cultured aplysia neuron on silicon wafer (image size 2.00 × 4.75 mm). Reproduced from Passerelli and Winograd,[45] with permission from Wiley.

184.07 fragment group and are abundant in neurons, the nitrogen group in the sphingosine moiety produces only odd molecular ion species.[46] Therefore, from the characteristic fragmentation peak and the odd–even parity, the authors were able to deduce that the m/z 746 is most likely a glycerophosphocholine lipid. Tandem MS of the m/z 768.5 peak produces an m/z 709.5 peak and confirms the fragmentation pathway theory. On the basis of information obtained, it was possible to reduce the possible lipid contributing to peak m/z 746.5 to the following lipids: a diacylglycerophospholipid with fatty acid constituents consisting of 33 carbons and one double bond (33a : 1 GPCho, 746.5779 Da), an ether glycerophospholipid (34e : 1 GPCho, 746.6143 Da), and a plasmalogen glycerophospholipid (34p : 0 GPCho, m/z 746.6143). This demonstrates the value and limitations of MS–MS. Without other separation methods such as ion mobility it will not be possible to distinguish the latter two molecules.

The first attempt to probe the 3D distribution of chemistry in a single cell also exploited a large cell system to make the study less challenging. The research focused on a freeze-dried *Xenopus laevis* oocyte (a frog's egg) using a 40 keV $C_{60}{}^+$ ion beam for both sputtering and analysis.[47] The frog's egg is approximately 1 mm in diameter and the size meant that topography effects were important in assessing the data generated. However, the study did demonstrate that the experimental approach is feasible and that molecular information continued to be accessible even after the removal of more than 100 μm of cell and that the spatial distribution of specific molecules in a cell could be described. Figure 8.11 shows the cell before and after depth profile analysis, together with the positive and negative ion spectra used to reconstruct the chemical images.

The study used a conventional pulsed ion beam ToF-SIMS system and, as a consequence, took many days of sputter etching and analysis. A total ion fluence of 10^{16} cm^{-2} was used. The intense peak at m/z 369 is attributed to the cholesterol $[M + H - H_2O]^+$. The m/z 540–700 envelop contains peaks at m/z 548, 574, and 576 that can be attributed to phosphatidylcholines with fatty acid side chains of composition C16 : 0, C18 : 2, and C18 : 1, respectively, without the phosphocholine head group. These assignments are corroborated by the detection of the corresponding fatty acid peaks in the negative ion spectra at m/z 255, 279, and 281 for fatty acid side chains of composition C16 : 0, C18 : 2, and C18 : 1, respectively. These lipids correspond to the three found most abundantly in analyses of biological cell extractions. There is a similar set of molecular ion peaks between m/z 800 and 1000. By generating a stack of 2D images for a particular m/z or range of m/z values, it is possible to put together a 3D image as shown on the left-hand side of Figure 8.12.

The right-hand side of Figure 8.12 shows the distribution of four lipid molecules within the volume of oocyte analyzed. Three-dimensional images, while useful in providing qualitative pictures of the distribution of specific chemistry are not always very useful in giving an idea of relative signal intensities as a function of location. One way to help assess relative intensities is to use the isosurface technique. The isosurface highlights regions of pixels with intensity values above a specified threshold. By selecting a high threshold, regions of high intensity can be clearly visualized. Low thresholds, just excluding noise, can be used to visualize the

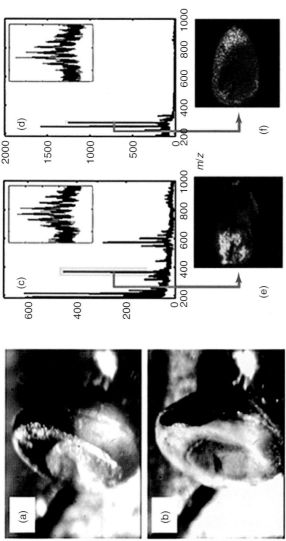

Figure 8.11 Optical micrograph of oocyte cell mounted on copper tape for ToF-SIMS analysis (a) before ion beam etching and (b) following 40 keV C_{60}^{+} etching. ToF-SIMS spectra (signal intensity [counts] *vs* m/z) of *X. laevis* oocyte (integrated over the scanned area), showing peaks above m/z 200 in the (c) positive and (d) negative ion modes, after an accumulated primary ion fluence of 1×10^{15} cm^{-2}. (e) and (f) are representative secondary ion images associated with m/z 369 (cholesterol $[M - H_2O + H]^{+}$) and m/z 281 (oleic acid $[M - H]^{-}$), respectively. Reproduced from Fletcher et al.,[47] with permission from ACS publications.

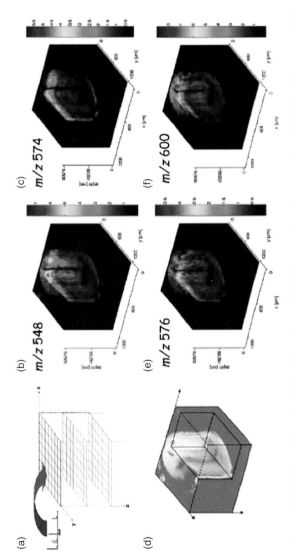

Figure 8.12 Three-dimensional imaging of part of the *Xenopus laevis* oocyte. (a and d) show how assembling a stack of 2D images enables a 3D image to be produced. (b, c, e, and f) show the distribution of four different lipids in the oocyte section. Reproduced from Fletcher et al.,[47] with permission from ACS publications.

sample as a whole. Thus, for this particular example, Figure 8.13 shows isosurface plots for the lipid head group m/z 184 with a low threshold so the whole cell is seen, the cholesterol fragment at m/z 369 with a median threshold, and the lipid less the head group fragments at m/z 540–600 with a high threshold showing that they are in high concentration in the outer membrane region. Although this example showed what is possible in principle, the cell was so large that spatial resolution was not challenged and yields were high for the molecules imaged. The topography of the cell was such that, although the cell-like structure is evident, no accuracy can be claimed in terms of the observable cell dimensions.

The frog egg study has been taken further using the J105 buncher-ToF-SIMS system and a DC C_{60}^{+} primary beam. The mode of operation of the J105 means that samples of complex geometry such as the frog's egg can be analyzed with relative ease. Figure 8.14 shows that the chemical changes during the cellular division associated with embryo development after fertilization can be followed in some detail (Tian et al., Unpublished Work and Reference 48). Furthermore, it is possible to focus on specific areas of the egg following cell division and the cells can be imaged and depth profiled (Figure 8.15). It can be seen that cellular lipid contents can be discerned that have been associated with Golgi structures. Linking these observations with the biology of embryo development is challenging; however, what is very clear is that ToF-SIMS imaging using C_{60}^{+} is an effective means for probing the chemistry of large single cells in three dimensions.

The studies described so far have been concerned with large cells. Good yields of molecules and fragments have been detected because pixel areas of many square micrometers have been sampled. However, most cells of interest are below 50 μm in diameter. There have been a number of studies that have explored the use of normal rat kidney (NRK) cells and HeLa cells as model systems to uncover the capability of SIMS in 2D and 3D imaging. Two approaches have been used: the first combines liquid metal cluster beams for analysis and 2D imaging and C_{60}^{+} for sputter etching on a pulsed beam ToF-SIMS. The second has used a DC C_{60}^{+} beam for sputtering with simultaneous collection of secondary ions to generate either 2D or 3D images using the J105 instrument (Figure 8.15).

A study by Breitenstein et al. illustrates the first approach.[49] NRK cells were studied whose cellular architecture is well known. The cells are approximately 20 μm in diameter. Six confluent layers of cells, grown on cover slips under ordinary cell-culture conditions, were analyzed using ToF-SIMS after the cells had been stabilized with chemical fixation. In this case, the layers were removed by 10 keV C_{60}^{+} sputtering and analysis was carried out using a metal cluster ion beam, 25 keV Bi_3^{+}. The authors were only able to clearly detect cholesterol, phospholipids, and immonium ions from the amino acids in proteins. The figure associated with this work was presented earlier in Chapter 1 (Fig. 1.8). Figure 1.8a–d shows the 2D xy distributions after the forty-fifth sputter cycle with amino acid fragments in red, phospholipids in green, and substrate peaks in blue. These are overlaid in Figure 1.8d. The depth profile experiment collects the data such that the observed xz distribution appears as a vertical inverse of the actual distribution in the cells sitting on a flat surface, as shown in Figure 1.8e. The authors carried out a mathematical transformation to provide a presentation that mirrors the real distribution,

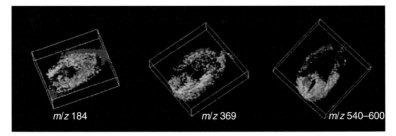

Figure 8.13 Isosurface representation of C_{60} depth profile of oocyte. See text for details. Figure reproduced from Fletcher et al. with permission from Springer.[50]

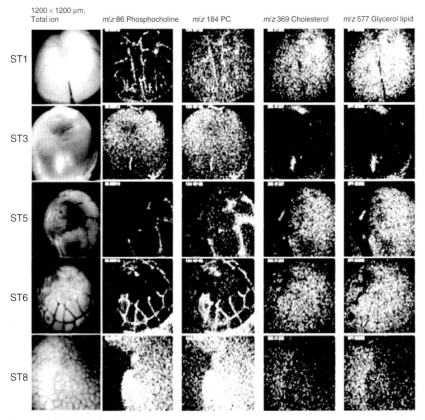

Figure 8.14 Biochemical changes at successive stages of embryo development after frog zygote fertilization as monitored by the J105 3D Chemical Imager using a DC 40 keV C_{60}^{+} primary beam. The images track the early stages of embryo development. The first column shows the total ion images at each stage. The morphographic changes are clearly seen here, from two cells to hundreds of cells on the animal side of the embryo. Some important biomolecules located on the surface are imaged. Phosphocholine lipids are mainly distributed at cell junctions. Cholesterol occurs all over the cell except at stage 3. At stage 8, phosphocholine lipids and cholesterol appear to be distributed complementarily. Reproduced from Vickerman,[48] with permission from the Royal Society of Chemistry.

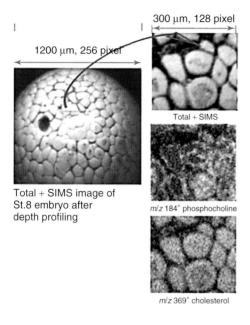

300 μm, 128 pixel

1200 μm, 256 pixel

Total + SIMS

Total + SIMS image of
St.8 embryo after
depth profiling

m/z 184+ phosphocholine

m/z 369+ cholesterol

Figure 8.15 Depth profiling of stage 8 embryo of *Xenopus laevis*. The 300 μm square region was profiled. The presented images show the accumulated secondary ions after a fluence of 4.26 × 10^{14} ions/cm^2. The m/z 184 phosphocholine fragment image shows Golgi-like structure. Reproduced with permission from Vickerman,[48] with permission from the Royal Society of Chemistry.

see Figure 1.8f. The fact that the lateral dimensions of the cell body and nucleus derived from this experiment correspond favorably with the values that were determined from light-microscopy images (diameter of the cell body 20 μm; diameter of the nucleus 10–12 μm) is very encouraging. Furthermore, the molecular information is provided without the need to label a subset of chemical species as, for instance, in fluorescence microscopy. From this study, the prospect of probing subcellular chemical information seems promising; however, the study did also show that detection of large molecular species was problematic. The use of the liquid metal ion beam did limit analysis to static conditions; however, the small pixel dimensions also limit the number of molecules available for analysis.

As mentioned in Section 8.2, the analysis of cells in the vacuum environment requires that some protocols have to be followed to seek to ensure that the biological integrity of the cells' physical and chemical structure is preserved. A study of the effects of sample handling on cellular structure illustrates the early use of the J105 analysis paradigm in cellular studies.[51,52] Three frequently used cellular preservation approaches were investigated: freeze-drying, in which the cells are rapidly frozen in liquid propane, stored in liquid nitrogen, and then slowly warmed to room temperature under vacuum to remove the water gradually while in theory maintaining the physical and chemical structure of the cell. The second method is to use formalin fixation that cross-links cellular protein to fix the chemical and physical structure. Finally the cells were rapidly frozen in liquid propane and then held in their frozen hydrated state under liquid nitrogen. They were then transferred to the reverse mousetrap fracturing device referred to earlier, at 100 K, and fractured to expose the inside of the cells. Samples of the cells treated in each of these three ways were loaded into the J105 instrument, analyzed and depth profiled with 40 keV C_{60}^+. A series of 2D chemical images of successive layers were acquired by accumulating the secondary ions generated between each layer. The 2D images were then combined to produce a 3D image. Unlike the situation with pulsed beam ToF-SIMS instruments, the vast majority of material sputtered from the cells was used in generating the analysis data enabling a higher ion yield per pixel (voxel). Figure 8.16 compares selected 2D images at three stages through each of the cell systems, top surface, after removal of several layers and then several further layers deeper in. Two ions have been used to image the cells, m/z 184 corresponding to the phosphocholine lipid head group that is mainly to be found in the cell membrane and m/z 120 corresponding to an amino acid, phenylalanine associated with protein in the cell nucleus. It should be remembered that the data shown represent secondary ions accumulated with ion fluences well beyond the static limit.

There is a clear difference between the three systems. The frozen hydrated freeze-fractured sample shows a ring of m/z 184 with m/z 120 in the center demonstrating that the cells had indeed been fractured. The membrane material is clearly separately located from the nucleus as the cell is profiled. The freeze-dried cells show a rather dispersed distribution of m/z 184 at the start and as the cells are profiled, m/z 120 does begin to appear, indicating that a membrane has been removed, but both ions are still quite dispersed and not as clearly located as they are for the frozen-hydrated sample suggesting that perhaps the sample pretreatment does

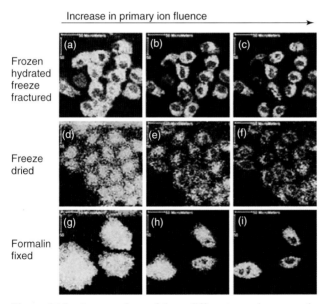

Figure 8.16 A comparison of three different sample preparations, freeze-fracture, frozen-hydrated (a–c), freeze-dried (d–f), and formalin fixed (g–i) for the m/z 184.06 (green) and m/z 120.10 (red) shown in color-overlay. The figure shows an increase in primary ion fluence from left to right with each layer obtained using a fluence of 3×10^{13} ions/cm^2 acquired over a field of view 256 × 256 μm^2 and 256 × 256 pixels. Reproduced from Rabbani et al.,[51] with permission from John Wiley & Sons.

not maintain the original cellular distribution as well as is hoped. Finally, the fixed cells seem to be much larger than the other two at the start. There are suggestions that the fixing process can expand the cell volume and the data would support this. As the cell is profiled, the membrane and nuclear components seem to be quite well localized however, and it is noteworthy that one cell shows evidence of mitosis giving rise to two nuclei. This imaging data have been complemented by assembling the data into depth profiles for the fixed- and frozen-hydrated samples, Figure 8.17. These show that for the frozen hydrated sample, as expected, membrane m/z 184 and nuclear material are evident at the surface. Nuclear material, in this case m/z 136 from adenine, is predominant in the center and membrane is evident just before reaching the substrate. The frozen-hydrated sample seems to preserve the physical and chemical state of the cell rather well. The profile for the fixed sample indicates some mixing of the nuclear material into the membrane in both the upper and lower regions of the cell, suggesting that the fixing chemistry may interfere with the original chemical distribution.

Overall, the data suggest that it may well be advantageous to carry out analyses in the frozen-hydrated state. The data assembled during this study has been used to produce a 3D representation of the cell structure and from these to produce movies to explore the structural distribution of cellular components. This 3D imaging data was presented in Chapter 6 (Fig. 6.26).[52] Such data present significant challenges in data handling as the raw data files that provide the information for the 3D images amount to many gigabytes.

A similar study of cell pretreatment for vacuum analysis has also been followed recently by Castner et al. using the liquid metal ion beam analysis with $C_{60}{}^{+}$ sputtering approach.[53] In this study, a somewhat different set of treatments were tried. The control set was a brief wash in ammonium acetate followed by drying in air. Figure 8.18 shows images of some of the positive ions detected immediately upon analysis, followed by the result of a sputter cycle 1 1 $\times$ 10^{14} 20 keV C_{60} ions. It can be seen that the C_{60} etch cleans up the substrate surface and also removes the phospholipid membrane (sum of peaks at m/z 58, 86, and 184) uncovering the nuclear area in the center. The K/Na ratio is now higher in the cell than outside it. When images are taken at higher resolution, the subcellular structure is clearly seen. In the negative ion images, ions associated with nucleic acids in the cell's nucleus are observed. Figure 8.19 shows a series of images of before and after C_{60} sputtering from samples prepared by other preparation protocols. It is interesting that in all cases the cell morphology is different from the control set, which seems to be closer to what might be expected. The authors suggest that the best result was obtained when the short ammonium rinse was followed by plunge-freezing in liquid ethane and freeze-drying, Figure 8.19e. They argue that this method resulted in an optimal K/Na ratio inside the cells, as well as the preservation of morphology with minimal membrane damage. It is not entirely clear why they believe it is better than their frozen hydrated sample Figure 8.19d. Certainly the above mentioned result seemed to suggest that the use of frozen hydrated samples was to be preferred. However, they do believe it is more important to observe that C_{60} "cleaning" always resulted in the removal of surface contamination, which made imaging of subcellular features possible.

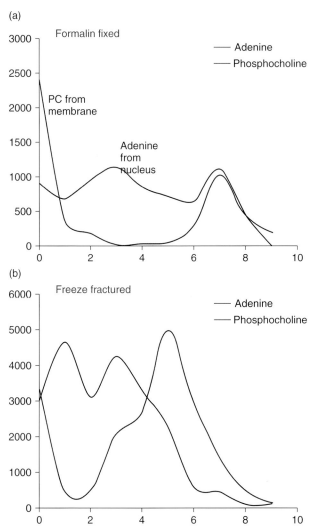

Figure 8.17 Plots showing the variation of phosphocholine and adenine signal (y-axis in arbitary units) as a function of depth (x-axis in arbitrary units) through the center of a single formalin fixed freeze-dried HeLa cell (a) and a freeze-fractured frozen hydrated HeLa cell (b). The freeze-fractured cell has a lower phosphocholine-to-adenine ratio at the start of the analysis and clear separation of the adenine and phosphocholine as the cell substrate interface is approached. Recreated from Fletcher et al. data,[52] with permission from Wiley.

8.5.5 Depth Scale Measurement

All the studies that exploit the depth profiling capability, whether it is simple removing contamination or exploring subsurface chemistry, or the construction of a 3D image would benefit from some measure of the z-scale. It is well known that different materials sputter at different rates, so pure sputtering time can at best only

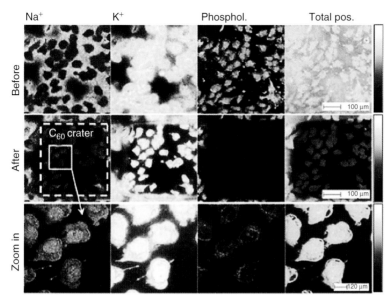

Figure 8.18   Positive and negative ion images of HeLa cells prepared by a wash-and-dry method. The images in the first row were acquired before C_{60}^{++} etching, the second and the third rows were acquired after etching with 1.03×10^{14} C_{60}^{++} cm^{-2}. Reproduced with permission from Brison et al.,[53] with permission from Wiley.

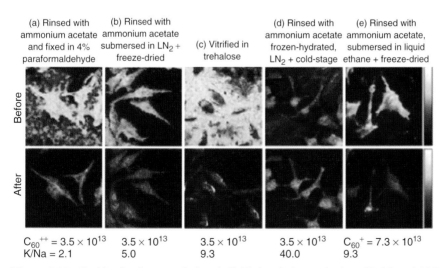

Figure 8.19   Positive ion images of phospholipid signals (sum of $m/z = 58, 86,$ and 184) before (first row) and after etching (second row). The methods used to prepare the cells are used to label the images. K^+ to Na^+ ratios shown in the bottom row were measured inside the cells after etching. Reproduced from Brison et al.,[53] with permission from Wiley.

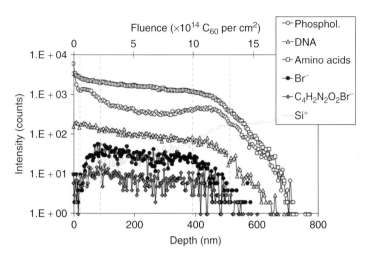

Figure 8.20 Depth profiles of HeLa cells acquired in the dual beam mode with 25 keV Bi_3^+ for analysis and 10 keV C_{60}^+ for etching. The depth profiles were reconstructed from a single cell nucleus region. The depth scales were calibrated using AFM. Reproduced from Brison et al.,[53] with permission from Wiley.

give a very crude estimate of depth. There is some evidence that organic molecular sputter rates do not differ too much, but nevertheless the study reported in by Cheng and Winograd in which peptides were incorporated into trehalose showed that the sputter rate of the principal matrix is influenced by guest molecules.[18] One way around the problem is to combine SIMS analysis with a technique that measures topography, such as AFM. Such an approach was pioneered on a model system in the Winograd laboratory (see Chapter 6).[54] The aim was to produce a model sample with components having a wide range of sputter yields under C_{60}^+ bombardment. Briefly, a 300 nm trehalose film was prepared doped with a 1% concentration of a small peptide GGYR on a silicon substrate. The film was bombarded with a focused Ga^+ beam to write a series of Ga lines and letters in the film. Ga was implanted and the sample topography changed. The resulting system was profiled using ToF-SIMS, and AFM images were collected at the start and end of the profile. Because of the topography effects and the fact that the gallium-implanted atoms and the silicon substrate sputtered much more slowly than the trehalose and peptide, if the raw profile was accepted, the apparent depth distribution of the gallium would be a gross distortion of actual distribution. A protocol was developed using the AFM data to successfully correct the images. This model sample starkly illustrated the potential problems involved in deriving 3D images from depth profiling data; however, it did include components having a very wide range of sputter yields. It is unlikely that biological materials would contain a high concentration of metallic species. The component sputter yields are likely to be much closer. However, particularly because the starting topography of biological systems is not likely to be flat, there is a real need to be able to monitor the z-scale as samples are sputtered. In the study by Castner's group referred to earlier (Figs. 8.19 and 8.20), the HeLa cells whose nucleus had been labeled with bromodeoxyuridine (BrdU) were sputtered to remove various components of the cell's structure, outer membrane, cytoplasm, nucleus, and lower membrane, and AFM used to measure the material removed at these different stages. They used ToF-SIMS peaks to determine when these components had been largely removed. This provided somewhat crude z-scale estimates as cells are not geometrically flat; however, they were able to estimate the sputter time for each cell component. For a 1 nA, 10 keV C_{60}^+ beam incident at $45°$ and rastered over 500×500 µm^2, 20 s of etching was necessary to remove the surface contamination ($\sim$20 nm). After approximately 75 s of etching, the top membrane was completely removed ($\sim$54 nm) with intracellular components visible. After 100 s of etching, the cytoplasm of the cells was completely sputtered away ($\sim$85 nm) and only the nucleus remained. Finally, the bottom membrane was reached after 403 s of etching, and the cell was completely sputtered after 515 s ($\sim$508 nm). Both the atomic and molecular species from BrdU were detected inside the cell's nucleus after the surface contamination and the top membrane were removed by C_{60} etching (see Figure 8.20).

This approach using AFM has been applied in a somewhat different mode for the analysis of a cheek cell.[55] Before analysis in a SIMS system, the topography of the cell was measured with AFM. Then the cell was analyzed using SIMS and a secondary ion image was generated. This image was rotated to correspond with the AFM image and then overlaid on that image to give the image in Figure 8.21. This image displays the topography in x, y, and z axes and the secondary ion intensity is provided as a gray scale. In this case, the z direction data is provided by the AFM. If the cell was then chemically profiled using SIMS, the z stack of images could be offset using the initial AFM image. Although this method of topography measurement provides a reasonable approximation, there are a number of disadvantages, particularly if the SIMS analysis is to be performed on cryogenically preserved samples. Also, if the sample is prone to considerable variations in sputter rate during the analysis, the only way of measuring these variations would be to remove the sample from the vacuum and remake the AFM measurement at regular intervals as in the example from the Castner group. This procedure would obviously be time

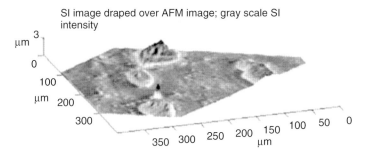

Figure 8.21 C_{60}^+ 3D secondary ion (SI) imaging of cheek cells — combined with AFM measurements to correct for topography and provide a z-scale measurement. The ion image 'draped' over the AFM data showing signal intensity (grey scale) and topography (z-axis). Reproduced from Fletcher et al.,[55] with permission from Elsevier.

consuming and in many cases may even be detrimental to sample integrity. There are a number of alternative methods of mapping topography. A possible solution that might prove more compatible with the UHV (ultra-high vacuum) conditions required for SIMS analysis is that of white light or laser interferometry. Often employed in metrological applications, a white light source is used with a beam splitter and one of a variety of interferometers. The systems are compatible with optical microscope type accessories, but produce an image of the interference pattern from the sample that can be transformed to provide a topographic map. Such a noncontact technique is much more applicable to UHV analysis, and if mounted in an appropriate manner could provide real-time mapping of topographic variation during a depth profile analysis. However, integrating such a system with a normal SIMS analysis arrangement could be technically challenging.

8.5.6 High Throughput Biomaterials Characterization

A somewhat different, although extremely important analysis direction is the application of cluster-based ToF-SIMS in high throughput analysis of combinatorial libraries of materials and drugs of potential use in medical therapeutic situations. Over a number of years Davies, Alexander, and coworkers have been exploring the use of polymer microarrays to discover new polymers with compositions that deliver favorable performance in cellular applications where the role of the material surface can have an important role in defining the nature and extent of the material–cellular interactions.[56,57] They have exploited methods of preparing polymer microarrays of varying composition using noncontact printing.[58] These arrays can then be characterized using a variety of surface-sensitive techniques, such as wettability using water contact angle (WCA), topography, roughness and hardness using AFM, elemental and chemical composition using XPS, chemical functionality using Raman, IR (infra-red spectroscopy), and ToF-SIMS. The polymer arrays can then be tested for the attachment of various proteins that mediate cell attachment. Following this the polymer protein combinations can be tested for cell attachment and activity. Such studies on microarrays that can number around 500 different sample spots generate enormous amounts of data. A ToF-SIMS spectrum from each spot with over 1000 significant peaks from the positive and negative spectrum obviously cannot be analyzed by "stare and compare" methods. Principal component analysis (PCA) and partial least squares (PLS) methods have been used to draw out the features of the analytical output that correlate with surface functionality that is of interest. PCA was introduced in Chapter 4. Briefly, PCA is a statistical technique that seeks to reduce the complexity of a large data set with multiple variables to a few principal components that describe the most significant trends and the largest differences in the data. From the output generated it is possible to determine which the key surface features are (e.g., ToF-SIMS ion fragments) that contribute to the chemical diversity of a large number of materials. PLS seeks to combine two different sets of experimental data, the so-called explanatory variables (e.g., surface chemical information from ToF-SIMS spectra) and the responses (cell adhesion or activity) to build a predictive model of the response data set. The degree of agreement between the experimental observations of, say, cell adhesion, and the prediction determine the degree of correlation between the explanatory and response functions. The PLS model also generates a set of regression coefficients for each explanatory variable from which key explanatory variables, for example, ToF-SIMS ion fragments controlling the response variable, such as cellular adhesion, can be extracted.

A particularly impressive study has been to develop new substrates that could be used to clonally expand human pluripotent stem cells on chemically-defined synthetic materials.[59] Human pluripotent stem cells hold great promise for regenerative medicine and human disease modeling and thus have attracted intense research interest for methodologies to expand these cell lines sufficiently for therapeutic use. The discovery of a material that supports the outgrowth of stem cells while maintaining their pluripotency would be an important step toward realizing the benefits of stem cells, and has been the focus of the authors' cell based-polymer microarray studies. Cell-compatible, biomaterial microarrays were prepared from 22 acrylate monomers with diversified hydrophobicity, hydrophilicity, and cross-linking densities. The arrays were prepared by copolymerization between each of 16 "major" monomers and each of six "minor" monomers at six different ratios [100 : 0, 90 : 10, 85 : 15, 80 : 20, 75 : 25, 70 : 30 (v/v)]. In this way, arrays with 496 different combinations were created, consisting of the major monomer (70–100%) and minor monomer (0–30%). These monomer mixtures were robotically deposited in triplicate on a noncell adhesive layer of poly(2-hydroxyethyl methacrylate) covering conventional glass slides (75 mm × 25 mm), and then polymerized with a long wave-length ultraviolet source. A prior study across the array of a range of different preadsorbed proteins had demonstrated that preadsorption of fetal bovine serum (FBS) most effectively promotes the growth of hES cells. Using FBS-coated arrays it was demonstrated that a moderate wettability (WCA ~70°) is associated with optimal frequency of hES cell-colony formation. Other factors such as polymer roughness or elastic modulus had no effect. To try to understand of how wettability modulates colony formation, a "secondary" group of 48 polymers with 12 replicates was assembled. This secondary array was designed to encompass the favorable range of WCA determined from the primary array. In good agreement with the primary array data, a moderate wettability (WCA ~70°) again effectively supported optimal hES cell clonal growth

In order to determine the polymer surface chemistry that provided favorable performance, a 25 keV Bi_3^+ primary beam probed each 300 μm diameter polymer spot on the 576 spot array under static conditions. An ION-TOF IV instrument was used. The cluster Bi_3^+ beam was rastered over a 100 × 100 μm area of each spot for 10 s acquisition time providing a good yield of ions in the mass range up to m/z 200. The instrument was operated in high mass resolution mode, $m/\Delta m$ ~6000. Using PLS on the ToF-SIMS spectra, the surface chemistry implied by the spectra was correlated to the colony-formation frequency observed on each polymer in the secondary array, Figure 8.22. A "leave one out" cross validation method was used for the PLS analysis. Both ToF-SIMS and hES cell data were mean-centered before analysis. The individual peak intensities were normalized to the total secondary ion count to remove the effect of primary ion beam fluctuation. The positive and negative ion intensity data was arranged into one concatenated

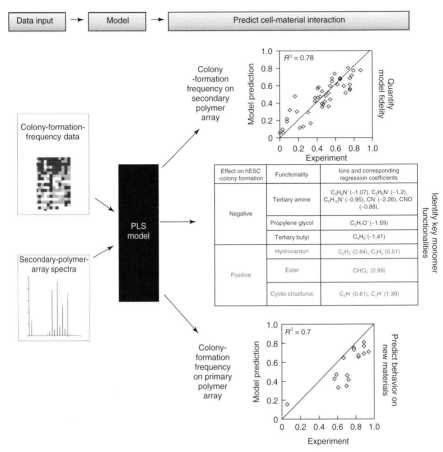

Figure 8.22 A multivariate PLS regression method was used to quantitatively analyze and predict the cell–material interactions by correlating ToF-SIMS spectra of polymer spots to their biological performance (colony-formation frequency). The fidelity of PLS models can be quantified by a linear correlation of predicted versus measured colony-formation frequency. Top: each point in the figure represents one of 48 different polymers in the secondary array, and the inserted line represents the ideal situation when predictions match experiments completely. Middle table showing functionalities and their associated characteristic ions supporting or inhibiting hES colony formation was generated as follows: Ions were identified by correlating ToF-SIMS spectra to hES colony formation using PLS regression. Each ion was designated with a regression coefficient, α, that characterizes the relative effect on hES cell-colony formation. Bottom: as in top plot, but the PLS model was developed on ToF-SIMS spectra from the secondary array and was used to predict the behavior of 16 homopolymers from the primary array. Reproduced with permission from Mei et al.,[59] with permission from Macmillan Publishers Ltd.

data matrix. A total of 181 positive and 43 negative ions were selected from a group of polymers from the array containing all 22 monomers to form the peak lists. The PLS model constructed from the training polymer samples produced a set of regression coefficients for each secondary ion. These regression coefficients were used to predict the hES cell colony formation on the test samples using their SIMS spectra. Good agreement between measured colony-formation frequency and that predicted from the ToF-SIMS spectra was found ($R^2 = 0.78$). Each secondary ion associated with functionalities in the polymer structures could be listed with its regression coefficient, α, a quantitative measure of its contribution to colony-formation frequency (Table in Fig. 8.22). The tertiary amine moiety (characteristic ions $C_3H_8N^+$, $C_2H_6N^+$, CN^-) and tertiary-butyl moiety ($C_4H_9^+$) were identified by the negative regression coefficient to be correlated most strongly with a low colony-formation frequency, and hydrocarbon ions ($C_2H_3^+$, $C_3H_3^+$), oxygen-containing ions (CHO_2^-, $C_3H_3O^+$, $C_2H_3O^+$) from esters, and ions from cyclic structures (C_6H^-, C_4H^-, C_2H^-) had the largest effect on promoting colony formation. The oxygen-containing ions and hydrocarbon ions can be attributed to the acrylate groups ($CH_2 = CHCO_2^-$) in each monomer, which form the backbone chain and the pendant ester groups after polymerization. Monomers with di- and triacrylates, which contain the most acrylate groups in the library, indeed showed the highest colony-formation frequencies.

The conclusions derived from "secondary" array provided an understandable view of the role of surface chemical structure in all the cell responses seen on the original larger array. To illustrate this, the hES cell colony formation of all 16 homopolymers in the original array was predicted entirely on their ToF-SIMS spectra using the PLS model from the "secondary" array (see lower regression plot in Fig. 8.22) suggesting that the model is able to predict the cell behavior of acrylate polymers outside the original training set of the model. Using this body of information, further FBS coated "hit" polymer arrays (polymers that support both robust growth and hES cell-colony formation) were developed that demonstrated capacity to maintain pluripotency of hES cells after more than 2 months of culture. The study was further extended to a more clinically relevant culture system for hES cells based on human serum (HS)-coated hit polymer arrays. It was shown that HS-coated hit polymer arrays supported the expansion of dissociated hES cells in a similar manner to arrays coated with FBS. Further, the HS-coated hit arrays could support long-term culture for more than 1 month, with robust expression of hES cell markers.

In contrast to the aspirations of cell and tissue analysis, this study demonstrates in a powerful way that accessing a relatively small mass range of fragment ions utilizing metal cluster initiated ToF-SIMS provides sensitivity to the detailed surface chemical functionality of complex systems that are important in determining their biological activity. Furthermore, because of its sensitivity and rapidity of analysis, this ToF-SIMS capability along with MVA techniques can be used to probe the impact of these surface properties on the biological activities of large libraries of polymeric materials in a high throughput mode of operation.

8.6 FINAL THOUGHTS AND FUTURE DIRECTIONS

As is clear from this chapter, the pace of progress in finding new applications for cluster SIMS in biology is picking up rapidly. So far, unfortunately, we have learned very little new biology. Researchers are still grappling with developing the appropriate instrumentation, sample preparation, and sensitivity, and it is likely that they will be important for years to come. As the sophistication with these issues deepens, it seems clear that by clever choice of the biological systems someone will hit a "home run" and the biologists will take notice.

Perhaps the most tantalizing development emerging from cluster bombardment is the emergence of 3D imaging protocols. This type of experiment will require a detailed understanding of molecular depth profiling and new strategies for calibrating the z-scale. We can look for both fundamental and applied studies to advance these aspects. Already, researchers are thinking about *in situ* AFM or laser interferometry as possible solutions. It is also clear that to resolve the sensitivity issue satisfactorily, the field will have to more efficiently detect the secondary ions. This means that either the instrumentation has to collect all the sputtered material, or that the ionization probability needs to be increased dramatically. We expect breakthroughs in both areas in the not too distant future.

ACKNOWLEDGMENTS

Results presented in this chapter were partially supported by the National Institutes of Health (grant no. 2R01 EB002016-19), the National Science Foundation (grant no. CHE-0908226), and the Department of Energy (grant no. DE-FG02-06ER15803).

REFERENCES

1. Fletcher, J. S.; Lockyer, N. P.; Vickerman, J. C. *Mass Spectrom. Rev.* **2011**, 30 (1), 142–174.
2. Carado, A.; Passarelli, M. K.; Kozole, J.; Wingate, J. E.; Winograd, N.; Loboda, A. V. *Anal. Chem.* **2008**, 80 (21), 7921–7929.
3. Fletcher, J. S.; Rabbani, S.; Henderson, A.; Blenkinsopp, P.; Thompson, S. P.; Lockyer, N. P.; Vickerman, J. C. *Anal. Chem.* **2008**, 80 (23), 9058–9064.
4. Colliver, T. L.; Brummel, C. L.; Pacholski, M. L.; Swanek, F. D.; Ewing, A. G.; Winograd, N. *Anal. Chem.* **1997**, 69 (13), 2225–2231.
5. Pacholski, M. L.; Donald, M. C.; Ewing, A. G.; Winograd, N. *J. Am. Chem. Soc.* **1999**, 121, 4716–4717.
6. Roddy, T. P.; Cannon Jr, D. M.; Meserole, C. A.; Winograd, N.; Ewing, A. G. *Anal. Chem.* **2002**, 74 (16), 4011–4019.
7. Parry, S.; Winograd, N. *Anal. Chem.* **2005**, 77 (24), 7950–7957.
8. Kurczy, M. E.; Piehowski, P. D.; Parry, S. A.; Jiang, M.; Chen, G.; Ewing, A. G.; Winograd, N. *Appl. Surf. Sci.* **2008**, 255 (4), 1298–1304.
9. Hoppert, M.; Holzenburg, A. *Electron Microscopy in Microbiology*; RMS Handbook Series, Bios Scientific Publishers Ltd., Oxford **1998**.
10. Severs, N. J.; Newman, T. M.; Shotton, D. M. A Practical Introduction to Rapid Freezing Techniques. In *Rapid Freezing, Freeze Fracture and Deep Etching*; Severs, N. J., Shotton, D. M., Eds.; Wiley-Liss: New York, **1995**.

11. Sjovall, P.; Johansson, B.; Lausmaa, J. *Appl. Surf. Sci.* **2006**, 252 (19), 6966–6974.

12. Jones, E. A.; Lockyer, N. P.; Vickerman, J. C. *Anal. Chem.* **2008**, 80 (6), 2125–2132.

13. Pacholski, M. L.; Winograd, N. *Chem. Rev.* **1999**, 99, 2977–3006.

14. Hill, R.; Blenkinsopp, P.; Thompson, S.; Vickerman, J.; Fletcher, J. S. *Surf. Interface Anal.* **2011**, 43 (1–2), 506–509.

15. Lanekoff, I.; Kurczy, M. E.; Hill, R.; Fletcher, J. S.; Vickerman, J. C.; Winograd, N.; Sjoêvall, P.; Ewing, A. G. *Anal. Chem.* **2010**, 82 (15), 6652–6659.

16. MacAleese, L.; Duursma, M. C.; Klerk, L. A.; Fisher, G.; Heeren, R. *J. Proteomics* **2011**, 74 (7), 993–1001.

17. Passarelli, M. K.; Winograd, N. *Biochim. Biophys. Acta Mol. Cell Biol. Lipids* **2011**, 1811 (11), 976.

18. Cheng, J.; Winograd, N. *Anal. Chem.* **2005**, 77 (11), 3651–3659.

19. Jones, E. A.; Lockyer, N. P.; Kordys, J.; Vickerman, J. C. *J. Am. Soc. Mass Spectrom.* **2007**, 18 (8), 1559–1567.

20. Cheng, J.; Wucher, A.; Winograd, N. *J. Phys. Chem. B.* **2006**, 110 (16), 8329–8336.

21. Cheng, J.; Kozole, J.; Hengstebeck, R.; Winograd, N. *J. Am. Soc. Mass Spectrom.* **2007**, 18 (3), 406–412.

22. McPhail, D. S.; Dowsett, M. G. In *Surface Analysis - The Principal Techniques*, 2nd ed.; Vickerman, J. C., Gilmore, I. S., Eds.; John Wiley & Sons, Ltd: Chichester, UK, **2009**; pp 207–254.

23. Tang, L.; Kebarle, P. *Anal. Chem.* **1991**, 63 (23), 2709–2715.

24. Amado, F. M. L.; Domingues, P.; Graca, S.-M. M.; Ferrer-Correia, A. J.; Tomer, K. B. *Rapid Commun. Mass Spectrom.* **1997**, 11, 1347.

25. King, R.; Bonfiglio, R.; Fernandez-Metzler, C.; Miller-Stein, C.; Olah, T. *J. Am. Chem. Soc.* **2000**, 11, 942.

26. Jones, E. A.; Lockyer, N. P.; Vickerman, J. C. *Appl. Surf. Sci.* **2006**, 252, 6727.

27. Piwowar, A. M.; Lockyer, N. P.; Vickerman, J. C. *Anal. Chem.* **2009**, 81 (3), 1040–1048.

28. Piwowar, A. M.; Fletcher, J. S.; Kordys, J.; Lockyer, N. P.; Winograd, N.; Vickerman, J. C. *Anal. Chem.* **2010**, 82 (19), 8291–8299.

29. Wucher, A.; Sun, S.; Szakal, C.; Winograd, N. *Anal. Chem.* **2004**, 76 (24), 7234–7242.

30. Conlan, X. A.; Lockyer, N. P.; Vickerman, J. C.. *Rapid Commun. Mass Spectrom.* **2006**, 20 (8), 1327–1334.

31. Lee, J. L. S.; Gilmore, I. S.. In *Surface Analysis - The Principal Techniques*, 2nd ed.; Vickerman, J. C., Gilmore, I. S., Eds.; John Wiley & Sons, Ltd: Chichester, UK, **2009**; pp 563–613.

32. Sjovall, P.; Lausmaa, J.; Johansson, B. *Anal. Chem.* **2004**, 76 (15), 4271–4278.

33. Benabdellah, F.; Seyer, A.; Quinton, L.; Touboul, D.; Brunelle, A.; Laprevote, O., *Anal. Bioanal. Chem.* **2010**, 396 (1), 151–162.

34. Briggs, D.; Hearn, M. J. *Vacuum* **1986**, 36 (11–12), 1005–1010.

35. Jones, E. A.; Lockyer, N. P.; Vickerman, J. C.. *Int. J. Mass Spectrom.* **2007**, 260 (2–3), 146–157.

36. Tahallah, N.; Brunelle, A.; De La Porte, S.; Laprevote, O. *J. Lipid Res.* **2008**, 49, 434–454.

37. Touboul, D.; Brunelle, A.; Halgand, F.; De La Porte, S.; Laprevote, O. *J. Lipid Res.* **2005**, 46, 1388–1395.

38. Touboul, D.; Roy, S.; Germain, D. P.; Chaminade, P.; Brunelle, A.; Laprevote, O. *Int. J. Mass Spectrom.* **2007**, 260 (2), 158–165.

39. Debois, D.; Bralet, M. P.; Le Naour, F.; Brunelle, A.; Laprevote, O. *Anal. Chem.* **2009**, 81 (8), 2823–2831.

40. Malmberg, P.; Borner, K.; Chen, Y.; Friberg, P.; Hagenhoff, B.; Mannsson, J. E.; Nygren, H. *Biochim. Biophys. Acta Mol. Cell Biol. Lipids* **2007**, 1771 (2), 185–195.

41. Brulet, M.; Seyer, A.; Edelman, A.; Brunelle, A.; Fritsch, J.; Ollero, M.; Laprevote, O. *J. Lipid Res.* **2010**, 51 (10), 3034–3045.

42. Camps, M.; Cortes, R.; Gueye, B.; Palacios, J. M. *Neuroscience* **1998**, (28), 275.

43. Willey, K. F.; Vorsa, V.; Braun, R. M.; Winograd, N. *Rapid. Commun. Mass Spectrom.* **1998**, 12 (18), 1253–1260.

44. Vaidyanathan, S.; Fletcher, J. S.; Goodacre, R.; Lockyer, N. P.; Micklefield, J.; Vickerman, J. C. *Anal. Chem.* **2008**, 80 (6), 1942–1951.

45. Passarelli, M. K.; Winograd, N. *Surf. Interface Anal.* **2011**, 43, 269–271.

46. Murphy, R. C.; Harrison, K. A. *Mass Spectrom. Rev.* **1994**, 13, 57.

47. Fletcher, J. S.; Lockyer, N. P.; Vaidyanathan, S.; Vickerman, J. C. *Anal. Chem.* **2007**, 79 (6), 2199–2206.

48. Vickerman, J. C. *Analyst* **2011**, 136 (11), 2199–2217.

49. Breitenstein, D.; Rommel, C. E.; Mollers, R.; Wegener, J.; Hagenhoff, B. *Angew. Chem. Int. Ed.* **2007**, 46 (28), 5332–5335.

50. Fletcher, J. S.; Vickerman, J. C. *Anal. Bioanal. Chem.* **2010**, 396 (1), 85–104.

51. Rabbani, S.; Fletcher, J. S.; Lockyer, N. P.; Vickerman, J. C. *Surf. Interface Anal.* **2011**, 43, 380–384.

52. Fletcher, J. S.; Rabbani, S.; Henderson, A.; Lockyer, N. P.; Vickerman, J. C. *Rapid Commun. Mass Spectrom.* **2011**, 25 (7), 925–932.

53. Brison, J.; Benoit, D. S. W.; Muramoto, S.; Robinson, M.; Stayton, P. S.; Castner, D. G. *Surf. Interface Anal.* **2011**, 43, 354–357.

54. Wucher, A.; Cheng, J.; Winograd, N. *Anal. Chem.* **2007**, 79 (15), 5529–5539.

55. Fletcher, J. S.; Henderson, A.; Biddulph, G. X.; Vaidyanathan, S.; Lockyer, N. P.; Vickerman, J. C. *Appl. Surf. Sci.* **2008**, 255, 1264–1270.

56. Davies, M. C.; Alexander, M. R.; Hook, A. L.; Yang, J.; Mei, Y.; Taylor, M.; Urquhart, A. J.; Langer, R.; Anderson, D. G. *J. Drug Target.* **2010**, 18 (10), 741–751.

57. Hook, A. L.; Anderson, D. G.; Langer, R.; Williams, P.; Davies, M. C.; Alexander, M. R. *Biomaterials* **2010**, 31 (2), 187–198.

58. Anderson, D. G.; Putnam, D.; Lavik, E. B.; Mahmood, T. A.; Langer, R. *Biomaterials* **2005**, 26 (23), 4892–4897.

59. Mei, Y.; Saha, K.; Bogatyrev, S. R.; Yang, J.; Hook, A. L.; Kalcioglu, Z. I.; Cho, S. W.; Mitalipova, M.; Pyzocha, N.; Rojas, F. *Nat. Mater.* **2010**, 9 (9), 768–778.

CHAPTER 9

FUTURE CHALLENGES AND PROSPECTS OF CLUSTER SIMS

Peter Williams and Christine M. Mahoney

9.1 INTRODUCTION

In this book, the benefits and challenges of cluster ion secondary ion mass spectrometry with reference to organic, polymeric, and biological sample analysis in three dimensions have been presented. A number of examples have demonstrated that the application of polyatomic or cluster primary ion sources is advantageous for analysis of organic, polymeric, and biological materials. Fundamental aspects have been described in detail, providing an understanding of both the physics and chemistry of sputtering of organic materials with cluster ion beams. Cluster secondary ion mass spectrometry (SIMS) has considerable promise for use as a mass spectrometric molecular imaging and depth profiling method. Despite the present limitations of SIMS for tissue analysis, the technique remains unrivaled in its ability to obtain high spatial resolution imaging information with finely focused ion beams, with spatial resolutions in the submicrometer range, as opposed to >20 μm for MALDI (matrix-assisted laser desorption/ionization) and even higher than that for DESI (desorption electrospray ionization). Some future challenges for cluster SIMS are readily identified as follows:

1. extend the molecular mass range accessible to focusable small cluster beams;
2. improve data rates for small clusters by moving away from pulsed primary beams;
3. develop imaging modalities for very large cluster beams;
4. develop understanding of parameters governing molecular ejection and ionization to allow improvements in useful ion yields;
5. develop understanding of matrix effects to allow quantitative analysis.

Cluster Secondary Ion Mass Spectrometry: Principles and Applications, First Edition.
Edited by Christine M. Mahoney.
© 2013 John Wiley & Sons, Inc. Published 2013 by John Wiley & Sons, Inc.

9.2 THE CLUSTER NICHE

Over roughly a 15 year period from the mid-1970s until about 1990, particle bombardment was the leading method to obtain intact gas-phase ions of biomolecules, such as peptides and proteins for mass spectrometric analysis. The (atomic) projectiles ranged from megaelectronvolt heavy ions, from particle accelerators or fission events, to kiloelectronvolt ions and neutrals that were shown in 1981 to generate stable sputtered molecular ion signals from samples presented in vacuum in glycerol solutions. This particle beam dominance ended abruptly around 1990 with the introduction of spray and laser ablation technologies—electrospray ionization (ESI) and MALDI. As this book demonstrates, there is still widespread interest in particle bombardment approaches for molecular analysis, dominantly in the form of cluster impact technologies, but applications now are focused on a niche centered on spatially resolved molecular and biomolecular analyses by depth profiling, two-dimensional imaging and the combination of both to yield three-dimensional molecular information. In all cases, the advantage conferred by clusters is the ability to detect stable molecular ion signals from solid targets while sputtering over an extended depth, that is, to sample a significant volume of the target—a voxel—rather than just a surface layer of molecules, together with the ability to focus some cluster beams to submicrometer dimensions. Molecular depth profiling using cluster bombardment typically samples fairly major components of laterally homogeneous materials so that ion formation efficiency often is not a major concern. For imaging applications, however, efficiency becomes crucial because one is always interested in finer detail, particularly in biological material, and concentrations of important components of a sample often are low, so that there may be very few analyte molecules in a voxel of interest. Efficiency is a central theme of this chapter; it can best be defined in terms of "useful ion yield," that is, the product of the fraction of target molecules consumed that are sputtered intact, the fraction of intact molecules that are ionized, and the fraction of ions that are transmitted and detected. It is not difficult to ensure high efficiency in transmission and detection, so that the major challenges concern sputtering and ionization efficiencies.

9.3 CLUSTER TYPES

We can rank particle beams for molecular analysis in the following rough hierarchy: atomic ions (e.g., Ar^+, Ga^+, Cs^+), small clusters (up to ~10 atoms, e.g., SF_5^+), medium size clusters (e.g., C_{60} or Ar_{100}), large clusters (e.g., Ar_{2500}), and massive clusters (e.g., $Glycerol_{100,000}^{100+}$). All of these species are capable of ejecting large biomolecular ions with some level of efficiency; the challenge is to achieve efficiencies (useful ion yields) high enough to produce useful signals from small analytical volumes.

The dominant competing techniques for biological tissue imaging are MALDI,[1,2] and more recently, DESI,[3] both of which are capable of obtaining information from quite massive proteins. The high mass range of MALDI and DESI imaging approaches is partly because of the relatively large volumes sampled and partly to the minimal damage imparted to the sample by the probe utilized for desorption (pulsed laser or water droplet). DESI is a particularly soft technique, comparable to ESI in that molecular ions result from solvent evaporation from charged solution droplets splashed off the sample surface. A significant added advantage of DESI is the ability to analyze samples at atmospheric pressure.

To date, most imaging work with cluster SIMS has used C_{60} ion beams and the data have rarely exceeded m/z values of ~ 1000, although the early detection of proteins with $m/z > 20$ kDa under 35 keV Cs^+ bombardment[4] and of $m/z > 29$ kDa by massive cluster impact (MCI)[5] suggests that there is no intrinsic physical limit to sputtering of high molecular weight protein molecular ions by fast particle impact. Part of the limitation of modern cluster SIMS is simply a result of the small analytical volumes sampled with the low duty cycles characteristic of time-of-flight SIMS (ToF-SIMS). Sample areas for MALDI imaging are 3–4 orders of magnitude larger than for cluster SIMS and the application of a liquid matrix solution probably leaches out sample should be singular from depths of some micrometers; DESI analytical volumes are even larger and DESI sampling uses DC beams. Similarly, the early Cs^+ and MCI data cited above used large area ion beams and DC measurement techniques.

Cluster types that show promise for efficient high molecular weight SIMS are large gas cluster ion beams (GCIBs),[6] which have been discussed in great detail throughout this book (see in particular, Chapters 2, 4, and 5) and electrosprayed massive clusters (MCI),[7] alternatively labeled *electrosprayed droplet impact (EDI)* by Hiraoka et al.[8] The low energy per cluster atom in large clusters should minimize sample damage and, indeed, both GCIB and MCI have already demonstrated the capability to desorb and ionize quite large proteins.[5,9] However, such sources can be difficult to focus and approaches other than microprobe imaging may be needed.

9.4 THE CHALLENGE OF MASSIVE MOLECULAR ION EJECTION

The requirements for obtaining an analytically useful gas-phase ion species from a condensed phase target are really quite stringent, especially for those techniques dependent on transient energy input—pulsed laser desorption, atomic and cluster ion impact. We require a gas-phase ion, that is (i) fully intact and (ii) fully separated from its surroundings in the condensed phase—that is, declustered, desolvated, and desalted so that it has a chemically specific mass. In other words, we must break multiple noncovalent interactions—hydrogen bonds, van der Waals, dipole–dipole, ion–dipole—amounting to a total energy that can be many electronvolts *and* have an ionization event, without breaking a *single* structural bond in the target molecule. If the ions are to be mass-analyzed using time-of-flight mass spectrometry (ToF-MS) with a DC acceleration field, which is not uncommon, desolvation and declustering must be completed within a few nanoseconds in order to achieve acceptable mass accuracy. This implies some degree of heating; however, to detect an intact molecular ion this heating must not result in the cleavage of a single structural bond in the molecule. It is this delicate balance

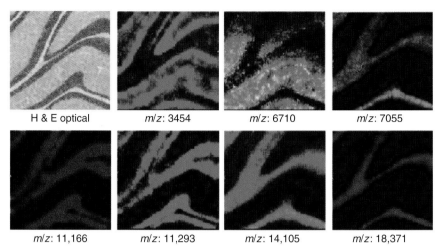

H & E optical   m/z: 3454   m/z: 6710   m/z: 7055

m/z: 11,166   m/z: 11,293   m/z: 14,105   m/z: 18,371

Figure 9.1 MALDI ion images of various proteins obtained from a 12 μm thick rat brain cerebellum section at 20 μm spatial resolution. Recreated from Yang and Caprioli,[11] with permission from ACS publications.

between heating/desolvation and molecular ion survival that transient energy impact desorption approaches strive, usually empirically, to achieve.

9.4.1 Comparing with MALDI: The Gold Standard

The surprising experimental simplicity and high spectral quality of the MALDI approach make it the gold standard for transient energy desorption techniques. MALDI has produced intact molecular ions of proteins in excess of 1 MDa molecular weight.[10] An example of MALDI imaging, which nicely illustrates the ability of the approach to desorb and detect large molecules is shown in Figure 9.1, which depicts intact protein imaging in rat brain tissue.[11] Here, proteins of up to ~20 kDa are imaged successfully; a feat not yet attainable using SIMS. In fact, with current SIMS imaging approaches, there is little useful signal beyond ~2000 Da, and counts are usually extremely low above ~1000 Da, making it difficult to image samples in a timely manner. Figure 9.2, for example, shows a direct comparison of the same phosphatidylcholine peak, m/z 772 imaged in a rat brain section using SIMS (Fig. 9.2a) and MALDI (Fig. 9.2b).[12]* As can be seen, in the case of SIMS, the signal is very low as compared to MALDI. Furthermore, the signals correlate with different features in the two images indicating that one or both techniques are subject to artifacts arising from different ionization mechanisms and matrix effects.

In a linear ToF-MS, MALDI mass spectra can be remarkably free of fragmentation, indicating that the majority of desolvated molecular ions survive at least

---

*The molecular images of the brain sample shown in Figure 9.2b are not from the same sample as shown in Figure 9.2a. MALDI images of the brain shown in Figure 9.2 were acquired at EMSL as well, with similar results, but are not displayed here as the results are not ready for publication.

(a)   (b)

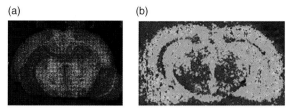

Figure 9.2   Positive ion SIMS (a) and MALDI (b) maps of the potassiated phosphatidylcholine 32 : 0 at m/z 772. The intensity scaling in (a) is 0–12 counts (0 for dark regions, 12 for brightest regions), while for (b) the scaling is 0–200 counts (0 for dark blue regions, 200 for red regions). Figure 9.2a was acquired at the Environmental Molecular Sciences Laboratory. Figure 9.2b is reproduced from Wang et al.,[12] with permission from Elsevier.

the acceleration interval intact. (In reflectron instruments, fragmentation is seen for ejected molecular ion species in excess of ∼10 kDa indicating that the ejected ions are vibrationally hot and can fragment over tens of microseconds following acceleration.) Estimates of temperatures in the ablated matrix plume are ∼1000 K.[13] This was argued by Chou et al.[14] to be consistent with resonant absorption of single UV photons by the matrix molecules followed by rapid equilibration over the vibrational degrees of freedom in the molecules. An energy of 3.7 eV (337 nm photon) per matrix molecule corresponds to maximum temperatures ∼1000–2000 K depending on the molecular complexity, broadly consistent with the measured plume temperatures and initial velocities of the matrix ions. Collisional equilibration should rapidly heat the embedded biomolecules to a similar temperature. Although it may sound surprising that biomolecules could survive such temperatures, in linear ToF instruments they only have to do so for the few microseconds needed to traverse the acceleration space. Note that such a temperature will very rapidly "boil off" any solvating species and this boil-off should act to cool and stabilize the analyte molecules.

9.4.2   Particle Impact Techniques

The central feature leading to the success of the MALDI process appears to be a localized explosion of a mass of heated gaseous material (the laser-excited matrix molecules) that can generate the motive force to eject large biomolecules into the vacuum, together with the nicely calibrated heating effect that results from the efficient absorption of a single resonant photon per matrix molecule and that therefore *clamps* the heating effect at ∼1000 K, sufficient for efficient desolvation but generally not resulting in rapid fragmentation. In relation to this picture, particle impact techniques start at a huge disadvantage. Multi-kiloelectronvolt atomic ion impacts are intensely damaging and even for C_{60} clusters the impact energies of the individual cluster atoms (250 eV per C atom for 15 keV C_{60}) are more than sufficient to break many few-electronvolt structural bonds, so that a significant fraction of molecules in the central impact site will be damaged and the observed intact molecular ejecta appear to originate largely around the rim of the impact crater, distant from the direct impact point. The advantages of C_{60} and other medium clusters are twofold: (i) much of the damaged material is ejected in the impact event so that succeeding cluster impacts sample largely undamaged material and (ii) sputter yields and erosion rates are high yielding acceptable data rates for imaging. However, the efficiency (useful yield) of the process is compromised by the collateral collisional damage.

For efficient ejection of intact molecules without accompanying fragmentation, it is desirable to produce an impact site in which molecules are rapidly heated to a temperature of 1000–2000 K, *without* high-energy collisions that can cause fragmentation. This was the premise that led to the first trials of MCI using electrosprayed glycerol clusters.[7] Subsequent calculation indicated that a massive, highly charged glycerol cluster (10^5–10^6 molecules, a few hundred charges) impacting at 6 km/s (kinetic energy $\sim$0.2 eV/nucleon) indeed could shock-heat an equivalent volume of a glycerol target sufficiently to vaporize much of that volume[15]; analyte molecules would be similarly heated and ejected in the ensuing vapor explosion resulting in desolvated gas-phase molecules. The total kinetic energy available in the cluster can be very large indeed—several megaelectronvolts—resulting in very large sputter yields, but the energy *density* in the impact site is very low and the maximum temperature excursion again is limited ("clamped") to not much more than $\sim$1000 K.[15] MCI has yielded the largest intact molecular ions yet desorbed by particle impact (carbonic anhydrase, 29 kDa[5]). A particular advantage of MCI over MALDI is that no intermediary matrix is required: molecular ion signals can be generated from neat protein samples and presumably from untreated tissue samples. There are experimental difficulties in coupling the MCI (EDI) approach to ToF-MS and to imaging systems, and so imaging work to date has used smaller clusters such as C_{60}, Au_3, and Bi_3 that can be focused to small spot sizes for imaging and pulsed for high resolution ToF measurements.

A promising recent development is the commercialization of Ar cluster technology (GCIB).[6] In principle, this approach can produce broadly tunable clusters of up to several thousand atoms with energy/nucleon approaching that of MCI. Thus, these clusters should allow efficient ejection of quite large molecules without fragmentation; the efficiency with which they can ionize the ejected molecules remains to be determined. Mochiji[9] has recently demonstrated that argon clusters can generate molecular ion signals from proteins as large as chymotrypsin (23 kDa). The heterogeneity of the gas clusters has posed problems for pulsing and imaging, but researchers at ION-TOF have recently used the momentum deflection mass filter (see Figure 3.11) on the ION-TOF to chop out well-defined size distributions (e.g., $\sim$2400–2600 atoms, full width) with correspondingly narrow time distributions allowing good mass resolving power $m/\Delta m \sim 60 - 120$.[16] However, imaging with gas clusters has yet to be demonstrated. For MCI/EDI, neither magnetic[5,7] or orthogonal ToF-MS[8] has been used to date and imaging has yet to be attempted.

9.5 IONIZATION

One of the more challenging problems in molecular sputtering concerns the mechanism(s) of ionization and possible approaches to improve ionization efficiency. For molecular imaging in particular, it is desirable not only to optimize the fraction of molecules sputtered intact, but also to maximize the efficiency with which those molecules are ionized. If ionization requires energetic collision events, these seem incompatible with molecular survival.

9.5.1 "Preformed" Ions

One theme in the ionization discussion has been speculation about the existence of "preformed" ions containing protonated, weakly basic groups such as amino groups. However, concepts such as basicity and acidity are developed for aqueous solutions where ions can be solvated and stabilized. With the exception of quaternary ammonium compounds that contain intrinsically charged cations, the extent to which protonated charge states observed in solution can persist in a dehydrated (or frozen) solid is not clear. Dissociation of weak acids or protonation of bases in aqueous solution is a dynamic equilibrium process in which ions are strongly stabilized by solvation and allowed to separate because of the high dielectric constant of water:

$$R_3NH^+(aq) + OH^-(aq) \rightleftharpoons R_3N(aq) + H_2O(l)$$

As solutions of these compounds are evaporated, the equilibrium shifts continuously to the right. What fraction of basic groups can remain protonated in a dried deposit in vacuum is not clear but it is probably low and dependent on the sample preparation details. Freezing a solution has a similar dehydrating effect as the growing ice crystals exclude the solute. It is not clear whether flash freezing approaches can operate rapidly enough to overcome this dehydration. Krueger et al.[17] have incorporated molecular pH probes in crystallized MALDI matrices and report that they retain their indicator hue from the solution, but suggest that in order to rationalize this finding it is necessary to assume that some water of solvation was incorporated into the matrix crystal.

9.5.2 Radical Ions and Ion Fragments

Are radical ion formation and/or polar dissociation of X−H bonds the initial ionization steps in cluster impact? Comparing C_{60} ionization with that produced by different-sized Ar cluster ions, Rabbani et al.[18] commented that ionization efficiency diminished as the atom count in the Ar cluster increased and the energy per atom decreased, which seems consistent with collisional ionization. Conlan et al.[19] observed water clusters centered on both H_2O^+ and H_3O^+ when sputtering water ice with Au_3 and C_{60}, demonstrating that radical ion formation can occur even with 20 keV C_{60} (only 333 eV per C atom). Ryan et al.[20] have simulated C_{60} bombardment of water ice and seen extensive dissociation of water molecules into H^+ and OH^- (delineated by an energy threshold of 17.2 eV, sufficient for ion-pair dissociation of a water molecule in vacuum). It is known that few hundred electronvolt atomic collisions in solids can lose ~10–15% of the collision energy into electronic excitation,[21] so that radical ionization certainly is possible, even likely. Still, 85–90% of the collision energy remains in atomic motion, so that the production of radical ions or dissociated protons and the survival of intact molecules in the same volume in a given event seem incompatible. The clue may lie in the observed delayed buildup of protonated signals with increasing ion dose: arguably this is the result of an ion residue left by prior impacts, as suggested by Willingham and colleagues.[22]

It is instructive to compare the ion residue picture for atomic and cluster projectiles. Both, atomic ions and C_{60} cluster ions can sputter intact molecular ions; in an early comparison of Ga^+ and C_{60}^+ depth profiling of histamine in ice it was estimated that the useful yield of histamine ions under C_{60} bombardment was not greatly different than for Ga^+ bombardment—perhaps a factor of 2–3 higher.[23] One expects that C_{60} should eject a significantly larger proportion of intact molecules so that this observation suggests that ionization efficiency may be higher for Ga^+, compensating for the more extensive molecular damage. An incident 15 keV Ga^+ can eject some fraction of intact molecules from the periphery of its impact site as it continues to penetrate deep within the solid and initiates both collisional damage and a significant amount of radical ionization. Much of this extensive subsurface ionization is wasted because subsurface molecules along the track are also largely destroyed, but the few surviving molecules may subsequently be sputtered with a fairly high ionization probability. In contrast, for 15 keV C_{60} projectiles, there is little penetration of the individual atoms or knock-ons beyond the bottom of the ejection crater. Even if radical ion, and subsequently protonated ion, formation is efficient in this shallow penetration zone, the result is that each following C_{60} impact scoops out a thin veneer of preionized material together with a much larger scoop of pristine, neutral molecules, suggesting that ionization efficiency may be less than optimal. This picture suggests that the ideal projectile might be a cluster whose ionization depth, on the order of the projected range of the component atoms, is roughly twice the depth of the individual impact craters. Then, a succeeding impact will scoop out a volume that is largely preionized, while the subsurface molecular damage is largely contained. Useful yields might therefore be maximized for smallish clusters, on the order of $\sim$10–20 atoms. Consistent with this picture, Gillen et al. have measured useful yields for a cocaine target under SF_5^+ bombardment that are a factor of $\sim$20 larger than for Ar^+, and arguably greater than for C_{60}, although this has yet to be demonstrated. It would be instructive to compare useful yields under C_{60} bombardment with these other projectiles.

9.5.3 Ionization Processes for Massive Clusters

As cluster sizes increase, the potential for electronic excitation must decrease, as already noted by Rabbani et al. for few thousand atom Ar clusters.[18] For truly massive clusters, $\sim$$10^6$–$10^7$ Da with energy content significantly less than 1 eV/nucleon, new ionization mechanisms must be invoked. One possibility is that the large amount of charge carried into the surface by the impacting massive cluster—typically several hundred excess protons—can contribute to ionization by attaching to analyte molecules. Rather surprisingly, Aksyonov and Williams[24] found that for high impact energies the cluster charge had little effect: either positive or negative clusters produced positive or negative analyte ions with similar efficiencies, a result later confirmed by Asakawa and Hiraoka.[25] These authors did observe that at low impact energies the cluster charge seems to play a role in ionization, however, at higher energies some more efficient process begins to dominate that they suggest is related to shock wave formation.

What mechanism could lead to shock-induced ionization? We can take water as a model as shock processes in water have been extensively studied. It is known, experimentally, that at high shock pressures the ionic conductivity of water increases enormously. At the $\sim$30 GPa shock pressure calculated for glycerol cluster impact[15] more than 10% of the water molecules are ionized.[26] This is due to the efficient autodissociation of water molecules into H_3O^+ and OH^- ions at the high shock temperature ($\sim$2000 K at a shock pressure of $\sim$30 GPa). Note that at this temperature and pressure water will be a dense supercritical fluid and because of ion stabilization by solvation the activation energy for ionic dissociation in the condensed phase is only $\sim$1 eV,[26] much lower than the $\sim$17 eV threshold for gas-phase ionic dissociation used by Ryan et al.[20] Thus, transiently, there can be an enormous concentration of dissociated charges in the impact zone of an ice target under cluster impact. Whether this effect extends to dissociation of other protic molecules has yet to be established, but the idea of considering ionization as a consequence of extreme superheating appears worth pursuing.

As the shock compression releases and the heated analyte volume explodes, do these hydronium ions efficiently find their counterions and reassociate, or do some fraction of them randomly persist to protonate the emerging analyte? The latter seems probable, but at present there appears to be no easy way to calculate an ionization probability. However, it is worth noting that Fabris et al.[5] observed extensive multiple charging of carbonic anhydrase and myoglobin sputtered by MCI from a thioglycerol matrix, suggesting that the fraction of molecules that remained uncharged was very small. In unpublished work, Hiraoka has estimated an ionization efficiency $\sim$10% for EDI of rhodamine B (Hiraoka, K. 2012, Personal communication). Careful measurements of useful yields with these cluster species are needed.

Ionization effects such as those discussed here must also come into play in sputtering by smaller clusters. Whether or not the overall effects can be characterized in terms of macroscopic shock wave phenomena, it is clear that there must be intense vibrational excitation (heating) within and around the sputtering site and so it is worth considering acid–base properties of condensed target materials at these transient elevated temperatures.

9.6   MATRIX EFFECTS AND CHALLENGES IN QUANTITATIVE ANALYSIS

One of the major challenges with SIMS as with any mass spectrometric imaging methods is that of quantification. As noted above, factors governing ionization efficiency are still largely unknown and the issue is further complicated by possibly different susceptibilities of different species to fragmentation. In addition, there is ample evidence, as discussed in Chapters 4 and 8, that the signal for a particular molecule can be suppressed and/or enhanced by the existence of other molecules in the local environment. Figure 9.2 shows an extreme example of an almost perfectly inverted contrast in similar samples using MALDI and cluster SIMS imaging of the same lipid species. In MALDI analysis of solution samples, such as blood samples, quantification issues can be resolved by the addition of internal standards, mass-shifted variants of the target proteins,[27,28] but such approaches are impossible in imaging analyses. Laser postionization approaches have long been advocated as a means to avoid ionization complications in the condensed phase, but molecular survival issues with this approach remain unquantified. It remains to be seen whether postionization efficiency can compete with direct ionization, and even if this is achieved, the cost and complication of laser systems and their relatively low data rates present significant barriers to their current use in imaging.

9.7 SIMS INSTRUMENTATION

Until very recently, molecular SIMS instrumentation has been largely based on ToF-MS designs. ToF-MS has significant advantages, in particular the ability to register an entire mass spectrum in each ion pulse, essentially unlimited mass range and compatibility with ion doses in the static SIMS range. However, imaging biological samples requires high data rates and also the ability to identify molecular species in what can be a highly complex mass spectrum. Data rates in ToF-MS are typically low, not only because of the low duty cycle of pulsed operation but also because detection is typically a multistop process that demands that a relatively low number of secondary ions be produced in each pulse—typically no more than one ion at any given mass. Molecular identification in complex samples calls for MS–MS approaches, involving collisional dissociation of chosen ion species followed by an orthogonal mass separation to identify the fragment ions and thus the parent. This is yet another step whose efficiency must be considered if high spatial resolution is to be achieved.

ToF-MS instrumentation for MALDI imaging has advanced to the point where pulse rates up to 5 kHz are possible and MS–MS analysis can be accomplished for multiple ion species in each spectrum.[29] While pulsed operation is enforced by the nature of the MALDI process, this is not the case in SIMS. Indeed, pulsed primary ion beams all begin as DC beams that are then chopped to produce the pulse. Cluster SIMS instrumentation is now starting to move toward DC operation,[30,31] with very significant throughput advantages, and this evolution is expected to continue. The mass spectrometry approaches have been either triple quadrupole or ToF-MS with postdesorption bunching. In either case, MS–MS capability will often be essential and high mass resolving power will be extremely useful in sorting through the complex mass spectra. Recently, a Fourier transform ion cyclotron resonance (FT-ICR) mass spectrometer has been developed for SIMS.[32] This instrument (described in Chapter 4) has been coupled to a C_{60} source, and shows mass resolving power >100,000. Other Fourier transform mass spectrometers (FT-MS), such as the Orbitrap, also appear promising.[33] With mass resolving power up to 50,000 and MS–MS capability, the Orbitrap offers great specificity at relatively low cost; spectrum acquisition time can be a fraction of a second so that a 10^4 pixel image could be acquired in about an hour. Finally, a continuing challenge for SIMS imaging of complex targets is the computing power needed to handle the large data loads.

There is continuing interest in the use of very large clusters for molecular SIMS—Ar gas clusters and massive electrosprayed clusters. Both types of clusters are difficult to focus to any reasonable beam size and so it may be useful to consider coupling such beams to direct-imaging mass spectrometers, an approach that has been pursued to good effect for MALDI imaging.[34] The Cameca IMS and 1280 magnetic instruments and the PHI TRIFT ToF-MS all allow direct imaging, and indeed the first cluster depth profile was performed on such an instrument.[35] With the present-day emphasis on ToF instrumentation it is easy to overlook the fact that the earliest MCI studies were performed using magnetic mass spectrometers and indeed the heaviest analyte species yet sputtered—carbonic anhydrase, 29 kDa—was detected (as multiply charged ions) on a magnetic instrument.[5] Although mass range is limited on magnetic instruments, the direct-imaging Cameca instruments can be operated up to at least ~1 kDa, which would allow detection of many lipids and small peptides. Lateral resolution is not dependent on the primary ion beam size, and can be as good as a few micrometers, depending on the ion signal. For molecular ions, which are sputtered with low initial energies, transmission at low mass resolving power can be several tens of percent. Transmission is compromised if high mass resolving power is needed, but on the other hand direct-imaging data rates are much faster than for microprobe approaches. For direct imaging on a TRIFT instrument, if the primary cluster beam cannot be pulsed with good time resolution the secondary beam can be pulsed instead. MS–MS analysis is not compatible with direct imaging, but the success to date of SIMS and MALDI (and DESI) approaches without MS–MS suggests that many useful analyses should be possible.

9.7.1 Massive Cluster Ion Source Technology

The potential advantage of MCI over smaller clusters is that MCI has been demonstrated to desorb molecular ions of quite large proteins while creating negligible surface damage, and it appears possible that ionization efficiency of the desorbed molecules can be quite high. A significant advantage over imaging MALDI is that MCI can desorb ions from untreated surfaces,[5] avoiding the difficulties of relocating or smearing analyte molecules by matrix addition that impose a limit on imaging resolution in MALDI. However, a robust and convenient massive cluster source technology does not yet exist. Glycerol was used in the initial MCI work[7] because it has a low vapor pressure and so is somewhat vacuum compatible. The downside is that electrospray produces a divergent beam that tends to coat and contaminate the ion optical column, and the involatile glycerol is not easily pumped away. It is attractive to think of spraying a more volatile liquid such as water, but then evaporative cooling is so rapid that the water freezes.

Freezing can be remedied by heating the emitter nozzle either electrically[24] or with an IR laser[36] but control of the spray is still problematic. Hiraoka earlier introduced the idea of electrospraying in the atmosphere and then introducing the charged droplets through an interface and called this approach EDI.[8] Although this approach removes the spraying difficulties and allows the use of a volatile spray liquid, it appears difficult with this approach to deliver significant current to an imaging target. Fujiwara and colleagues[37] are exploring the electrospraying of ionic liquids. These are significantly less volatile even than glycerol, but the authors argue that any liquid deposits on electrodes should continue to conduct. None of these alternatives seem ideal and future improvements are hoped for.

9.8   PROSPECTS FOR BIOLOGICAL IMAGING

The great promise for cluster imaging in biology is the possibility of surpassing the image resolution achievable by MALDI, into the submicrometer regime that could allow imaging with subcellular resolution. Human cells have typical diameters on the order of 10 μm, so image voxel sizes of 1 μm or smaller might allow visualization of intracellular molecular features. The most intriguing imaging targets are proteins; however, the number of proteins of a given species—the copy number—varies enormously within the cell. In a human cell line protein copy numbers vary over at least seven decades down from $\sim$20,000,000 copies per cell,[38] with many of the more important proteins being present at very low levels. If molecular useful yields under cluster bombardment are typically only $\sim$10^{-4} then much of the cell proteome will be undetectable. It seems essential to achieve a sputtering solution that does not fragment the analyte molecules and that ionizes them with high efficiency.

An added concern is that many proteins undergo extensive processing, post-translational modification, and metabolic alteration so that the dominant species may bear little relation to the intact protein predicted from the gene sequence. Figure 9.3 shows an example of this in the MALDI spectrum of parathyroid hormone (PTH) extracted from blood plasma by a polyclonal antibody that targets different epitopes along the sequence.[39] Strikingly, only a tiny fraction of the captured material is the full-length protein. Various enzymatic processing steps yield a variety of fragments, several of which are biologically active and none of which would be recognizable by molecular weight alone. At a minimum, MS–MS techniques will be required to identify many peptide or protein signals, as was realized many years ago by Todd and coworkers.[40]

Lipids appear to be a more promising imaging target. They are fairly abundant and are conveniently concentrated in membranes; molecular weights are mostly below $\sim$1 kDa. Intermolecular binding in lipid membranes is relatively weak, so that a large fraction of molecules should be sputtered intact,[41] although ionization efficiency may not be high. Striking differences have been found in the composition of certain lipids between cancer and normal cells, so that imaging lipids has the potential to be of diagnostic value.

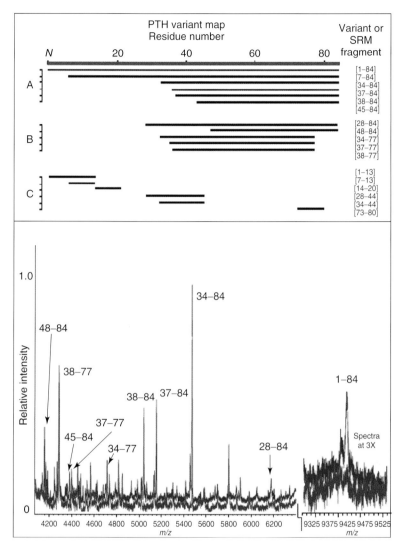

Figure 9.3 Variant map and MALDI mass spectra of PTH affinity-isolated from serum from healthy (red) and renal failure (blue) patients. In healthy patients, the full-length protein (1–84) is barely detectable. From Lopez et al.,[39] with permission from American Association for Clinical Chemistry.

9.9 CONCLUSIONS

Cluster SIMS is still a fairly young technique. Apart from the early ventures into MCI technology, the modern field dates from the striking demonstration in 1998 by Gillen and Roberson[35] that SF_5^+ clusters allowed depth profiling of organic thin film samples, and the later development of commercial C_{60} ion sources. There is an exceptionally rich variety of cluster species to be explored, with differing merits as to ionization efficiency, data rate, focusing capability, and fundamental physics. It seems clear that large clusters are essential for efficient ejection of very large molecules, either argon clusters or massive electrosprayed clusters. For smaller analytes, the situation is not yet clear, and we have suggested here that the optimum cluster size for best useful yield may actually be in the ~10 atom range. Similarly, the optimum mass spectrometric configuration arguably has yet to be achieved, and different configurations may be needed to couple to different cluster species and different applications. On the theoretical front, computer simulations have provided invaluable insights into details of molecular motion under bombardment by different types and sizes of clusters. Still, ionization phenomena are hard to model and new insights into ionization are needed. In particular, careful quantitative comparisons of useful ion yields for different cluster species and analytes should help illuminate the cluster sputtering phenomenon. It seems clear that a rich future remains to be explored.

REFERENCES

1. Hillenkamp, F.; Karas, M. The MALDI Process and Method. In *MALDI-MS: A Practical Guide to Instrumentation, Methods and Applications*; Hillenkamp, F., Peter-Katalinic, J., Eds.; Wiley-VCH Verlag GmbH & Co.: Weinheim, Germany, **2007**, Chapter 1.
2. Schwamborn, K.; Caprioli, R. M. *Nat Rev. Cancer* **2010**, 10 (9), 639–646.
3. Wiseman, J. M.; Ifa, D. R.; Song, Q.; Cooks, R. G. *Angew. Chem. Int. Ed.* **2006**, 45 (43), 7188–7192.
4. Siegel, M. M.; Tsao, R.; Doelling, V. W.; Hollander, I. J. *Anal. Chem.* **1990**, 62 (14), 1536–1542.
5. Fabris, D.; Wu, Z.; Fenselau, C. C. *J. Mass Spectrom.* **1995**, 30 (1), 140–143.
6. Ninomiya, S.; Nakata, Y.; Ichiki, K.; Seki, T.; Aoki, T.; Matsuo, J. *Nucl. Instrum. Methods Phys. Res. B.* **2007**, 256 (1), 493–496.
7. Mahoney, J. F.; Perel, J.; Ruatta, S. A.; Martino, P. A.; Husain, S.; Cook, K.; Lee, T. D. *Rapid Commun. Mass Spectrom.* **1991**, 5 (10), 441–445.
8. Hiraoka, K.; Mori, K.; Asakawa, D. *J. Mass Spectrom.* **2006**, 41 (7), 894–902.
9. Mochiji, K. *J. Anal. Bioanal. Tech. S2* **2011**, 2, 2.
10. Nelson, R. W.; Dogruel, D.; Williams, P. *Rapid Commun. Mass Spectrom.* **1995**, 9 (7), 625.
11. Yang, J.; Caprioli, R. M. *Anal. Chem.* **2011**, 83 (14), 5728–5734.
12. Wang, H. Y. J.; Post, S. N.; Woods, A. S. *Int. J. Mass Spectrom.* **2008**, 278 (2), 143–149.
13. Jaskolla, T. W.; Karas, M. *J. Am. Soc. Mass Spectrom.* **2008**, 19 (8), 1054–1061.
14. Chou, C. W.; Nelson, R. W.; Williams, P. *Eur. J. Mass Spectrom.* **2009**, 15 (2), 305.
15. Mahoney, J. F.; Perel, J.; Lee, T. D.; Martino, P. A.; Williams, P. *J. Am. Soc. Mass Spectrom.* **1992**, 3 (4), 311–317.
16. Kayser, S.; Rading, D.; Moellers, R.; Kollmer, F.; Niehuis, E. Surface spectrometry using large argon clusters. *Surf. Interface Anal.* **2013**, 45, 131–133.
17. Krueger, R.; Pfenninger, A.; Fournier, I.; Gluckmann, M.; Karas, M. *Anal. Chem.* **2001**, 73 (24), 5812–5821.
18. Rabbani, S.; Barber, A. M.; Fletcher, J. S.; Lockyer, N. P.; Vickerman, J. C. *Anal. Chem.* **2011**, 83 (10), 3793–3800.
19. Conlan, X. A.; Fletcher, J. S.; Lockyer, N. P.; Vickerman, J. C. *J. Phys. Chem. C.* **2010**, 114 (12), 5468–5479.
20. Ryan, K. E.; Wojciechowski, I. A.; Garrison, B. J. *J. Phys. Chem. C.* **2007**, 111 (34), 12822–12826.
21. Jones, K. W.; Kraner, H. W. *Phys. Rev. A.* **1975**, 11 (4), 1347–1353.
22. Willingham, D.; Brenes, D. A.; Wucher, A.; Winograd, N. *J. Phys. Chem. C.* **2010**, 114 (12), 5391–5399.
23. Wucher, A.; Sun, S.; Szakal, C.; Winograd, N. *Anal. Chem.* **2004**, 76 (24), 7234–7242.

24. Aksyonov, S. A.; Williams, P. *Rapid Commun. Mass Spectrom.* **2001**, 15 (21), 2001–2006.

25. Asakawa, D.; Hiraoka, K. *Rapid Commun. Mass Spectrom.* **2011**, 25 (5), 655–660.

26. Hollenberg, K. *J. Phys. D.* **1983**, 16, 385.

27. Nelson, R. W.; Krone, J. R.; Bieber, A. L.; Williams, P. *Anal. Chem.* **1995**, 67 (7), 1153–1158.

28. Niederkofler, E. E.; Tubbs, K. A.; Gruber, K.; Nedelkov, D.; Kiernan, U. A.; Williams, P.; Nelson, R. W. *Anal. Chem.* **2001**, 73 (14), 3294–3299.

29. Vestal, M. L. *J. Am. Soc. Mass Spectrom.* **2011**, 22 (6), 953–959.

30. Carado, A.; Passarelli, M. K.; Kozole, J.; Wingate, J. E.; Winograd, N.; Loboda, A. V. *Anal. Chem.* **2008**, 80 (21), 7921–7929.

31. Fletcher, J. S.; Rabbani, S.; Henderson, A.; Blenkinsopp, P.; Thompson, S. P.; Lockyer, N. P.; Vickerman, J. C. *Anal. Chem.* **2008**, 80 (23), 9058–9064.

32. Smith, D. F.; Robinson, E. W.; Tolmachev, A. V.; Heeren, R.; Pasa-Tolic, L. *Anal. Chem.* **2011**, 83, 9552–9556.

33. Makarov, A.; Denisov, E.; Kholomeev, A.; Balschun, W.; Lange, O.; Strupat, K.; Horning, S. *Anal. Chem.* **2006**, 78 (7), 2113–2120.

34. Luxembourg, S. L.; Mize, T. H.; McDonnell, L. A.; Heeren, R. M. A. *Anal. Chem.* **2004**, 76 (18), 5339–5344.

35. Gillen, G.; Roberson, S. *Rapid Commun. Mass Spectrom.* **1998**, 12, 1303–1312.

36. Ninomiya, S.; Chen, L. C.; Sakai, Y.; Suzuki, H.; Hiraoka, K. Development of a high-performance electrospray droplet beam source. *Surf. Interface Anal.* **2013**, 45, 126–130.

37. Fujiwara, Y.; Saito, N.; Nonaka, H.; Ichimura, S. Emission characteristics of a charged-droplet beam source using vacuum electrospray of an ionic liquid. *Surf. Interface Anal.* **2013**, 45, 517–521.

38. Beck, M.; Schmidt, A.; Malmstroem, J.; Claassen, M.; Ori, A.; Szymborska, A.; Herzog, F.; Rinner, O.; Ellenberg, J.; Aebersold, R. *Mol. Syst. Biol.* **2011**, 7 (549), 1–8.

39. Lopez, M. F.; Rezai, T.; Sarracino, D. A.; Prakash, A.; Krastins, B.; Athanas, M.; Singh, R. J.; Barnidge, D. R.; Oran, P.; Borges, C. *Clin. Chem.* **2010**, 56 (2), 281–290.

40. Todd, P. J.; Short, R. T.; Grimm, C. C.; Holland, W. M.; Markey, S. P. *Anal. Chem.* **1992**, 64 (17), 1871–1878.

41. Paruch, R.; Rzeznik, L.; Czerwinski, B.; Garrison, B. J.; Winograd, N.; Postawa, Z. *J. Phys. Chem. C.* **2009**, 113 (14), 5641–5648.

INDEX

Cluster Secondary Ion Mass Spectrometry: Principles and Applications, First Edition.
Edited by Christine M. Mahoney.
© 2013 John Wiley & Sons, Inc. Published 2013 by John Wiley & Sons, Inc.

WILEY SERIES ON MASS SPECTROMETRY

Series Editors

Dominic M. Desiderio
Departments of Neurology and Biochemistry
University of Tennessee Health Science Center

Nico M. M. Nibbering
Vrije Universiteit Amsterdam, The Netherlands

Joseph A. Loo
Department of Chemistry and Biochemistry
UCLA

John R. de Laeter • *Applications of Inorganic Mass Spectrometry*
Michael Kinter and Nicholas E. Sherman • *Protein Sequencing and Identification Using Tandem Mass Spectrometry*
Chhabil Dass • *Principles and Practice of Biological Mass Spectrometry*
Mike S. Lee • *LC/MS Applications in Drug Development*
Jerzy Silberring and Rolf Eckman • *Mass Spectrometry and Hyphenated Techniques in Neuropeptide Research*
J. Wayne Rabalais • *Principles and Applications of Ion Scattering Spectrometry: Surface Chemical and Structural Analysis*
Mahmoud Hamdan and Pier Giorgio Righetti • *Proteomics Today: Protein Assessment and Biomarkers Using Mass Spectrometry, 2D Electrophoresis, and Microarray Technology*
Igor A. Kaltashov and Stephen J. Eyles • *Mass Spectrometry in Structural Biology and Biophysics: Architecture, Dynamics, and Interaction of Biomolecules, Second Edition*
Isabella Dalle-Donne, Andrea Scaloni, and D. Allan Butterfield • *Redox Proteomics: From Protein Modifications to Cellular Dysfunction and Diseases*
Silas G. Villas-Boas, Ute Roessner, Michael A.E. Hansen, Jorn Smedsgaard, and Jens Nielsen • *Metabolome Analysis: An Introduction*
Mahmoud H. Hamdan • *Cancer Biomarkers: Analytical Techniques for Discovery*
Chabbil Dass • *Fundamentals of Contemporary Mass Spectrometry*
Kevin M. Downard (Editor) • *Mass Spectrometry of Protein Interactions*
Nobuhiro Takahashi and Toshiaki Isobe • *Proteomic Biology Using LC-MS: Large Scale Analysis of Cellular Dynamics and Function*
Agnieszka Kraj and Jerzy Silberring (Editors) • *Proteomics: Introduction to Methods and Applications*
Ganesh Kumar Agrawal and Randeep Rakwal (Editors) • *Plant Proteomics: Technologies, Strategies, and Applications*
Rolf Ekman, Jerzy Silberring, Ann M. Westman-Brinkmalm, and Agnieszka Kraj (Editors) • *Mass Spectrometry: Instrumentation, Interpretation, and Applications*
Christoph A. Schalley and Andreas Springer • *Mass Spectrometry and Gas-Phase Chemistry of Non-Covalent Complexes*
Riccardo Flamini and Pietro Traldi • *Mass Spectrometry in Grape and Wine Chemistry*
Mario Thevis • *Mass Spectrometry in Sports Drug Testing: Characterization of Prohibited Substances and Doping Control Analytical Assays*